Concepts of

HUMAN

ANATOMY

&

physiology

LABORATORY MANUAL

HUMAN

Concepts of

ANATOMY

&

physiology
LABORATORY MANUAL

fourth edition

Kent M. Van De Graaff

Stuart Ira Fox
Pierce College

WCB **Wm. C. Brown Publishers**

Dubuque, IA Bogota Boston Buenos Aires Caracas Chicago
Guilford, CT London Madrid Mexico City Sydney Toronto

Book Team

Editor *Colin H. Wheatley*
Developmental Editor *Kristine Noel*
Production Editor *Jane E. Matthews*
Art Editor *Renee Grevas*
Photo Editor *John C. Leland*
Permissions Coordinator *Gail I. Wheatley*

Wm. C. Brown Publishers
A Division of Wm. C. Brown Communications, Inc.

Vice President and General Manager *Beverly Kolz*
Vice President, Publisher *Kevin Kane*
Vice President, Director of Sales and Marketing *Virginia S. Moffat*
Vice President, Director of Production *Colleen A. Yonda*
National Sales Manager *Douglas J. DiNardo*
Marketing Manager *Craig S. Marty*
Advertising Manager *Janelle Keeffer*
Production Editorial Manager *Renée Menne*
Publishing Services Manager *Karen J. Slaght*
Royalty/Permissions Manager *Connie Allendorf*

Wm. C. Brown Communications, Inc.

President and Chief Executive Officer *G. Franklin Lewis*
Senior Vice President, Operations *James H. Higby*
Corporate Senior Vice President, President of WCB Manufacturing *Roger Meyer*
Corporate Senior Vice President and Chief Financial Officer *Robert Chesterman*

Copyedited by Moira Urich

Cover Credit:
Hands and ears: © Wm. C. Brown Communications, Inc.
Mechanical gears: Photone-Letraset.
Running and jumping series: Courtesy George Eastman House.
Female model: © Butch Martin/The Image Bank.
Illustration of male: Courtesy Francis A Countway Library of Medicine.
Eye: © Digital Stock

The credits section for this book begins on page 431 and
is considered an extension of the copyright page.

Some of the laboratory experiments included in this text may be
hazardous if materials are handled improperly or if procedures are
conducted incorrectly. Safety precautions are necessary when you are
working with chemicals, glass test tubes, hot water baths, sharp
instruments, and the like, or for any procedures that generally require
caution. Your school may have set regulations regarding safety
procedures that your instructor will explain to you. Should you have any
problems with materials or procedures, please ask your instructor for
help.

Printed in the United States of America by Wm. C. Brown Communications, Inc.,
2460 Kerper Boulevard, Dubuque, IA 52001

10 9 8 7 6 5 4 3 2 1

Contents

Preface

Laboratory Manual for Concepts of Human Anatomy and Physiology provides the laboratory exercises needed to support and enrich lecture material in a human anatomy and physiology course. This manual is designed to be used in conjunction with the lecture textbook *Concepts of Human Anatomy and Physiology,* 4th ed. The laboratory manual is organized into the same units and topic sequences as the textbook, and specific textbook chapters and topics are referenced at the beginning of each exercise. Referenced information in the textbook provides the introductory concepts required to understand the exercises, and students will need this information to complete the laboratory reports. This correlation of lecture textbook and laboratory manual eliminates redundant information and encourages the students' more thorough integration of the lecture and laboratory portions of the course.

The laboratory exercises in this manual help students become actively involved in learning the principles and applications of anatomy and physiology. Active learning is required to perform the exercises and to complete the laboratory report at the end of each exercise. Through these activities, students interact with the subject matter, other students, and the instructor in a more personal way than they could in lecture alone. In order to help reach this goal, a wide variety of exercises is provided to support most of the learning objectives of the course.

Clinically oriented laboratory exercises, along with those that relate to exercise physiology, are provided where possible to heighten student interest and demonstrate the health applications of anatomy and physiology. Human anatomy is emphasized by inclusion of figures to be labeled and colored while referring to the textbook. Because many laboratories do not have facilities for the dissection of human cadavers, preserved cats are often used. This dissection experience is supported by exercises that provide detailed directions for dissection, descriptions of cat anatomy, and excellent line drawings and color photographs. The organization of these exercises helps students correlate cat anatomy with human anatomy.

The exercises in this manual are organized in the following manner:

1. Each exercise begins by noting the chapters and sections of the textbook (*Concepts of Anatomy and Physiology,* 4th ed.) that provide the background information for understanding the principles behind each exercise (and that provide labeled figures for reference).

2. Each exercise begins with an **introduction** that contains a concise summary of the concepts presented. This short section does not, however, provide all of the information needed to understand the principles of the exercise.

3. **Objectives** are listed after the introduction so that students can guide their learning while performing the exercise.

4. **Materials** required for the laboratory exercise are listed next (before the exercise itself) to make setup easier.

5. Many exercises are divided into **sections** that break large subject areas into logical subdivisions. This assists students in performing the exercises and allows instructors to be flexible in laboratory assignments.

6. In many exercises, **procedures** are set off by a heading and are presented in a different typeface. This helps students locate and use the appropriate information as a step-by-step guide.

7. A **laboratory report** follows each exercise. Students enter data here, when appropriate, and answer questions. The questions in the laboratory report begin with the simplest forms (objective and fill-in-the-blank questions) and progress to essay questions. Instructors may require students to hand in these reports, they may use them as a basis for laboratory quizzes, or the students may use them as study guides for the laboratory.

We are indebted to our colleagues and students for their suggestions and encouragement in the development of the laboratory exercises. Special thanks go to Brian E. Day and L. Nick Crockett for their assistance in the cat dissections, and to Dr. Sheril D. Burton for photographing the dissected specimens. Dr. Lawrence Thouin and Mr. Edmont Katz made many useful suggestions for updating some of the laboratory procedures.

Finally, we wish to state our appreciation to our long-suffering wives, Ellen and Karen, and to our children for their continued support and sacrifice during the preparation of this manual.

Laboratory Safety

Most of the reagents (chemicals) and equipment in an anatomy and physiology laboratory are potentially dangerous, so it is important that certain safety rules be followed. Unfortunately, rules of behavior that seem obvious when read at the beginning of the course may be forgotten in the hustle (and, it is hoped, excitement) of the laboratory exercises. Therefore, the rules of laboratory safety should not simply be memorized but should become a habit.

1. Assume that all reagents are poisonous, and act accordingly. **Do not** ingest any reagents; **do not** eat, drink, or smoke in the laboratory; **do not** carry reagent bottles around the room; and **do not** pipette anything by mouth unless specifically told to do so by your instructor.
   Do wash your hands thoroughly before leaving the laboratory; stopper all reagent bottles when they are not in use; thoroughly clean up spills; wash spilled reagents off yourself and your clothing; and, if you accidentally get any reagent in your mouth, immediately inform the instructor and rinse your mouth thoroughly.
2. Follow the written procedure instructions or the instructor's modifications of these instructions completely. **Do not** improvise unless the instructor specifically approves it.
3. In order to protect yourself from the possible danger of contact with blood infected with the AIDS virus, use only your own blood for these exercises. If you must stick someone else with a lancet, wear disposable gloves. Place used lancets and other objects that contain blood in an appropriate container that can later be sterilized.
4. Clean glassware at the end of each exercise so that residue from one exercise does not carry over to the next.
5. Keep your work area clean, neat, and organized. This will reduce the chances of error and will help make your work safer and more accurate.
6. Study the theory and procedures for the laboratory exercises *before* coming to the laboratory. This will increase your understanding, enjoyment, and safety during exercises. Confusion can be dangerous.
7. **Do not** operate any equipment until you are instructed in its proper use. If you are unsure of the procedures, ask the instructor.
8. Be careful about open flames in the laboratory. **Do not** leave a flame unattended, **do not** light a Bunsen burner near any gas tank or cylinder, and **do not** move a lit Bunsen burner around on the desk. Make sure that long hair is well out of the way of the flame.
9. Always make sure that gas jets are off when you are not operating the Bunsen burner.
10. Handle hot glassware with test-tube clamps or tongs.
11. Note the nearest location of an emergency first aid kit, eyewash bottle, and fire extinguisher. Report all accidents to the instructor immediately.
12. Wear safety glasses during exercises in which glassware and solutions are heated with a Bunsen burner.

Orientation and Organization of the Human Body

The following exercises are included in this unit:

1. Body Organization and Terminology
2. Microscope and Metric System
3. Homeostasis and Negative Feedback
4. Measurement of Plasma Glucose, Protein, and Cholesterol
5. Cell Structure and Cell Division
6. Genetic Control of Metabolism
7. Diffusion, Osmosis, and Tonicity
8. Histology

These exercises are based on information presented in the following chapters of *Concepts of Human Anatomy and Physiology,* 4th ed.:

1. Introduction to Human Anatomy and Physiology
2. Chemical Composition of the Body
3. Cell Structure and Genetic Regulation
4. Enzymes, Energy, and Metabolism
5. Membrane Transport and the Membrane Potential
6. Histology

These exercises provide laboratory support for the basic molecular and cellular concepts needed for an understanding of anatomy and physiology. The "wet" laboratory exercises—such as measurements of plasma glucose, protein, and cholesterol—should give you a feeling for the importance of these basic concepts in clinical diagnosis. Microscopic examination of cells provides an introduction to the use of the microscope. It also adds a visual dimension to understanding concepts that otherwise might seem abstract. For example, when you see the swelling and shrinking of red blood cells in different solutions, the practicality of *osmosis* and *tonicity* is obvious. Although the concepts covered in this unit of the text and laboratory manual are basic, they are not simple. This laboratory work should help you understand these principles better and should strengthen the foundation for comprehending the concepts introduced later in the course.

Body Organization and Terminology

Before coming to class, review the following sections in chapter 1 of the textbook: "Body Organization," "Anatomical Position and Directional Terms," "Body Regions," "Body Cavities and Membranes," and "Planes of Reference and Descriptive Terminology."

Introduction

Because anatomy is a descriptive science, the meanings of certain terms that describe body parts and positions must be known. These basic terms describe levels of organization (cells, tissues, organs, and systems), body regions, body cavities, planes of reference, and relative directional position.

Objectives

Students completing this section will be able to:

1. Identify the structures of a cell, a tissue, an organ, and a system; and explain the relationship between these structures.
2. List the principal body regions and body cavities.
3. Describe the planes of reference of the body.
4. Define the descriptive terms used to convey information about position and direction in anatomy.

Materials

1. Anatomical charts and illustrations
2. Human torso model
3. Reference text
4. Colored pencils

A. Body Organization

Study the descriptions and definitions of the terms *cells, tissues, organs,* and *systems* in chapter 1 of the text, and answer the questions in section A of the laboratory report.

B. Body Regions

Identify and label the following structures and regions of the body in figure 1.1:

Abdomen	Lower extremity
Antebrachium	Lumbar region
Axillary region	Neck
Brachium	Palmar surface
Buttock	Patellar region
Calf	Plantar surface
Cervical region	Popliteal fossa
Cubital fossa	Pubic region
Deltoid region	Thigh
Elbow	Thorax
Head	Upper extremity

C. Body Cavities

1. From the list below, identify and label the structures and cavities in figures 1.2, and 1.3.
2. Using colored pencils, color-code the structures identified in the figures. Red is suggested for the heart, for example, and pink for the lungs. Neural tissue is traditionally shown in a yellow color, so yellow might be used for the brain and spinal cord.

Anterior (ventral) cavity
    Abdominal cavity
    Diaphragm
    Heart within the pericardial cavity
    Large intestine (transverse colon)
    Liver
    Lung within the pleural cavity
    Mesentery
    Omentum (greater and lesser)
    Pancreas
    Pelvic cavity
    Rectum
    Small intestine (duodenum)
    Stomach
    Thoracic cavity
    Urinary bladder
Posterior (dorsal) cavity
    Brain
    Cranial cavity
    Spinal cord
    Vertebral canal
Abdominopelvic cavity
Anterior cavity
Posterior cavity

D. Descriptive Terminology

1. Study the *anatomical position,* as shown in chapter 1 of the text; all directional terms that describe the relationship of one body part to another are made in reference to the anatomical position.

2. Referring to chapter 1 of the text, write the definitions of the directional terms listed in part D, number 2, of the laboratory report.

3. Study figure 1.4 and compare the use of directional terms and planes of reference for a cat, which is a quadruped (walks on four legs), with those for a human, which is a biped (walks on two legs).

Figure 1.1   Major body regions and localized areas.

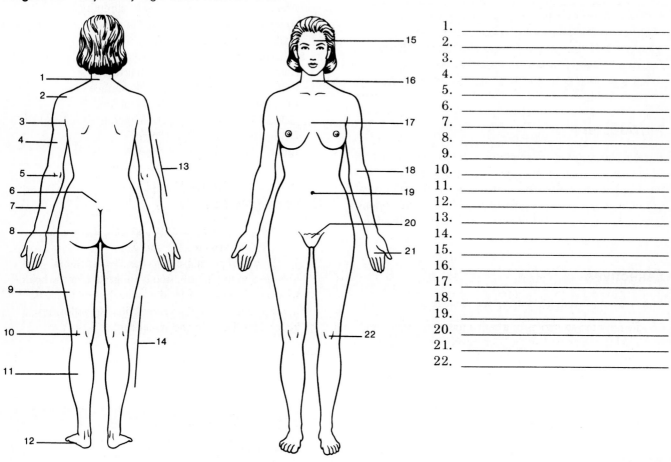

1. _____
2. _____
3. _____
4. _____
5. _____
6. _____
7. _____
8. _____
9. _____
10. _____
11. _____
12. _____
13. _____
14. _____
15. _____
16. _____
17. _____
18. _____
19. _____
20. _____
21. _____
22. _____

Figure 1.2 An anterior view of body cavities.

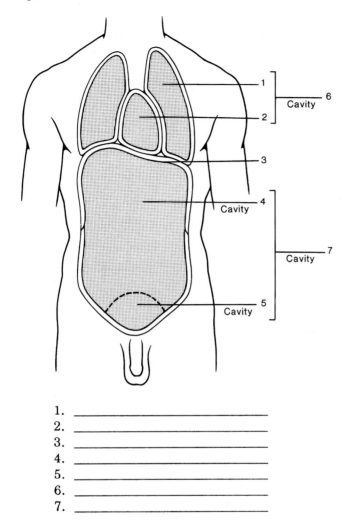

1. _____
2. _____
3. _____
4. _____
5. _____
6. _____
7. _____

Figure 1.3 A medial view of body cavities.

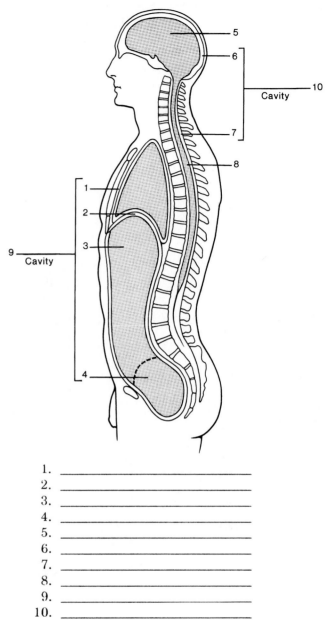

1. _____
2. _____
3. _____
4. _____
5. _____
6. _____
7. _____
8. _____
9. _____
10. _____

Figure 1.4 Directional terminology in (*a*) bipedal (human) and (*b*) quadrupedal (cat) animals.

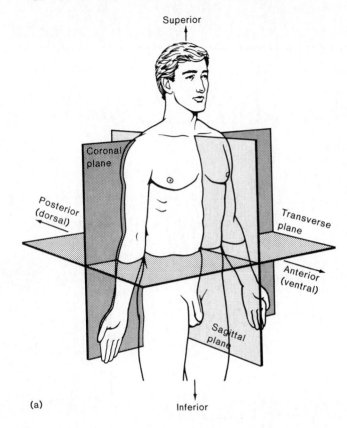

(a)

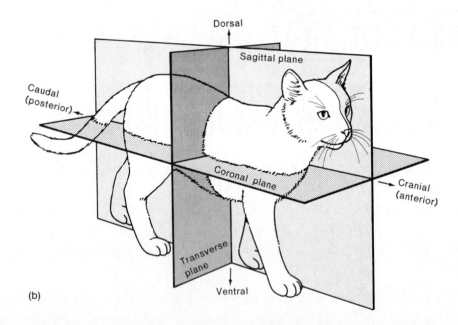

(b)

Laboratory Report 1

Name _____

Date _____

Section _____

Body Organization and Terminology

Read the assigned sections in the textbook before completing the laboratory report.

A. Body Organization

1. The basic structural and functional unit of any living organism is the _____ .

2. List the four primary tissues:

   _____

   _____

   _____

   _____

3. Distinguish between the terms *tissue* and *organ*.

4. Distinguish between the terms *organ* and *system*.

B. Body Regions

Provide the *anatomical* term for each common expression listed below:

Skull _____

Hand _____

Navel _____

Back of knee _____

Surface of hand _____

Small of back _____

Armpit _____

Shoulder blade _____

Upper arm _____

Neck _____

Surface of foot _____

Upper leg _____

C. Body Cavities

Identify the body cavity in which each of the following organs is found, and identify the body system to which the organ belongs:

Organ	Cavity	System
Spinal cord		
Heart		
Lung		
Liver		
Small intestine		
Brain		

D. Descriptive Terminology

1. Complete the following sentences by filling in the correct directional terms:

   a. The thorax is _____ to the abdomen.

   b. The heart is _____ to the lungs.

   c. The hand is _____ to the elbow.

   d. The navel is on the _____ side of the body.

   e. The skin is _____ to the muscles.

   f. The legs are _____ to the trunk.

   g. The buttocks are _____ to the pelvis.

   h. The ears are on the _____ sides of the head.

   i. The knee is _____ to the foot.

   j. The brain is _____ to the cranium.

2. Write in the definition of each directional term listed below:

Directional Term	Definition
Superior (cranial, cephalic)	
Inferior (caudal)	
Anterior (ventral)	
Posterior (dorsal)	
Medial	
Lateral	
Internal (deep)	
External (superficial)	
Proximal	
Distal	
Visceral	
Parietal	

Microscope and Metric System

Before coming to class, read the material about the microscope in this exercise, as well as "Units of Measurement" in chapter 1 of the textbook.

Introduction

The interaction of structure and function in the study of anatomy and physiology can best be appreciated by observing organs under a microscope. In addition, microscopic examination is clinically valuable when distinguishing normal from pathological (disease) conditions. This exercise is an introduction to the use of the microscope. Because the size of microscopic objects must often be estimated, this exercise also includes an overview of the use of the metric system, which is the universally used system of scientific measurement.

Objectives

Students completing this exercise will be able to:

1. Identify the parts of a microscope, and demonstrate proper care of this instrument.
2. Focus the microscope using different objective lenses, and move the slide to observe different regions.
3. Estimate the size of microscopic objects, using the metric system.
4. Convert from one metric unit to another, using the technique of dimensional analysis.

Materials

1. Compound microscope
2. Prepared microscope slides (with the letter *e*)
3. Lens paper

A. Examination of the Microscope

Carefully carry the microscope from the storage cabinet to the table by firmly grasping the *arm* (fig. 2.1) with one hand and supporting the *base* with the other hand. Place the microscope gently on the table, and identify the following parts as they are described.

1. The *base* is the solid foundation of the microscope that rests on the table.
2. The *arm* is the angular portion of the frame that extends upward from the base.
3. The *stage* is the adjustable platform upon which a microscope slide is placed for examination. A circular opening in the center of the stage admits the light that illuminates the object being viewed.
4. The *stage clips* are the metal clips that secure the slide. Some microscopes have a *mechanical stage,* a more elaborate mechanical clamp that allows slow-motion, systematic scanning of the slide.
5. The *condenser* and *iris diaphragm* are attached directly under the opening in the stage, and they regulate the intensity of the light reaching the slide.
6. The *condenser adjustment knob,* or *substage adjustment,* raises or lowers the condenser and alters the illumination.
7. A *substage lamp* is found in microscopes without a built-in *mirror.* The adjustable mirror reflects the beam of light from its source to the condenser.
8. The *coarse adjustment knob* is used initially to bring an object into view.
9. The *fine adjustment knob* is then used to sharpen the focus. It is usually located below the coarse adjustment knob, but some microscopes have a combined coarse and fine adjustment knob.
10. The *revolvable nosepiece* is a circular attachment on the body tube: on the revolvable nosepiece are commonly three *objective lenses:*
    a. A *low-power objective,* which is a scanning lens with a magnification of 10× on most instruments.
    b. A *high-power objective,* which usually has a magnification of 43× or 45×.
    c. An *oil-immersion objective* with a magnification of 100×. This lens is usually identified by an etched red dot and is used with a drop of immersion oil on the microscope slide.
11. The *eyepiece* with *ocular lens* is what one looks directly into at the upper end of the body tube. The ocular usually has a magnification of 10× and can be removed for cleaning. It may contain a built-in pointer, which is helpful when working with another person. A *binocular* microscope has two body tubes and oculars that can be separately focused and adjusted for optimal imaging.

Figure 2.1 The parts of a compound microscope.

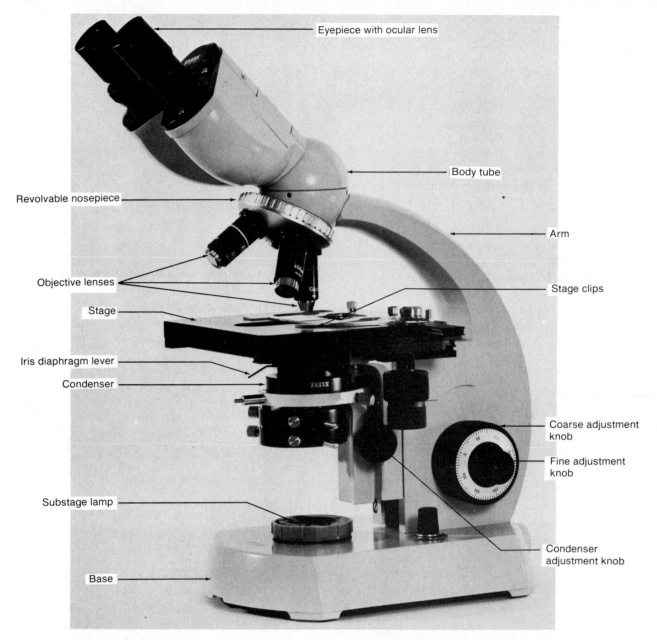

Eyepiece with ocular lens

Body tube

Revolvable nosepiece

Arm

Objective lenses

Stage clips

Stage

Iris diaphragm lever

Condenser

Coarse adjustment knob

Fine adjustment knob

Substage lamp

Condenser adjustment knob

Base

B. Use of the Microscope

Carefully remove the ocular, and clean the lens with special lens paper. Do not moisten the lens paper, and do not use paper towels or cloth to clean the lens; this may scratch the delicate glass surfaces. Replace the ocular, and clean the condenser and mirror.

1. Obtain a specially prepared slide of a newsprint letter *e*. Turn the revolvable nosepiece of the microscope so that the low-power objective clicks into place.
2. Before placing the slide on the stage, use the coarse adjustment knob to raise the body tube to its highest position so that the lenses are as far as possible from the stage.

3. Secure the slide on the stage with the stage clamps, and position the letter directly over the condenser.
4. Watching from the side of the microscope, carefully use the coarse adjustment to lower the body tube as far as possible. Note that on low magnification, the lens does not come in contact with the slide. Be careful never to focus the lens downward toward the slide, because this could shatter the slide or damage the lens.
5. Look through the ocular, and slowly raise the body tube with the coarse adjustment until an image is clearly seen. When using a microscope with a single ocular, it is best to make the observations with both eyes open. This technique may be difficult at first, but when you adapt to it, eyestrain is significantly reduced.

Table 2.1 SI Unit Prefixes

Multiplication Factor			Prefix	Symbol	Term
1,000,000	=	10^6	Mega	M	One million
1,000	=	10^3	Kilo	k	One thousand
100	=	10^2	Hecto	h	One hundred
10	=	10^1	Deka	da	Ten
0.1	=	10^{-1}	Deci	d	One-tenth
0.01	=	10^{-2}	Centi	c	One-hundredth
0.001	=	10^{-3}	Milli	m	One-thousandth
0.000001	=	10^{-6}	Micro	μ	One-millionth
0.000000001	=	10^{-9}	Nano	n	One-billionth
0.000000000001	=	10^{-12}	Pico	p	One-trillionth
0.000000000000001	=	10^{-15}	Femto	f	One-quadrillionth

Table 2.2 Sample Metric Conversions

To Convert from	To	Factor	Move Decimal Point
Meter (liter, gram)	Milli-	× 1000 (10^3)	3 places to right
Meter (liter, gram)	Micro-	× 1,000,000 (10^6)	6 places to right
Milli-	Meter (liter, gram)	÷ 1000 ($\times 10^{-3}$)	3 places to left
Micro-	Meter (liter, gram)	÷ 1,000,000 ($\times 10^{-6}$)	6 places to left
Milli-	Micro-	× 1000 (10^3)	3 places to right
Micro-	Milli-	÷ 1000 ($\times 10^{-3}$)	3 places to left

6. Use the fine adjustment to complete the focusing. Everything that can be viewed through a stationary ocular is referred to as a *field,* or a *field of vision.*
7. The slide can now be slowly maneuvered from side to side as the scanning technique is practiced. Observe the illumination of the object on the slide as the condenser and the amount of light that is transmitted are adjusted. The light should be adjusted until the image is easy to view.
8. To observe an object on high magnification, the following steps must be taken:
   a. Place the letter *e* in the center of the field while observing under low power.
   b. Bring the letter into sharp focus, and then turn the revolvable nosepiece until the high-power objective clicks into place. If the microscope is "parfocal," the image should be in focus under all magnifications, and coarse adjustment should not be necessary. If the microscope is not parfocal, carefully switch to high power while observing the stage from the side. Slowly lower the objective until it is nearly touching the slide. Look through the ocular, and focus upward until the object comes into focus. Refine the focus, using the fine adjustment. With increased magnification, more illumination may be necessary to see structures clearly.

9. If the microscope has an oil-immersion objective, the laboratory instructor may give instructions on how it should be used. Often it is used for demonstration only.

C. Metric System and Dimensional Analysis

The definitions for the metric units of length, mass, volume, and temperature are as follows:

meter (m)—unit of length equal to 1,650,763.73 wavelengths of the orange-red spectrum line of krypton-86 in a vacuum

gram (g)—unit of mass based on the mass of 1 cubic centimeter (cm^3) of water at the temperature of its maximum density

liter (L)—unit of volume equal to 1 cubic decimeter (dm^3) or 0.001 cubic meter (m^3)

Celsius (C)—temperature scale on which 0° is the freezing point of water and 100° is the boiling point of water (equivalent to the centigrade scale)

Conversions between different orders of magnitude in the metric system are based on powers of 10 (table 2.1). Therefore, you can convert from one order of magnitude to another simply by moving the decimal point the correct number of places to the right (for multiplying by whole numbers) or to the left (for multiplying by decimal fractions). This is illustrated in table 2.2.

Dimensional Analysis

If you are unsure about the proper factor to use in making a metric conversion, you can use the technique called *dimensional analysis*. This technique is based on two principles:

1. Multiplying a number by 1 does not change the number.
2. A number divided by itself is equal to 1.

These principles can be used to change the units of any measurement.

Example

Since 1 m is equivalent to 1,000 millimeters (mm),

$$\frac{1 \text{ m}}{1000 \text{ mm}} = 1 \text{ and } \frac{1000 \text{ mm}}{1 \text{ m}} = 1.$$

Suppose you want to convert 0.032 m to mm:

$$0.032 \text{ m} \times \frac{1000 \text{ mm}}{1 \text{ m}} = 32.0 \text{ mm}.$$

Notice that in dimensional analysis the problem is set up so that the unwanted units cancel each other. This technique is particularly useful for complex conversions or when some conversion factor cannot be recalled.

Example

Suppose you want to convert 0.1 milliliter (ml) to microliters (µl). If you remember that 1 ml = 1,000 µl, you can set up the problem as follows:

$$0.1 \text{ ml} \times \frac{1000 \text{ µl}}{1 \text{ ml}} = 100 \text{ µl}.$$

If you can remember only that a milliliter is one-thousandth of a liter and that a microliter is one-millionth of a liter, you can set up the problem this way:

$$0.1 \text{ ml} \times \frac{1.0 \text{ L}}{1000 \text{ ml}} \times \frac{1000000 \text{ µl}}{1.0 \text{ L}} = 100 \text{ µl}.$$

D. Estimation of Microscopic Size

If the magnification power of your ocular lens is 10× and you use the 10× objective lens, the total magnification of the visual field is 100×. At this magnification, the diameter of the visual field is approximately 1,600 micrometers (µm).

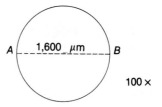

You can estimate an object's size by comparing it with the diameter of the visual field.

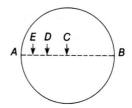

How long is line *AC* in µm? _____

How long is line *AD* in µm? _____

How long is line *AE* in µm? _____

The diameter of the field of vision using the 45× objective lens (total magnification 450×) is approximately 356 µm. Using the same diagram and procedure that you used for the 100× power, answer the following questions:

How long is line *AC* in µm? _____

How long is line *AD* in nanometers (nm)? _____

Laboratory Report 2

Name _____

Date _____

Section _____

Microscope and Metric System

Read the assigned section in the textbook before completing the laboratory report.

A. Examination of the Microscope

What is the function of each of the following components of the microscope?

1. Stage clips _____
2. Oil-immersion objective _____
3. Arm _____
4. Fine adjustment knob _____
5. Iris diaphragm _____
6. Ocular _____

B. Use of the Microscope

1. What is the procedure for focusing the microscope? What precautions must be taken when using the high-power objective lens?

2. What is the power of each of your objective lenses?
    a. Scanning lens _____
    b. Low-power lens _____
    c. High-power dry lens _____
    d. Oil-immersion lens _____

3. What is the total magnification when one uses each of the above lenses?
    a. _____
    b. _____
    c. _____
    d. _____

C. Metric System and Dimensional Analysis

1. Give the metric units for the following:

    a. The weight of one cubic centimeter of water at its maximum density _____

    b. The temperature at which water freezes _____

    c. The unit of volume equal to 0.001 cubic meter _____

2. Match the following equivalent measurements.

    _____ (1) 100 ml   (a) 100 µl

    _____ (2) 0.10 ml   (b) 0.00001 L

    _____ (3) 0.0001 ml   (c) 1.0 dl

    _____ (4) 0.01 ml   (d) 100 nl

D. Estimation of Microscopic Size

Estimate the size of an object (in micrometers) when the following conditions occur:

1. It occupies one-fourth of the visual field when the 10× objective is used:

    _____

2. It occupies one-sixth of the visual field when the 10× objective is used:

    _____

3. It occupies one-third of the visual field when the 45× objective is used:

    _____

4. It occupies three-fourths of the visual field when the 45× objective is used:

    _____

Homeostasis and Negative Feedback

Before coming to class, review the following sections in chapter 1 of the textbook: "Homeostasis and Feedback Control" and "Neural and Endocrine Regulation."

Introduction

Although body structures are functional, the study of body function involves much more than studying structure. The extent to which each organ performs its genetically programmed functions is determined by regulatory mechanisms that coordinate body functions that, in turn, serve the entire organism. Controlled by *negative feedback* mechanisms, these regulatory systems maintain a state of *homeostasis:* the dynamic constancy of the internal environment.

Objectives

Students completing this exercise will be able to:

1. Define the term *homeostasis.*
2. Explain how the negative feedback control of effectors helps maintain homeostasis.
3. Explain how sensors, integrating centers, and effectors interact in a negative feedback loop.
4. Explain how a normal range of values for the internal environment is obtained, and explain the significance of these values.

Materials

1. Watch or clock with a second hand
2. Constant-temperature water bath

A. Negative Feedback in a Constant-Temperature Water Bath

When homeostasis is disturbed—for example, by an increase or decrease in body temperature or some other parameter—a *sensor* detects the change. The sensor then activates an *effector,* which induces changes opposite to those that activated the sensor. Activation of the effector thus compensates for the initial disturbance. Homeostasis is therefore a state of ever-changing, or dynamic, constancy, rather than of absolute constancy.

The negative feedback control of a heating unit (the effector) does not allow the water bath temperature to rise or fall too far from the *set point.* The temperature of the water is at the set point only in passing; the set point is in fact only the *average* value within a *range* of fluctuating temperatures. The *sensitivity* of this particular negative feedback system is measured by the deviation in temperature from the set point that is required to activate the compensatory response (the heater turning on or off).

Procedure

1. The temperature of the water bath is set by the instructor somewhere between 40°C and 60°C.
2. A red indicator light goes on when the heating unit goes on, and it goes off when the heater turns off. In the spaces provided in the laboratory report, record the temperature of the water when the light first goes on and when the light first goes off.
3. Determine the temperature range, the set point, and the sensitivity of the water bath to deviations from the set point.
4. Record your data in the laboratory report.

B. Negative Feedback Control of Pulse Rate

The rate of the heartbeat, and thus the pulse rate, is largely controlled by the antagonistic actions of two types of nerves that innervate (serve) the heart. Impulses through the *sympathetic nerve* stimulate the heart to beat faster, and impulses through the *parasympathetic nerve* cause the heart to beat more slowly. The sensors are blood pressure receptors in the arteries, and the integrating center is in the medulla oblongata of the brain. In this exercise, each of you will demonstrate that pulse rate is in a state of dynamic constancy (implying negative feedback control) and will determine the set point of the pulse rate as the average value of the measurements taken.

Procedure

1. Gently press the index and middle fingers (not the thumb) against the radial artery in the wrist until you feel a pulse. Alternatively, the carotid pulse in the neck may be used for these measurements.
2. The pulse rate is usually expressed as pulses per minute. However, only the number of pulses per 15 seconds need be measured; multiplying this by 4 gives the number of pulses per minute. Record the number of pulses per 15-second interval in the data table provided.
3. Pause 15 seconds, and then take the pulse again in the next 15-second interval. Repeat this procedure over a 5-minute period. A total of 10 measurements of pulses per minute will thus be obtained by taking the pulse once every half-minute for 5 minutes.
4. Graph your results on the grid in the laboratory report: place a dot at the point corresponding to the pulse rate for each measurement, and then connect the dots.

C. Normal Values

Questions that students often ask include, "How does my pulse rate compare with others' pulse rates?" and "Is my measurement normal?" Keep in mind that normal values are those that normal (that is, healthy) people have. Since the values from healthy people may differ to some degree, "normal" is usually expressed as a range that spans the measurements of most healthy people. A normal range is thus a statistical determination, an estimate subject to statistical error and to the interpretation of the term *healthy*.

Health, in this context, means the absence of known cardiovascular disease. However, this applies to both endurance-trained athletes (who usually have lower-than-average pulse rates) and sedentary people (who have higher-than-average pulse rates). Therefore, normal ranges will vary depending on the proportion of each group in the sample tested; the average value and range of values of a particular group of students will probably differ somewhat from those of the general population.

Procedure

1. Determine your own average pulse rate from the data in part B, either by finding the arithmetic average or by simply observing the average value of the fluctuations in the graph. Record your own average in the laboratory report.
2. Record the number of students whose average pulse rates fall within the categories in the table in the laboratory report. Also, calculate the percentage of the class within each category, and record these in the laboratory report.
3. Divide the class into two groups: those who exercise regularly (at least three times a week) and those who do not. Determine the average pulse rate and range of values in each of these groups, and enter this information in the table in the laboratory report.

Laboratory Report 3

Name _____

Date _____

Section _____

Homeostasis and Negative Feedback

Read the assigned sections in the textbook before completing the laboratory report.

A. Negative Feedback in a Constant-Temperature Water Bath

1. Enter your data in the spaces below:
   _____ Temperature at which light goes on and heater is activated
   _____ Temperature at which light and heater go off
   _____ Temperature range permitted by negative feedback mechanism
   _____ Set point of constant-temperature water bath
   _____ Sensitivity of water bath to temperature deviations

2. Define *homeostasis,* and describe how it is maintained by negative feedback mechanisms. Draw a flow diagram, using arrows to indicate cause and effect, and to show how constant temperature is maintained in a water bath.

3. Imagine a constant-temperature water bath with two antagonistic effectors: a heater and a cooler. Using a flow diagram, show how a constant temperature would be maintained in this system.

B. Negative Feedback Control of Pulse Rate

1. Pulse rate measurements:

Measurement number	1	2	3	4	5	6	7	8	9	10
Beats per 15 seconds										
Beats per minute										

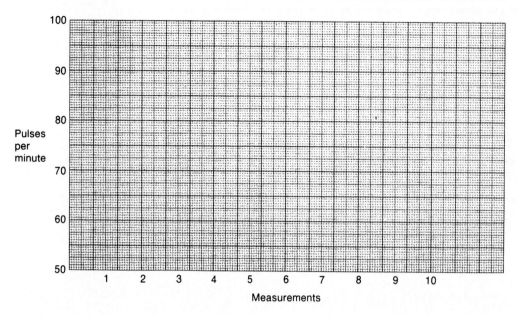

2. Explain why your graph of pulse rate measurements suggests the presence of negative feedback control mechanisms. Using the previously described effects of sympathetic and parasympathetic nerve stimulation on the heart rate, draw a flow diagram to show how these antagonistic effectors maintain dynamic constancy of the resting pulse rate.

C. Normal Values

1. Your average pulse rate: _____ beats per minute.
2. Enter the pulse rate values for the class in the tables below:

Pulse Rate (beats per minute)	Number of Students	Percentage of Total
Over 100		
90–100		
80–90		
70–80		
60–70		
50–60		
Under 50		

Averages for exercise and nonexercise groups:

	Exercise Group	Nonexercise Group
Range of pulse rates		
Average of pulse rates		

3. Describe how the normal range for a given measurement is obtained. Explain why published values for normal ranges may differ and why these values must continually be checked and updated.

Measurement of Plasma Glucose, Protein, and Cholesterol

Before coming to class, study "Carbohydrates," "Lipids," and "Proteins" in chapter 2 of the textbook.

Introduction

The plasma concentrations of various molecules, including glucose, protein, and cholesterol, are maintained within a narrow range by physiological regulatory mechanisms. Concentration measurements outside normal ranges are clinically useful when diagnosing disease states.

Objectives

Students completing this exercise will be able to:

1. Explain how Beer's law can be used to determine the concentration of molecules in solution.
2. Use the formula method and the graphic method to determine the concentration of molecules in serum samples.
3. Describe the chemical categories and characteristics of glucose, protein, and cholesterol.
4. Explain the physiological and clinical significance of measurements of plasma glucose, protein, and cholesterol.

Materials

1. Test tubes, mechanical pipettors, microliter pipettes (40 μl, 50 μl, and 100 μl capacities)
2. Constant-temperature water bath, set at 37°C
3. Colorimeter and cuvettes
4. Cholesterol kit (Medical Analysis Systems, Inc., available from Curtin-Matheson Scientific, Inc.)
5. Protein human albumin standards (purchased or previously diluted) of the following concentrations: 2, 4, 6, 8, and 10 g/dl
6. Glucose kit ("glucose liquicolor" from Stanbio, available from Curtin-Matheson Scientific, Inc.)
7. Biuret reagent: add 45 g of sodium potassium tartarate and 15 g of $CuSO_4$-$5H_2O$ to a 1.0-liter volumetric flask; fill three-quarters full with 0.2N NaOH; shake to dissolve; add 5.0 g of potassium iodide, and fill to volume with 0.2N NaOH
8. Artificial serum with known values (available from supply houses), or serum from cats or dogs (available from veterinarians)

A. Introduction to the Colorimeter

The colorimeter is a device used in physiology and in clinical laboratories to measure the concentration of a substance in solution. **Beer's law** states that the concentration of a substance in solution is directly proportional to the amount of light *absorbed* by the solution, and inversely proportional to the logarithm of the amount of light *transmitted* by the solution.

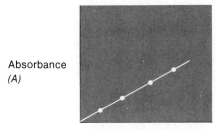

Absorbance (A)

Concentration

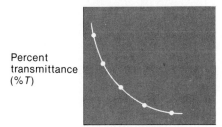

Percent transmittance (%T)

Concentration

Beer's law will be followed only if the *incident light* (the light entering the solution) is monochromatic; that is, light composed of a single wavelength. *White light* is composed of light with many wavelengths between 380 and 750 nanometers (nm), or millimicrons (mμ). The rods and cones in the eyes are stimulated by the light waves, and the brain interprets these different wavelengths as different colors.

380–435 nm violet
435–480 nm blue
480–580 nm green
580–595 nm yellow
595–610 nm orange
610–750 nm red

By means of a prism or diffraction grating, the colorimeter separates white light into its component wavelengths. The operator of this device can select incident light of any wavelength by turning the appropriate dial to that wavelength. This light enters the *cuvette,* a tube that contains the test solution. Some of the incident light is absorbed by the solution (the amount absorbed depends on the concentration of the solution), and the remainder of the light—the *transmitted* light—passes through the cuvette. The transmitted light generates an electric current in a photoelectric cell, and the amount of this current is registered on a galvanometer.

A pointer indicates the **percent transmittance (%T).** The amount of light that goes into and that leaves the solution are known, and a ratio of the two indicates the light **absorbance (A)** of that solution. In the colorimeter, the absorbance scale is adjacent to the percent transmittance scale. In this exercise, the absorbance scale is used because absorbance is directly proportional to concentration. This relationship can be described by a simple formula:

$$\frac{\text{Concentration}_1}{\text{Absorbance}_1} = \frac{\text{Concentration}_2}{\text{Absorbance}_2}$$

where 1 and 2 are different solutions.

One solution might be a sample of plasma in which the concentration of the test substance (glucose, for example) is unknown. The second solution might be a *standard,* which contains a known concentration of the test substance. If the absorbances of both solutions are determined, the unknown concentration of the substance in plasma can easily be calculated:

$$C_x = C_{std} \times \frac{A_x}{A_{std}}$$

where x = the "unknown" plasma
std = the standard solution
A = the absorbance
C = the concentration.

Suppose there are three standards. Standard 1 has a concentration of 1 gram per milliliter (1 g/ml). Standards 2 and 3 have concentrations of 2 g/ml and 4 g/ml, respectively. Since standard 2 has twice the concentration of standard 1, it should (according to Beer's law) have twice the absorbance. The second standard, similarly, should have half the absorbance of the third standard, because it has half the concentration. Unfortunately, experimental error makes this very unlikely. It is therefore necessary to average the answers obtained for the unknown concentration when different standards are used. This can be done arithmetically or by means of a graph called a **standard curve,** which is the line that "best fits" the data obtained from standards with different concentrations.

Absorbance	Concentration
0.25	30 mg/100 ml
0.38	50 mg/100 ml
0.41	60 mg/100 ml
0.57	70 mg/100 ml

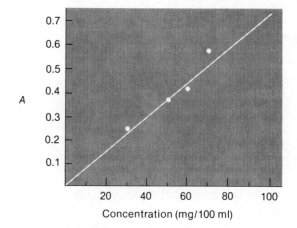

Suppose that a solution of unknown concentration has an absorbance of 0.35. The standard curve previously prepared can be used to determine its concentration.

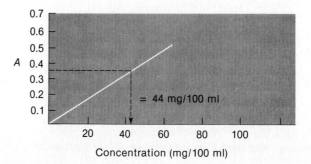

Figure 4.1 A typical colorimeter (spectrophotometer): (*a*) the sample holder; (*b*) the power switch/zero control; (*c*) the 100% *T* control; (*d*) the wavelength control (monochromator dial).

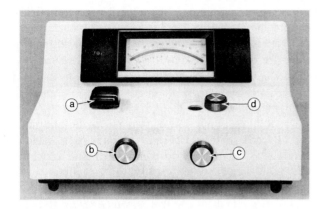

Standardizing the Colorimeter

The following procedure is intended specifically for the Spectronic 20 (Bausch & Lomb). Although the procedure is similar for all colorimeters, details may vary with different models.

1. Turn on the colorimeter by rotating the power switch/zero control (knob *b* in fig. 4.1) to the right.
2. Set the monochromator dial (knob *d* in fig. 4.1) so that the desired wavelength in nanometers is lined up with the indicator in the window. (The appropriate wavelength is provided with each exercise.)
3. When there is no cuvette in the sample holder (*a* in fig. 4.1), the light source is off. The pointer should thus read zero transmittance or infinite absorbance (at the left end of the scale). Turn knob *b* until the pointer is aligned with the left end of the scale.
4. Place in the sample holder the **blank,** which is the cuvette that contains all the reagents in the other tubes except the test substance (glucose, for example). The concentration of test substance in the blank is zero. The blank should therefore have zero absorbance or 100% transmittance (the needle should be at the right end of the scale). Set the pointer to the right end of the scale using the 100% *T* control (knob *c* in fig. 4.1).
5. Repeat steps 3 and 4 to confirm settings.
6. Place each of the other cuvettes that contain the standard solutions and the unknown in the sample holder. Close the hatch and read the absorbance of each solution.

☞ **Note:** *Before placing each cuvette in the chamber, wipe it with a lint-free, soft paper towel. If the cuvette has a white indicator line, place the cuvette so that this line is even with the line in the front of the cuvette holder.*

B. Measurement of Plasma Glucose Concentration

The brain's major energy source is glucose in the blood; consequently, numerous physiological mechanisms ensure a constancy of the blood glucose concentration. If the blood glucose levels should drop too low, glycogen (a polysaccharide of glucose) in the liver would be broken down to glucose, which would then be secreted into the blood. Following carbohydrate meals, conversely, the hormone *insulin* promotes the uptake of blood glucose into liver cells, as well as the conversion of that glucose into glycogen and fat.

Clinical Significance

The most important regulator of blood glucose is the hormone **insulin,** produced by the **pancreatic islets.** This hormone promotes the entry of glucose into the cells of the body; hence, it lowers the blood glucose level. An elevated blood glucose level (*hyperglycemia*) results from insufficient insulin secretion. This disease is called **diabetes mellitus.** Low blood glucose (*hypoglycemia*) can result from excessive insulin secretion.

Procedure

1. Obtain three test tubes and number them 1 through 3.
2. Measure 5.0 ml of glucose reagent into each tube, using the mechanical pipettor.
3. Pipette the following solutions into each of the indicated test tubes. Be sure to use different pipettes for each solution.

	Unknown	Standard	Blank
a. Sample (serum)	40 µl	—	—
b. Standard (100 mg/dl)	—	40 µl	—
c. Water	—	—	40 µl

Note: *All tubes contain equal volumes of solution. The blank consists of the reagent plus water.*

4. Mix the tubes and allow them to stand at room temperature for 10 minutes.
5. Set the monochromator (wavelength) dial at 500 nm, and standardize the colorimeter using the blank.
6. Record the absorbance values in the laboratory report.
7. Determine the concentration of the unknown plasma sample. Enter the value for the glucose concentration of the unknown sample in the laboratory report.

The normal fasting range of glucose in the plasma is 70–100 mg/dl.

C. Measurement of Plasma Cholesterol Concentration

Steroids, along with triglycerides and phospholipids, are in the category known as *lipids,* because they are nonpolar and thus not water soluble. Cholesterol is produced by various organs in the body, and it is the precursor (parent molecule) of the steroid hormones. An excessively high blood concentration of cholesterol, however, is a risk factor for the development of cardiovascular disease.

Clinical Significance

There is evidence that high blood cholesterol, together with other risk factors, such as hypertension and cigarette smoking, contributes to *atherosclerosis.* In atherosclerosis, deposits of cholesterol, other lipids, and calcium salts build up in the walls of arteries and reduce blood flow. These deposits—called *atheromas*—also serve as sites for the production of *thrombi* (blood clots), which further occlude blood flow. The reduction in blood flow through the artery may result in heart disease or cerebrovascular accident (stroke). It is generally believed that blood cholesterol levels—and thus the risk of atherosclerosis—may be significantly lowered by diets restricted in cholesterol and saturated fats.

Procedure
1. Using a mechanical pipette, add 5.0 ml of cholesterol reagent to each of three test tubes.
2. Pipette each of the following into the indicated tube:

	Unknown	Standard	Blank
a. Sample (serum)	50 µl	——	——
b. Standard (200 mg/dl)	——	50 µl	——
c. Water	——	——	50 µl
d. Reagent	5.0 ml	5.0 ml	5.0 ml

3. Mix the contents of each tube by gentle tapping, and allow the tubes to stand at room temperature for 15 minutes.
4. Transfer the solutions to three cuvettes. Standardize the spectrophotometer at 500 nm, using solution 3 as the blank.
5. Record the absorbances of solutions 1 and 2 in the laboratory report.
6. Calculate the cholesterol concentration in the unknown plasma sample. Enter this value in the laboratory report.

$$C_{plasma} = \frac{A_{plasma}}{A_{std}} \times C_{std}$$

Normal values for plasma cholesterol are 130–250 mg/dl.

D. Measurement of Plasma Protein Concentration

Proteins in the plasma serve a variety of functions: enzymes, hormones, carrier molecules (which transport lipids, iron, some hormones, and antibodies). The plasma proteins are divided into classes based on their behavior during biochemical separation procedures. These classes are the *albumins,* the *alpha* and *beta globulins* (synthesized mainly in the liver), and the *gamma globulins* (antibodies produced by the lymphoid tissue).

The total concentration of proteins in the plasma is physiologically important because proteins exert osmotic pressure, called the **colloid osmotic pressure,** which "pulls" fluid from the tissue spaces into the capillary blood. This compensates for the continuous filtration of fluid from the capillaries into the tissue spaces, which results from the hydrostatic pressure of the blood (fig. 4.2).

Clinical Significance

An abnormally low concentration of total blood protein (*hypoproteinemia*) may result from inadequate production by a diseased liver (e.g., cirrhosis or hepatitis), or from urinary loss of protein (*albuminuria*) caused by kidney diseases. Hypoproteinemia decreases the colloid osmotic pressure of the blood and may lead to **edema,** the accumulation of fluid in the tissue spaces.

An abnormally high concentration of total plasma protein (*hyperproteinemia*) may be due to dehydration or to increased production of the plasma proteins. An increased production of gamma globulins (antibodies), for example, is characteristic of many infections (e.g., pneumonia) and parasitic diseases (e.g., malaria).

Procedure
1. Obtain seven clean test tubes, and label them 1 through 7.
2. Using a mechanical pipettor, put 5.0 ml of biuret reagent in each of the tubes.
3. Pipette the following solutions into each of the indicated tubes.
   1—0.10 ml of distilled water (this is the blank)
   2—0.10 ml of 2.0 g/dl protein standard
   3—0.10 ml of 4.0 g/dl protein standard
   4—0.10 ml of 6.0 g/dl protein standard
   5—0.10 ml of 8.0 g/dl protein standard
   6—0.10 ml of 10.0 g/dl protein standard
   7—0.10 ml of unknown serum sample
4. Let the tubes stand at room temperature for 30 minutes.
5. Transfer the solutions to seven clean cuvettes.
6. Standardize the spectrophotometer at 555 nm, using solution 1 as the blank.
7. Record the absorbance of each solution in the laboratory report.

Figure 4.2 The circulation of fluid between the blood plasma in a capillary and the tissues. Arrows pointing away from the capillary indicate the force exerted by the blood pressure, whereas arrows pointing toward the capillary indicate the force exerted by the colloid osmotic pressure of the plasma proteins. The heavy arrow indicates the direction of blood flow along the capillary from arteriole to venule.

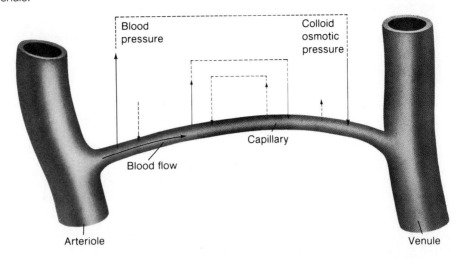

8. Using the absorbances of solutions 2 through 6, plot a standard curve.
9. Determine the concentration of protein in the unknown serum sample, and enter this value in the laboratory report.

> The normal fasting protein level is 6.0–8.0 g/dl.

Laboratory Report *4*

Name _____

Date _____

Section _____

Measurement of Plasma Glucose, Protein, and Cholesterol

Read the assigned sections in the textbook before completing the laboratory report.

A. Introduction to the Colorimeter

1. The concentration of a solution is _____ to its absorbance.
2. The above relationship is described by _____ law.
3. What does the blank tube contain, and what is its function in a colorimetric assay?

4. Why does one always draw a linear standard curve (absorbance vs. concentration) even though one's experimental values may deviate slightly from a straight line? Why must the line one draws intersect the origin (zero concentration = zero absorbance)?

B. Measurement of Plasma Glucose Concentration

1. Enter the absorbance data in the following table:

Tube Number	Glucose Concentration (mg/dl)	Absorbance
1	0 (*blank*)	0
2	100	
3	Unknown (*serum sample*)	

2. Calculate the glucose concentration of the unknown serum sample. Enter this value in the space below:
   _____ mg/dl
3. Is the glucose concentration of the serum sample normal or abnormal? Explain.
4. Glucose is in the general chemical category known as _____ ; more specifically, it is a(n) _____.
5. Hyperglycemia is characteristic of the disease _____.
6. Hyperglycemia may be caused by a deficiency in the hormone _____.
7. Although the ingestion of carbohydrates fluctuates greatly, the blood glucose concentration remains within a fairly narrow range. How is this accomplished?

C. Measurement of Plasma Cholesterol Concentration

1. Record the absorbance data in the following table:

Tube Number	Glucose Concentration	Absorbance
1	0 (*blank*)	0
2	200 mg/dl	
3	Unknown (*serum sample*)	

2. Calculate the cholesterol concentration of the unknown serum sample, using the formula shown in step 6 of the procedure. Enter this concentration in the space below:

   _____ mg/dl

3. Is the cholesterol concentration of the serum sample normal or abnormal? Explain.

4. Cholesterol is in the general chemical category known as _____ ; more specifically, it is a(n) _____.

5. High blood cholesterol is a contributing factor in the disease _____.

6. All fats are lipids, but not all lipids are fats. Explain.

D. Measurement of Plasma Protein Concentration

1. Record the absorbance of each solution in the table below:

Tube Number	Protein Concentration (g/dl)	Absorbance
1	0 (*blank*)	0
2	2.0	
3	4.0	
4	6.0	
5	8.0	
6	10.0	
7	Unknown (*serum sample*)	

2. Use the graph paper at the end of the exercise to plot a standard curve.

3. Using this standard curve, determine the protein concentration of the unknown serum sample, and enter this value in the space below:

_____ g/dl

4. Is the protein concentration of the serum sample normal or abnormal? Explain.

5. Proteins are composed of subunits called _____ , which are joined by _____ bonds.

6. The specific folding and bending of a protein is known as its _____ structure.

7. Most of the plasma proteins are produced by the _____ .

8. Low plasma protein concentration can produce a physical condition known as _____ .

Cell Structure and Cell Division

Before coming to class, review the structure and function of cellular organelles, as well as the discussion of mitosis and meiosis in chapter 3 of the textbook.

Introduction

The cell is the basic unit of structure and function in the body. The structure of tissues and organs results from the structure and organization of the different types of cells that exist within these larger units. Similarly, the functions of the body are ultimately the result of cellular activities. A knowledge of cell structure and function is thus a prerequisite to the study of the anatomy and physiology of the human body.

Objectives

Students completing this exercise will be able to:

1. Identify the major parts of a cell.
2. Describe the functions of the different organelles within a cell.
3. List the different stages of mitosis, and describe the characteristics of each stage.
4. Describe the stages of meiosis, and understand the significance of meiosis and mitosis.

Materials

1. Compound microscope
2. Prepared slides of whitefish blastula
3. Lens paper
4. Reference text
5. Colored pencils

A. Structure of a Cell

Identify the following cellular structures in figure 5.1 by labeling and coloring with colored pencils. You might use blue-to-purple for chromatin and pink for cytoplasmic structures, because that is how cells appear when standard mixtures of the hematoxylin and eosin (H & E) stain are used for microscope slide preparation.

Cytoplasm
    Centriole
    Endoplasmic reticulum
    Golgi apparatus
    Lysosome
    Microtubules
    Mitochondria
    Ribosomes
    Small vacuole
Nucleus
    Chromatin
    Nuclear membrane
    Nucleolus
Cell membrane

B. Cell Division

1. In figure 5.2, label the stages of mitosis—interphase, prophase, metaphase, anaphase, and telophase. Also label the following: spindle fibers, daughter cells, nucleus, centrioles, and centromeres.
2. Color and label the nuclear membrane, centrioles, and spindle fibers. Also, color-code the homologous chromosomes, using different colors for each pair.
3. Obtain a slide of a whitefish blastula (an early embryonic stage of development), and observe the different stages of mitosis.
4. Label the different stages of meiosis in figure 5.3. Color-code the homologous chromosomes as you did for figure 5.2.

Figure 5.1 The cell.

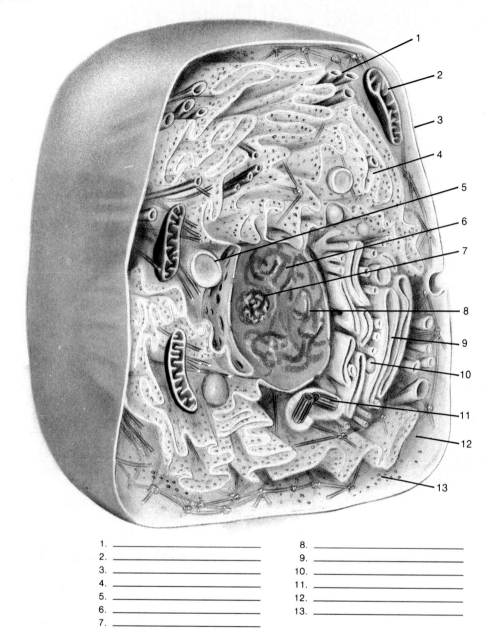

1. _____	8. _____
2. _____	9. _____
3. _____	10. _____
4. _____	11. _____
5. _____	12. _____
6. _____	13. _____
7. _____	

Figure 5.2 The sequence of mitosis. Label the stages using the text as a guide.

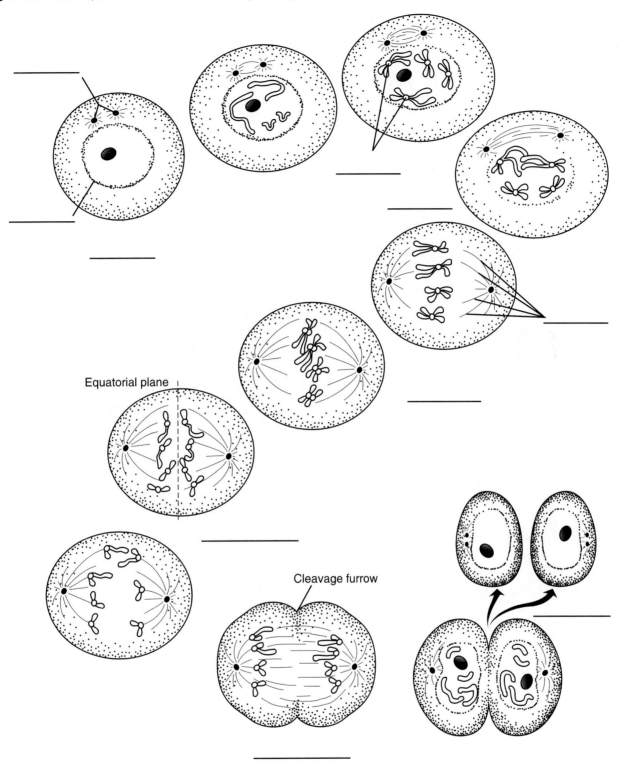

Equatorial plane

Cleavage furrow

Figure 5.3 The process of meiosis. Label the stages using the text as a guide.

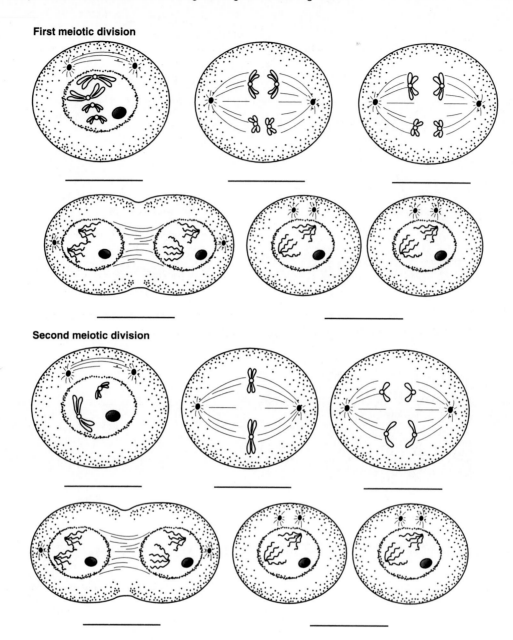

First meiotic division

Second meiotic division

Laboratory Report 5

Name _____

Date _____

Section _____

Cell Structure and Cell Division

Read the assigned sections in the textbook before completing the laboratory report.

A. Structure of a Cell

Identify the organelle that fits each of the following descriptions:

1. A prominent role in cell division _____
2. The major site of energy production in the cell _____
3. A system of membranous sacs in the cytoplasm _____
4. The location of genetic information _____
5. The vesicle that contains digestive enzymes _____
6. The site of protein synthesis _____
7. Regulates the entry of substances into the cell _____

B. Cell Division

1. Match the following events of mitosis with the names of the stages:

____ (1) The nuclear membrane disappears; spindles appear

____ (2) Chromosomes line up at the equator of the cell

____ (3) Duplicated chromosomes separate and are pulled toward the centrioles

____ (4) Chromosomes elongate into threads; nuclear membranes reappear

(a) metaphase
(b) telophase
(c) anaphase
(d) prophase

2. Compare the processes of mitosis and meiosis, using the following criteria:

	Mitosis	Meiosis
Organs where process occurs	_____	_____
Number of chromosomes in end product	_____	_____
Number of stages or steps involved	_____	_____

3. What functions do mitosis and meiosis serve in the human organism?

Genetic Control of Metabolism

Before coming to class, review "Genetic Transcription—RNA Synthesis" and "Formation of a Polypeptide" in chapter 3 of the textbook. Also review "Metabolic Pathways" and "Metabolic Disturbances" in chapter 4 of the textbook.

Introduction

The metabolism of the body is controlled to a large degree by the activity of many enzymes within each cell. These enzymes can produce sequential chemical changes that make up metabolic pathways. Because the production of each enzyme is coded by a gene specific for that enzyme, a defective gene can cause a specific disruption in a metabolic pathway. Such inborn errors of metabolism can be detected by tests for particular enzyme products.

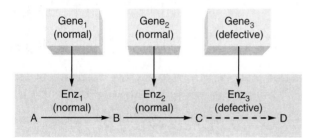

Objectives

Students completing this exercise will be able to:

1. Define the terms *genetic transcription* and *genetic translation.*
2. Describe how enzymes cooperate to produce metabolic pathways.
3. Describe the cause of inborn errors of metabolism.
4. Explain the causes of phenylketonuria (PKU) and albinism.

Materials

1. Test tubes, Pasteur pipettes (droppers), urine collection cups
2. Urine samples
3. Phenistix (Ames), silver nitrate (3 g/dl), 10% ammonium hydroxide, 40% sodium hydroxide

A. Phenylketonuria

A person with *phenylketonuria* (*PKU*) lacks the enzyme needed to convert the amino acid phenylalanine into the amino acid tyrosine. The accumulated phenylalanine is converted instead to phenylpyruvic acid, which is excreted in the urine. High blood levels of phenylpyruvic acid in an infant can affect its developing central nervous system and produce mental retardation. These effects can be avoided by placing the child with PKU on a diet low in phenylalanine.

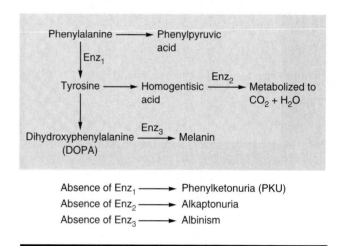

Clinical Significance

Since we cannot at present remove defective genes from people and replace them with good genes, the only course available involves early, correct diagnosis and prevention. For this reason, newborn babies are routinely tested for PKU and placed on low phenylalanine diets if they have this disease. Unfortunately, many other inborn errors of metabolism cannot be treated by dietary restrictions; the only way to prevent some of these diseases is by genetic counseling of potential parents who are carriers of the disease.

When there is a defective enzyme in a metabolic pathway, the substrate molecule of that enzyme accumulates. Such inborn errors of metabolism are not restricted to enzymes involved in amino acid metabolism. In *Tay–Sachs disease,* for example, the defect is in the enzyme that breaks down a complex lipid in the axon sheaths around neurons, which results in lipid accumulation in the brain and retina. Found predominately in Ashkenazi Jews, this disease results in death by age 4.

There are also errors in carbohydrate metabolism: in the **glycogen storage diseases,** for example, an enzyme that breaks down glycogen is defective. This results in the excessive accumulation of glycogen, which causes liver disease.

Procedure

1. Dip the test end of a Phenistix[1] strip into a sample of urine.
2. Compare the color of the strip with the color chart provided.

B. Alkaptonuria

A person with alkaptonuria excretes large amounts of homogentisic acid, which reacts with silver to form a black precipitate. This condition does not appear to have any adverse effects on the health of the individual.

Procedure

1. Add 10 drops of urine to a test tube containing 5 drops of 3% silver nitrate ($AgNO_3$).
2. Add 5 drops of 10% ammonium hydroxide (NH_4OH) to the tube and mix.
3. A positive test is indicated by the presence of a black precipitate.

[1]From Ames Laboratories.

Laboratory Report 6

Name _____

Date _____

Section _____

Genetic Control of Metabolism

Read the assigned sections in the textbook before completing the laboratory report.

1. Define the following terms:

    a. Genetic transcription _____

    b. Genetic translation _____

    c. Metabolic pathway _____

2. Explain, in general terms, how a genetic defect can result in the production of a defective enzyme and thus in an inborn error of metabolism.

3. Draw a schematic illustration to show the metabolic pathways by which the amino acid phenylalanine is broken down. Using this diagram, explain how PKU, alkaptonuria, and albinism are produced.

Diffusion, Osmosis, and Tonicity

Before coming to class, review the following sections in chapter 5 of the textbook: "Diffusion," "Diffusion through the Cell Membrane," and "Osmosis."

Introduction

The cell membrane is selectively permeable; it allows some substances to pass through but prevents the passage of other substances. In some cases, the degree of permeability varies according to physiological conditions. If the membrane is permeable to a given molecule or ion, the direction of the net movement of that molecule or ion through the membrane will be determined by the concentration gradient. The net diffusion of water through a membrane, called **osmosis,** follows the general laws governing the diffusion of any substance.

If two solutions of different concentrations are separated by a membrane that is impermeable to the solutes but permeable to water, there will be a net diffusion of water—osmosis—from the more dilute to the more concentrated solution. This is illustrated in figure 7.1.

Objectives

Students completing this exercise will be able to:

1. Define the terms *diffusion* and *passive transport.*
2. Define the terms *osmosis* and *osmolality,* and describe the meaning of *osmotic pressure.*
3. Calculate the osmolality of solutions when the concentration (in weight per volume) and molecular weight of a solute are known.
4. Describe *isotonic, hypertonic,* and *hypotonic solutions,* and explain how cells respond to solutions of these tonicities.

Materials

1. Test tubes, thistle tubes, dialysis tubing
2. Beakers, ring stands, burette clamps
3. Lancets, alcohol swabs
4. Microscopes, slides, cover slips, and Pasteur pipettes (droppers)
5. Sucrose (30 g/dl) and sodium chloride solutions (0.20 g, 0.45 g, 0.85 g, 3.5 g, and 10 g per dl each)

Figure 7.1 A model of osmosis. The solution on the left is more dilute than the one on the right, causing osmosis to occur from left to right across the semipermeable membrane.

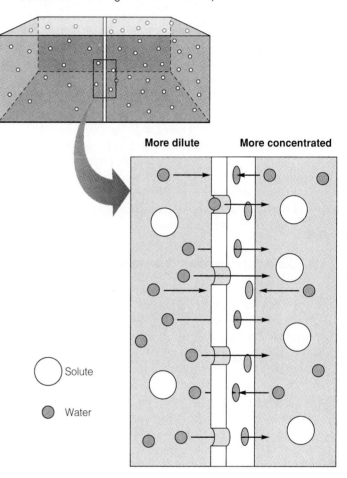

More dilute   **More concentrated**

○ Solute

● Water

A. Osmosis across a Synthetic Semipermeable Membrane

1. Cut a 2½-inch piece of dialysis tubing. Soak this piece in tap water until the layers separate. (You can speed this process by rotating the tubing between two fingers.) Slide one blade of a scissors between the two sides, and cut along the margin, producing a single rectangular sheet of dialysis tubing.

☞ **Note:** *Dialysis tubing is a plastic porous material that can be used to separate molecules on the basis of their size. It is a synthetic semipermeable membrane. Molecules that are larger than the pore size remain inside the tubing; smaller molecules (including water) can diffuse through the membrane. This technique of biochemical separation is called* dialysis.

2. Hold a thistle tube vertically with the mouth upward. While one person closes the lower opening with one finger, another pours the 30% sucrose solution into the mouth of the tube until the solution is about to overflow the tube.

3. Place the rectangular piece of dialysis tubing tightly over the mouth of the thistle tube so that no air is trapped between the dialysis tubing and the sucrose. Keeping the dialysis tubing taut, secure it to the thistle tube with several wrappings of a rubber band.

4. Invert the thistle tube and check for leaks. If leaks are observed, remove the dialysis tubing and repeat step 3.

5. With the thistle tube inverted, immerse it in a larger beaker of water (fig. 7.2). Secure the thistle tube in this position, using a ring stand and a burette clamp. A folded wad of paper towel will enable you to safely clamp the narrow part of the thistle tube.

6. Mark the meniscus of the 30% sucrose solution with a marking pen and, in the laboratory report, record the change in the level of this meniscus (in centimeters) every 15 minutes for 1 hour.

B. Osmosis across the Red Blood Cell Membrane

Osmosis depends on the difference in solute concentrations of two solutions separated by a membrane. The units that best describe concentration are *molal* and *osmolal*. A 1-molal (M) solution contains 1 mole of solute dissolved in 1 liter of water. If the molal concentrations of different solutes in a given solution are added, the total solute concentration (total moles per liter of water) is given in units of osmolality (Osm). In physiology and medicine, these units are usually expressed in units of milliosmolality (mOsm), 1 Osm = 1000 mOsm. Plasma, for example, has a total solute concentration of approximately 290 mOsm.

The red blood cell (RBC) has the same osmolality and thus the same osmotic pressure as blood plasma. When a RBC is placed in a hypotonic solution, it will expand or perhaps even burst (*hemolysis*) because of the influx of water, extruding hemoglobin into the solution. When a RBC is placed in a hypertonic solution, it will shrink (*crenate*) as a result of the efflux of water (fig. 7.3).

Red blood cells can thus be used as *osmometers* to determine the osmolality of blood plasma, since the cells will neither expand nor shrink in an isotonic solution.

Figure 7.2 A thistle tube setup for the osmosis exercise.

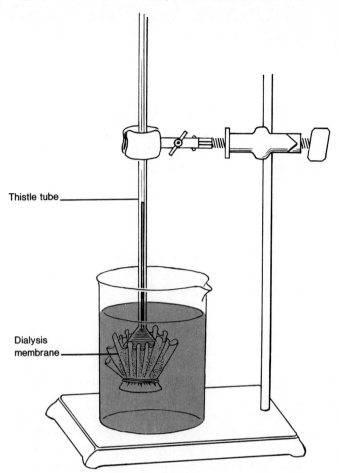

Thistle tube _____

Dialysis membrane _____

Figure 7.3 A scanning electron micrograph of a crenated red blood cell attached to a fibrin thread. Note the notched or scalloped appearance. From Kessel, R. G., and Kardon, R. H.: *Tissues and Organs: A Text-Atlas of Scanning Electron Microscopy.* © W. H. Freeman and Company, 1979.

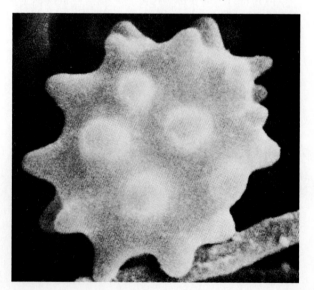

Clinical Significance

Osmolality determines the distribution of fluids between the intracellular compartment and the extracellular compartment of the body. An accumulation of fluid in the tissues (*edema*), for example, can result from increased osmolality of the tissue spaces due to protein accumulation. Blood volume and, therefore, blood pressure are maintained by an osmotic equilibrium between the blood plasma and the tissue fluids. Intravenous fluids infused to maintain blood volume and pressure must be isotonic to prevent the expansion or crenation of the body cells.

Procedure

1. Measure 2.0 ml of the solutions indicated in part B of the laboratory report into each of five numbered test tubes.
2. Wipe the tip of your finger with alcohol and, using a sterile lancet, prick your finger to draw a drop of blood.

☞ **Note:** *Caution must be exercised when handling blood, because of the possibility of contracting the AIDS virus and other infectious agents. Handle only your own blood, and discard all objects containing blood into the receptacles provided by the instructor.*

3. Allow the drop of blood to drain down the side of test tube 1. Mix the blood with the saline (salt) solution by inverting the test tube a few times.
4. Repeat the above procedure for test tubes 2 through 5. Additional drops of blood can be obtained by milking the finger.
5. Using a transfer pipette, place a drop of solution 1 on a slide, and cover it with a cover slip. Observe the cells using the 45× objective.
6. Repeat step 5 for the other solutions, and record your observations in the laboratory report.

Laboratory Report 7

Name _____

Date _____

Section _____

Diffusion, Osmosis, and Tonicity

Read the assigned sections in the textbook before completing the laboratory report.

A. Osmosis across a Synthetic Semipermeable Membrane

1. Enter your data in the table below:

Time	Distance Meniscus Has Moved
15 minutes	
30 minutes	
45 minutes	
60 minutes	

2. Explain your results.

3. Define the term *osmosis*. _____

4. Explain why osmosis is considered to be passive membrane transport. What factors regulate the rate of osmosis across a membrane?

B. Osmosis across the Red Blood Cell Membrane

1. Enter your data in the table below:

Tube and Contents	Molality	Milliosmolality	Appearance of RBC	Diameter of RBC (micrometers)
10 g/dl NaCl				
3.5 g/dl NaCl*				
0.85 g/dl NaCl	0.145m	290 mOsm		
0.45 g/dl NaCl				
0.20 g/dl NaCl				

*Approximately the concentration of seawater.

2. Which solution is isotonic? _____

3. Which solutions are hypotonic? _____

4. Which solutions are hypertonic? _____

5. Define the term *isotonic*. _____

6. Red blood cells _____ in a hypertonic solution.

7. A 0.10M NaCl solution is _____ (iso-/hypo-/hypertonic) to a 0.10M glucose solution.

8. Suppose a salt and a glucose solution are separated by a membrane that is permeable to water but not to the solute. The NaCl solution has a concentration of 1.95 g/250 ml (molecular weight = 58.5). The glucose solution has a concentration of 9.0 g/250 ml (molecular weight = 180). Calculate the molality, millimolality, and milliosmolality of both solutions. Describe the direction in which osmosis occurs (if it does), and explain your answer.

9. Suppose a man crashes on a desert island. Trace the course of events leading up to his sensation of thirst. Can he satisfy his thirst by drinking seawater? Explain.

10. Before the invention of refrigerators, pioneers preserved meat by salting it. Explain why this procedure can preserve meat.

Histology

Before coming to class, review the following sections in chapter 6 of the textbook: "Epithelial Tissues," "Connective Tissues," "Muscle Tissues," and "Nervous Tissues."

Introduction

Even though all cells have certain basic similarities, they differ considerably in size, structure, and function. An aggregation of cells that are similar in structure and that function together is called a **tissue.** *Histology* is the study of tissue structure and function. Tissues are classified into four principal types based on cell arrangement, shape, and function. The four principal types are *epithelial tissues, connective tissues, muscular tissues,* and *nervous tissues.*

Objectives

Students completing this exercise will be able to:

1. Define a *tissue,* and explain how tissues are related to organs.
2. List the four principal types of tissues and describe the structural and functional characteristics of each.

Materials

1. Microscopes and lens paper
2. Prepared slides of the different tissues to be observed
3. Histology atlas
4. Colored pencils

A. Epithelial Tissues

1. Obtain microscope slides of different types of epithelia. Observe these, using the 10× objective (total magnification = 100×). Each slide can be examined with the high-dry (45×) lens after it has been observed with the low objective.

2. Identify and label the following types of epithelia in figure 8.1:

Ciliated columnar epithelium
Columnar epithelium
Cuboidal epithelium
Pseudostratified ciliated columnar epithelium
Simple squamous epithelium
Stratified squamous epithelium
Transitional epithelium

B. Connective Tissues

1. Obtain prepared slides of different types of connective tissues and observe these slides by first using the 10× and then the 45× objective lens.
2. Identify and label the following types of connective tissues and their structures in figures 8.2 through 8.6. Where applicable, color the ground substance, fibers, and cells so that they can be more easily distinguished.

Mesenchymal tissue: homogenous matrix, mesenchymal cell
Connective tissue (proper): adipose, dense fibrous, elastic, loose connective (areolar), reticular
Cartilage tissue: elastic (chondrocyte, elastic fibers), fibrocartilage (lacuna, collagenous bundles), hyaline (lacuna, chondrocyte)
Bone tissue: canaliculi, central canal, lacuna, lamella, osteocyte within a lacuna
Vascular tissue: erythrocyte (red blood cell), leukocyte (white blood cell: eosinophil, monocyte, erythrocyte, basophil, neutrophil, thrombocyte, lymphocyte), thrombocyte (platelet)

C. Muscular Tissues

1. Obtain microscope slides of muscle tissues and observe them under first the 10× and then the 45× objective lens.
2. Identify and label the following types of muscle tissues in figure 8.7:

Cardiac: intercalated disc, nucleus
Skeletal: nucleus, striations
Smooth: nucleus

D. Nervous Tissues

1. Obtain slides of nervous tissues and observe them under first the 10× and then the 45× objective lens.
2. Identify the following in figure 8.8:

   Neuroglial cells
   Neuron
   Axon

   Branches of an axon
   Cell body
   Dendrite
   Nucleus
   Schwann cell (neurolemmocyte)

3. Examine figure 8.9 and study the different tissues depicted.

Figure 8.1 Epithelial tissues.

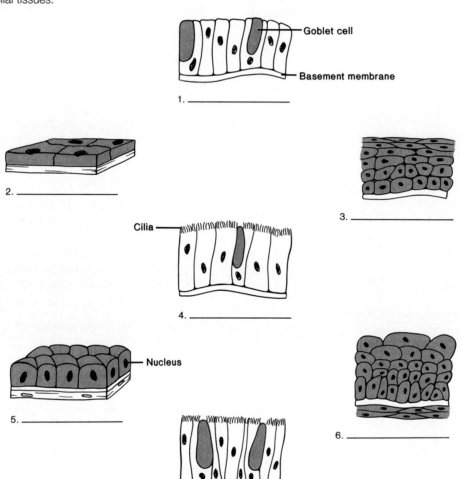

Goblet cell

Basement membrane

1. _____

2. _____

3. _____

Cilia

4. _____

5. _____

Nucleus

6. _____

7. _____

Figure 8.2 Mesenchymal tissue.

1. _____
2. _____

Figure 8.3 Connective tissue proper.

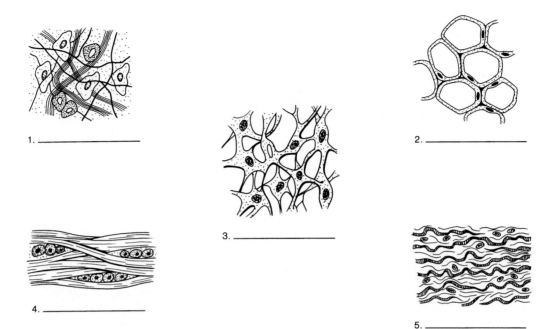

1. _____

2. _____

3. _____

4. _____

5. _____

Figure 8.4 Cartilage tissue.

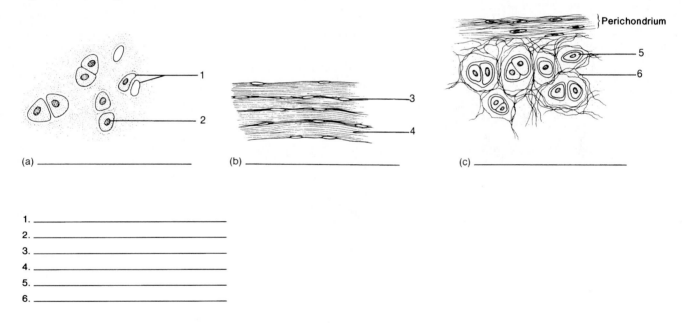

Perichondrium

(a) _____

(b) _____

(c) _____

1. _____
2. _____
3. _____
4. _____
5. _____
6. _____

Figure 8.5 Bone tissue.

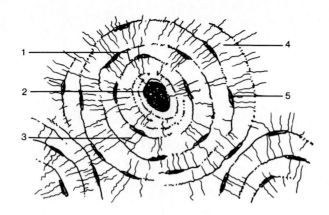

1. _____
2. _____
3. _____
4. _____
5. _____

Figure 8.6 Vascular tissue.

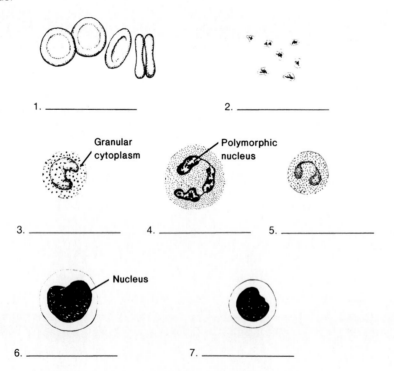

1. _____   2. _____

Granular cytoplasm   Polymorphic nucleus

3. _____   4. _____   5. _____

Nucleus

6. _____   7. _____

Figure 8.7 Muscular tissue.

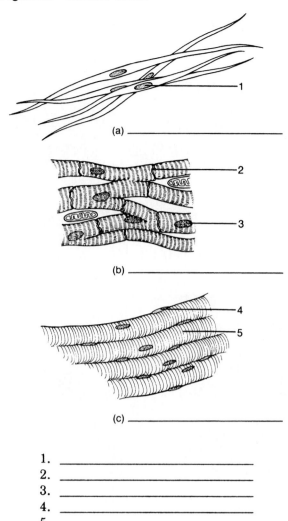

(a) _____

(b) _____

(c) _____

1. _____
2. _____
3. _____
4. _____
5. _____

Figure 8.8 Nervous tissue.

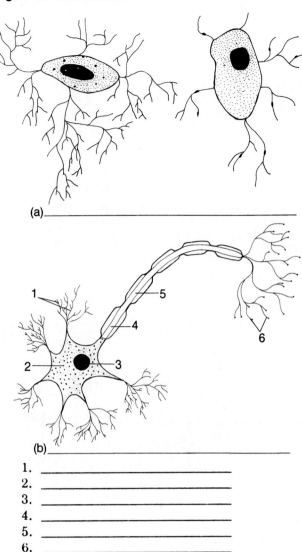

(a) _____

(b) _____

1. _____
2. _____
3. _____
4. _____
5. _____
6. _____

Figure 8.9 Histological examples of common types of tissues: (*a*) simple cuboidal epithelium; (*b*) pseudostratified ciliated columnar epithelium; (*c*) simple columnar epithelium; (*d*) loose connective tissue; (*e*) dense connective tissue; (*f*) bone (osseous) tissue; (*g*) stratified squamous epithelium; (*h*) a cross section of nerve; (*i*) blood (*j*) cardiac muscle.

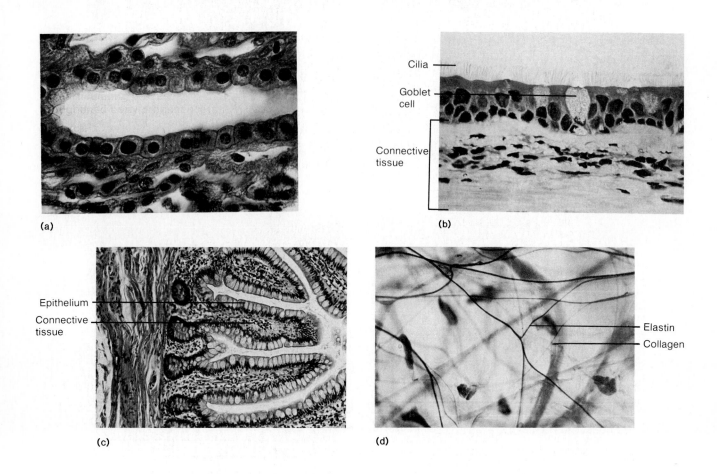

(a)

Cilia

Goblet cell

Connective tissue

(b)

Epithelium

Connective tissue

(c)

Elastin

Collagen

(d)

Figure 8.9 Continued

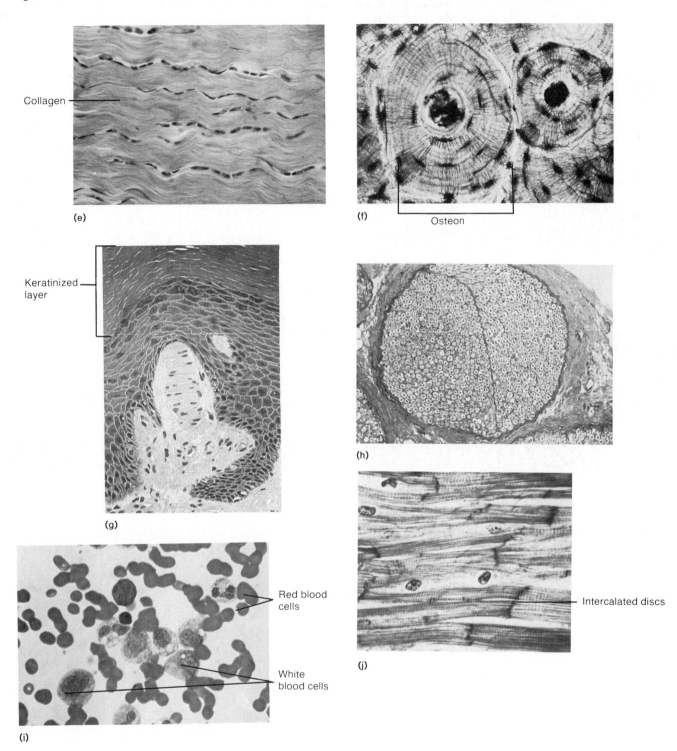

Collagen

(e)

Osteon

(f)

Keratinized layer

(g)

(h)

Red blood cells

White blood cells

(i)

Intercalated discs

(j)

Laboratory Report 8

Name _____

Date _____

Section _____

Histology

Read the assigned sections in the textbook before completing the laboratory report.

A. Epithelial Tissues

1. Define *tissue,* and discuss the relationship between tissues and organs.

2. Discuss how cell shape and cell layering are used as criteria in the classification of epithelial membranes.

B. Connective Tissues

1. What is the distinguishing feature of all connective tissues? How does this differ from epithelial tissues?

2. List the major categories of connective tissues.

3. List the subcategories of cartilage, and indicate at least one body location where each is found.

4. Explain why blood is classified as a connective tissue. What is the intercellular matrix of blood? What are the blood cells?

C. Muscular Tissues

1. List the three types of muscle tissue, and indicate where in the body each is found.

2. Compare and contrast the structure of the three types of muscle tissue.

D. Nervous Tissues

1. What are the two types of cells that form the nervous system? Which cell type is more abundant? Which type produces and conducts nerve impulses?

2. Define a *cell process*. Name the two types of processes that extend from a neuron cell body, and describe their function.

Support and Movement of the Human Body

The following exercises are included in this unit:

9. Integumentary System
10. Structure of Bone Tissue
11. Skeletal System: The Axial Skeleton
12. Skeletal System: The Appendicular Skeleton
13. Articulations
14. Neural Control of Skeletal Muscle Contraction
15. Summation, Tetanus, and Fatigue
16. Muscles of the Head and Neck
17. Muscles of the Pectoral Girdle and Upper Extremity
18. Muscles of the Trunk, Pelvic Girdle, and Lower Extremity
19. Cat Musculature

These exercises are based on information presented in the following chapters of *Concepts of Human Anatomy and Physiology,* 4th ed.:

7. The Integumentary System
8. Skeletal System: Bone Tissue and Development
9. Skeletal System: The Axial Skeleton
10. Skeletal System: The Appendicular Skeleton
11. Articulations
12. Muscle Tissue and Muscle Physiology
13. Muscular System

The skin, the largest organ in terms of area, offers protection as the outer boundary of the body. It also has functions that are less obvious; for example, it produces vitamin D, and it helps maintain a constant core body temperature. The skeletal system and the skeletal muscle system function together to produce most of the movements of the human body. Unfortunately, the physiology of these systems is difficult to study experimentally in a student laboratory. Aside from some basic muscle physiology exercises, unit 2 of this manual is primarily devoted to the anatomy of these systems.

In this unit, you will study in detail the anatomy of the skin, the skeleton, and the system of skeletal muscles. Use the figures and photographs to aid your dissection and examination of the cat and the human cadaver (if available). Also, use the textbook to help you "fill in the blanks" of the line diagrams in these exercises. Actually writing in the labels will help you learn and retain the new terms, and it will help you study laboratory material when you are not in the laboratory.

Integumentary System

Before coming to class, review the following sections in chapter 7 of the textbook: "The Integument as an Organ," "Layers of the Integument," "Functions of the Integument," and "Epidermal Derivatives."

Introduction

The skin and its epidermal structures (*hair, glands,* and *nails*), constitute the integumentary system. In certain body areas it is adaptively modified and has protective or metabolic functions. The skin (integument) is a dynamic interface between the constantly changing external environment and the body's internal environment, and it helps maintain homeostasis.

Objectives

Students completing this exercise will be able to:

1. Classify the skin as an organ and a component of the integumentary system.
2. List the basic functions of the skin.
3. Identify the layers of skin by discussing their location, structure, and function.
4. Describe the accessory structures of the integumentary system, such as hair, glands, and nails.

Materials

1. Microscope
2. Prepared slides of human skin (showing hair and associated glands) and a longitudinal section of fingertip
3. Magnifying glass
4. Reference text
5. Drawing pencils and colored pencils
6. Ink pad and black ink (washable)

A. Layers of the Skin

1. Identify and label the following structures in figure 9.1:

   Epidermis
   Stratum basale
   Stratum corneum
   Stratum granulosum
   Stratum lucidum (not shown)
   Stratum spinosum
   Dermis
   Hypodermis
   Papillary layer
   Reticular layer

2. Using colored pencils, color-code the epidermis, dermis, and hypodermis.
3. Examine a prepared microscope slide of the skin, first on low and then on high power; identify as many structures as possible.
4. Record an ink print of the thumb and middle finger in the laboratory report. (Firmly press the thumb and middle finger on an ink pad and then, with moderate pressure, roll the surface of the thumb and middle finger on the paper.)

B. Epidermal Derivatives

1. Identify and label the following items in figures 9.2, 9.3, and 9.4:

   Hair
      Arrector pili muscle
      Bulb
      Hair follicle
      Root
      Shaft
      External root sheath
      Internal root sheath
      Matrix
      Dermal papilla

Nail
    Body
    Eponychium
    Hyponychium
    Matrix
    Hidden border
Glands
    Apocrine sweat gland
    Eccrine sweat gland
    Sebaceous gland
       Duct of sebaceous gland

2. Using colored pencils, color-code the sweat glands, sebaceous glands, and arrector pili muscles in figures 9.1 and 9.4. Color-code the external and internal root sheaths in figure 9.2.
3. Examine a fingernail with a magnifying glass, and identify as many structures as possible.
4. Examine a microscope slide of a longitudinal section of a fingertip that shows the nail and associated structures.

Figure 9.1 A section of skin.

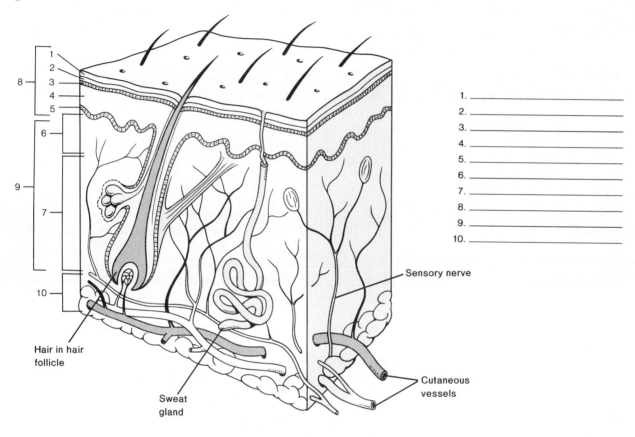

1. _____
2. _____
3. _____
4. _____
5. _____
6. _____
7. _____
8. _____
9. _____
10. _____

Sensory nerve

Hair in hair follicle

Sweat gland

Cutaneous vessels

Figure 9.2 A diagram of a hair, a hair follicle, a sebaceous gland, and an arrector pili muscle.

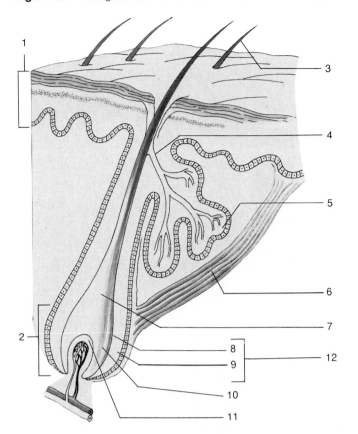

1. _____
2. _____
3. _____
4. _____
5. _____
6. _____
7. _____
8. _____
9. _____
10. _____
11. _____
12. _____

Figure 9.3 Photomicrograph of a fingernail of a human neonate.

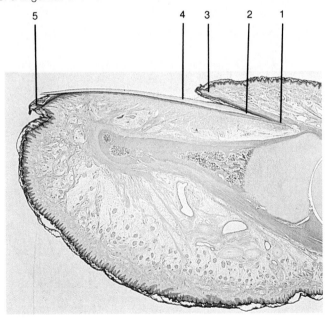

1. _____

2. _____

3. _____

4. _____

5. _____

Figure 9.4 Types of skin glands.

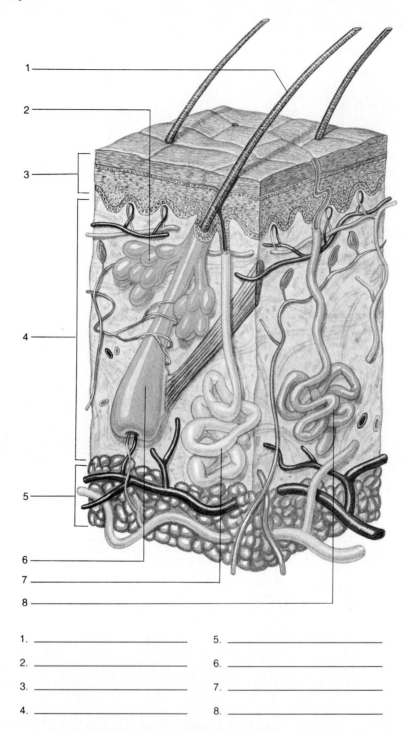

1.
2.
3.
4.
5.

6.
7.
8.

1. _____ 5. _____

2. _____ 6. _____

3. _____ 7. _____

4. _____ 8. _____

Laboratory Report *9*

Name _____

Date _____

Section _____

Integumentary System

Read the assigned sections in the textbook before completing the laboratory report.

A. Layers of the Skin

1. Make an ink print of your middle finger and thumb in the space provided.

2. Examine the fingerprints of other students. How many fingerprint types can be identified?

3. Describe the structure of thick skin and thin skin. On what parts of the body is the skin thickest? Why is thick skin located in these areas?

4. Explain the functions of (a) the stratum basale and (b) the stratum corneum and keratin.

5. Discuss the process of suntanning. What causes suntans to occur, and of what value are they? In what ways can excessive exposure to sunlight be harmful?

B. Epidermal Derivatives

1. Make a simple sketch of a hair follicle, showing its relationship to the arrector pili muscle and sebaceous glands.

2. Contrast sebaceous glands and sweat glands in terms of structure, function, and location in the body.

3. What is acne?

Structure of Bone Tissue

Before coming to class, review the following sections in chapter 8 of the textbook: "Organization of the Skeletal System," "Functions of the Skeletal System," "Development of the Skeletal System," and "Bone Growth."

Introduction

Bone contains living osteocytes, blood vessels, and nerves; these are specifically arranged to enable them to remain alive—despite the fact that they are entombed within a calcified matrix.

Objectives

Students completing this exercise will be able to:

1. State the principal functions of the skeleton.
2. Diagram and label the histological structure of a bone.
3. Describe the different surface features of bones.
4. Discuss the process of ossification and the mechanisms of bone growth.

Materials

1. Histological slides of bone
2. Longitudinally sectioned femur showing compact and spongy bone
3. Colored pencils
4. Reference text

Structure of Bones

1. Identify and label the following structures and regions in figures 10.1, 10.2, and 10.3, as well as on the slides and specimens provided.

    Canaliculi
    Compact bone
    Diaphysis
    Diploe
    Distal epiphysis
    Endosteum
    Epiphyseal plate
    Central canal
    Inner compact bone
    Lacuna
    Lamellae (matrix)
    Medullary cavity
    Nutrient foramen
    Nutrient vessel
    Osteocyte
    Osteon
    Outer compact bone
    Periosteum
    Proximal epiphysis
    Spongy bone
    Suture

2. Using colored pencils, color-code the compact bone, spongy bone, and diploe in figures 10.1 and 10.2.
3. Write a description of each of the terms listed in table 10.1.

Figure 10.1 A section through the skull showing diploe.

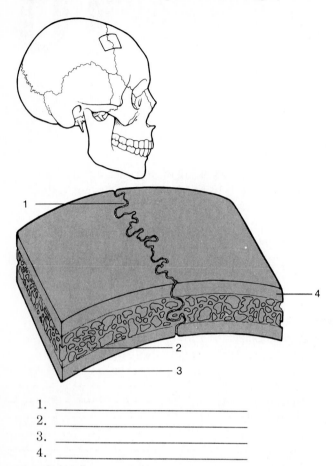

1. _____
2. _____
3. _____
4. _____

Figure 10.2 A diagram of a long bone shown in longitudinal section.

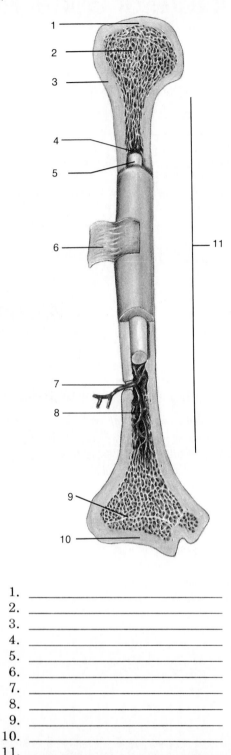

1. _____
2. _____
3. _____
4. _____
5. _____
6. _____
7. _____
8. _____
9. _____
10. _____
11. _____

Figure 10.3 The histological structure of bone tissue: (*a*) a section cut out of the femur; (*b*) an enlarged section of bone showing compact and spongy bone tissue; (*c*) an osteon (haversian system).

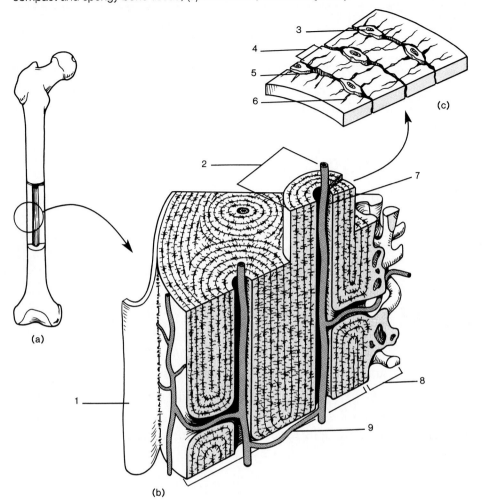

1. _____
2. _____
3. _____
4. _____
5. _____
6. _____
7. _____
8. _____
9. _____

Table **10.1** Surface features of bones

Feature	Description
Articulating surface	
Condyle	_____
Head	_____
Facet	_____
Nonarticulating process (roughened bony prominence)	
Tubercle	_____
Tuberosity	_____
Trochanter	_____
Spine	_____
Crest	_____
Epicondyle	_____
Depressions and openings	
Fossa	_____
Fissure	_____
Meatus	_____
Foramen	_____
Sinus	_____

Laboratory Report **10**

Name _____

Date _____

Section _____

Structure of Bone Tissue

Read the assigned sections in the textbook before completing the laboratory report.

Structure of Bones

1. Discuss the four principal kinds of bones according to shape, and give an example of each.

2. Briefly explain how osteocytes within an osteon receive nutrients and oxygen.

3. Each term in the following pairs of terms has some specific relationship to the other term in its pair but also a notable difference. State each in the space provided.

	Specific relationship	Specific difference
Membranous and endochondral bone formation	_____	_____
Diaphysis and epiphysis	_____	_____
Fossa and foramen	_____	_____
Compact and spongy bone tissue	_____	_____
Tubercle and trochanter	_____	_____

Skeletal System: The Axial Skeleton

Before coming to class, review the following sections in chapter 9 of the textbook: "Development of the Skull," "Skull," "Vertebral Column," and "Rib Cage."

Introduction

Familiarity with the bones of the body is important for many reasons. For example, many muscles are named for the bones to which they are attached. Learning the bones will be easier and more meaningful if you palpate (feel with firm pressure) the bones of your own body.

Handle bone specimens with care because they are expensive and become brittle as they dry. Avoid touching bones with the tip of a pen or pencil. Refer to the articulated skeleton to see the actual position of individual, disarticulated bones.

Objectives

Students completing this section will be able to:

1. Identify the individual bones of the axial skeleton from an articulated or disarticulated skeleton.
2. Identify the principal sutures, processes, and foramina of the skull.

Materials

1. Articulated and disarticulated skeletons
2. Skull of a fetus showing fontanels
3. Colored pencils
4. Reference text

A. Skull

Cranial Bones and Features

1. Identify the structures and processes listed below on the skulls in the laboratory and label them in figures 11.1 through 11.6.
2. Using colored pencils, color-code the bones in the figures in a consistent manner. For example, if you use blue for the temporal bone in one figure, the temporal bone should also be colored blue in the other figures.

Structures of the fetal skull
    Anterior fontanel
    Anterolateral fontanel
    Posterior fontanel
    Posterolateral fontanel
Structures of the cranium
    Coronal suture
Teeth
    Premolar teeth
    Molar teeth
    Canine teeth
    Incisor teeth
Ethmoid bone
    Ethmoidal sinuses
    Cribriform plate of ethmoid bone
    Crista galli of ethmoid bone
    Nasal conchae
    Perpendicular plate of ethmoid bone
Frontal bone
    Frontal sinus
    Supraorbital foramen
    Supraorbital margin
    Zygomatic process of frontal bone
Lambdoidal suture
Occipital bone
    External occipital protuberance
    Foramen magnum
    Occipital condyle
    Superior nuchal line

Parietal bone
Sagittal suture
Sphenoid bone
    Foramen ovale
    Foramen spinosum
    Foramen rotundum
    Foramen lacerum
    Greater wing of sphenoid bone
    Lesser wing of sphenoid bone
    Optic foramen
    Pterygoid process of sphenoid bone
    Sella turcica
    Sphenoidal sinus
    Superior orbital fissure
Squamosal suture

Temporal bone
    Mandibular fossa
    Mastoid part of temporal bone
    Petrous part of temporal bone
        Carotid canal
        Jugular foramen
    Stylomastoid foramen
    Styloid process of temporal bone
    Tympanic part of temporal bone
        External acoustic meatus
        Internal acoustic meatus
    Zygomatic arch
    Zygomatic process
Zygomatic bone

Figure 11.1 The fetal skull.

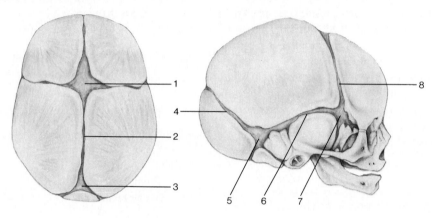

1. _____
2. _____
3. _____
4. _____

5. _____
6. _____
7. _____
8. _____

Figure 11.2 A lateral view of the skull.

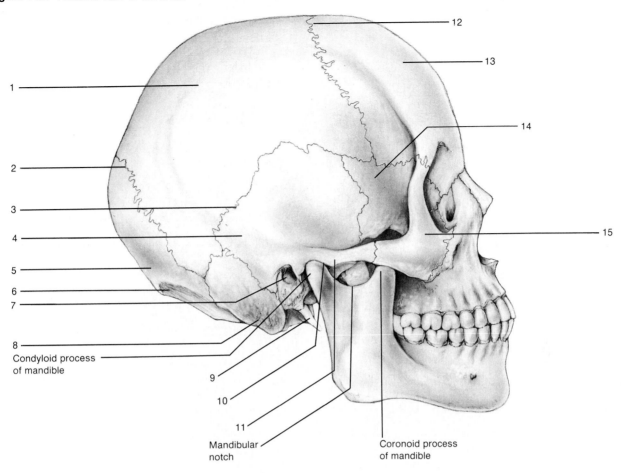

Condyloid process
of mandible

Mandibular
notch

Coronoid process
of mandible

1. _____
2. _____
3. _____
4. _____
5. _____
6. _____
7. _____
8. _____

9. _____
10. _____
11. _____
12. _____
13. _____
14. _____
15. _____

Figure 11.3 An inferior aspect of the skull (with the lower jaw removed).

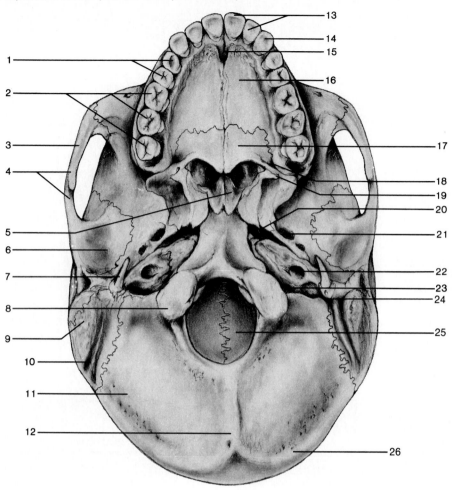

1. _____
2. _____
3. _____
4. _____
5. _____
6. _____
7. _____
8. _____
9. _____
10. _____
11. _____
12. _____
13. _____

14. _____
15. _____
16. _____
17. _____
18. _____
19. _____
20. _____
21. _____
22. _____
23. _____
24. _____
25. _____
26. _____

Figure 11.4 The floor of the cranial cavity.

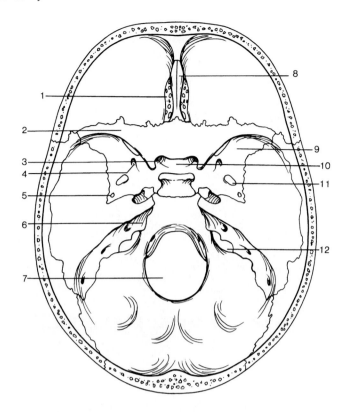

1. _____	7. _____
2. _____	8. _____
3. _____	9. _____
4. _____	10. _____
5. _____	11. _____
6. _____	12. _____

Figure 11.5 Radiographs of the skull showing paranasal sinuses: (*a*) an anteroposterior view; (*b*) a right lateral view.

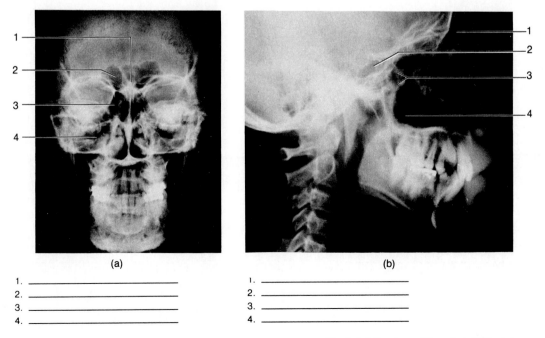

(a)   (b)

1. _____	1. _____
2. _____	2. _____
3. _____	3. _____
4. _____	4. _____

Figure 11.6 An anterior view of the skull.

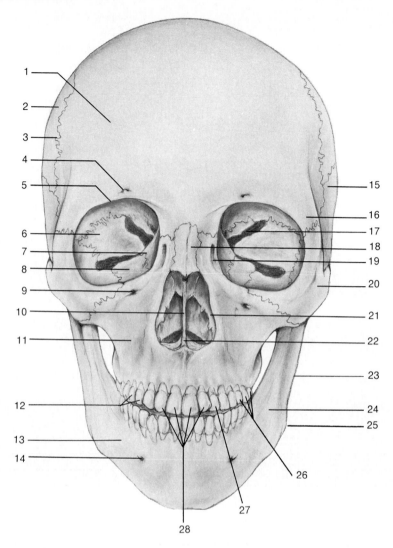

1. _____
2. _____
3. _____
4. _____
5. _____
6. _____
7. _____
8. _____
9. _____
10. _____
11. _____
12. _____
13. _____
14. _____

15. _____
16. _____
17. _____
18. _____
19. _____
20. _____
21. _____
22. _____
23. _____
24. _____
25. _____
26. _____
27. _____
28. _____

Facial Bones and Features

Identify the following structures and processes on the sample skulls and label them in figures 11.3, 11.6, and 11.7. Color-code the bones in the figures as you did before.

Lacrimal bone
    Nasolacrimal canal
Mandible
    Alveolar margin
    Angle of mandible
    Body of mandible
    Condyloid process of mandible
    Coronoid process of mandible
    Mandibular foramen

Mandibular notch of mandible
Mental foramen
Ramus of mandible
Maxilla
    Alveolar process
    Incisive foramen
    Infraorbital foramen
    Maxillary sinus
    Palatine process
Nasal bone
Palatine bone
    Greater palatine foramen
Vomer
Zygomatic bone

Figure 11.7 The mandible.

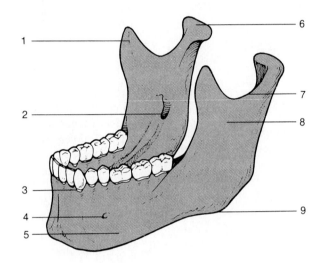

1. _____   6. _____

2. _____   7. _____

3. _____   8. _____

4. _____   9. _____

5. _____

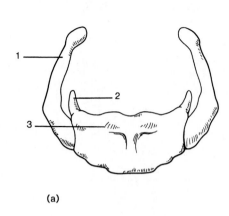

(a)

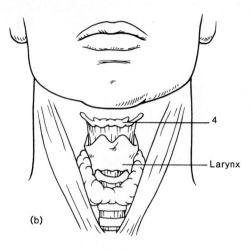

(b)

1. _____

2. _____

3. _____

4. _____

B. Hyoid Bone and Auditory Ossicles

Identify and label the bones and structures listed below in figures 11.8 and 11.9. Using colored pencils, color the hyoid bone and larynx in different colors.

Hyoid bone
    Body of hyoid bone
    Greater cornua
    Lesser cornua
Auditory ossicles
    Malleus
    Incus
    Stapes

C. Vertebral Column

Identify the following bones and structures in your samples and label these bones and structures in figures 11.10 through 11.14:

General structure
    Body
    Inferior articular process
    Intervertebral discs
    Intervertebral foramen
    Lamina
    Pedicle
    Spinous process
    Superior articular process
    Transverse process and foramina
    Vertebral foramen

Regional characteristics
    Cervical curvature
    Cervical vertebrae
        Atlas
            Anterior arch of atlas
        Axis
        Bifid spinous process
        Dens (odontoid process)
        Superior articular surface for occipital condyles
        Transverse foramen
    Coccyx
        Coccygeal cornua
    Lumbar curvature
    Lumbar vertebrae
    Pelvic curvature
    Sacrum
        Auricular surface
        Inferior portion of sacral canal
        Posterior sacral foramen
        Median sacral crest
        Pelvic foramen
        Sacral canal
        Sacral promontory
        Sacral tuberosity
        Superior articular processes
        Superior portion of sacral canal
        Transverse lines
    Thoracic curvature
    Thoracic vertebrae
        Fovea

Figure 11.9 The three auditory ossicles within the middle ear cavity.

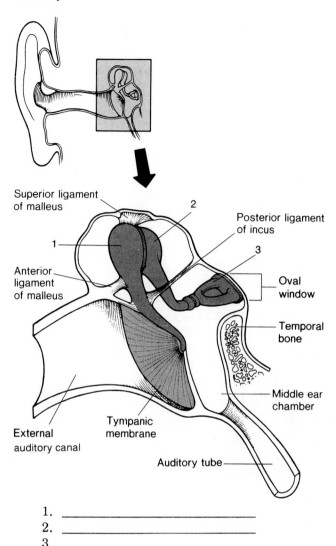

1. _____
2. _____
3. _____

Figure 11.10 The vertebral column of an adult, showing curvatures and regions.

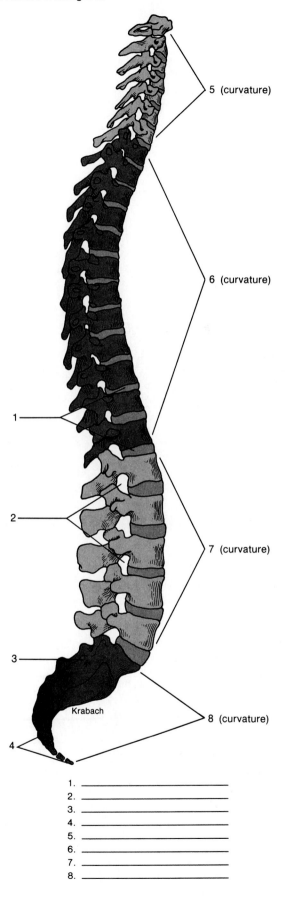

Krabach

1. _____
2. _____
3. _____
4. _____
5. _____
6. _____
7. _____
8. _____

Skeletal System: The Axial Skeleton   81

Figure 11.11 Cervical vertebrae: (*a*) a typical cervical vertebra; (*b*) the atlas and axis as they articulate.

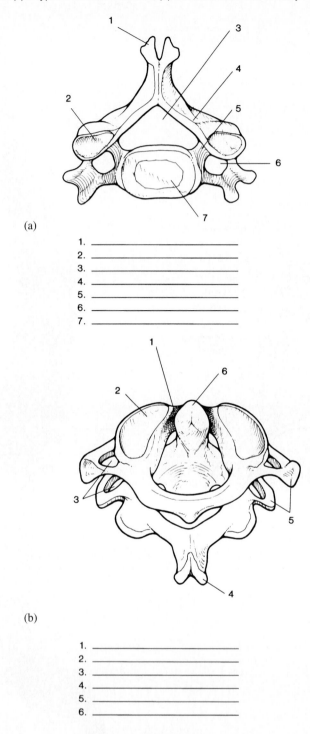

(a)

1. _____
2. _____
3. _____
4. _____
5. _____
6. _____
7. _____

(b)

1. _____
2. _____
3. _____
4. _____
5. _____
6. _____

Figure 11.12 Thoracic vertebrae: (*a*) a right lateral view; (*b*) a superior view.

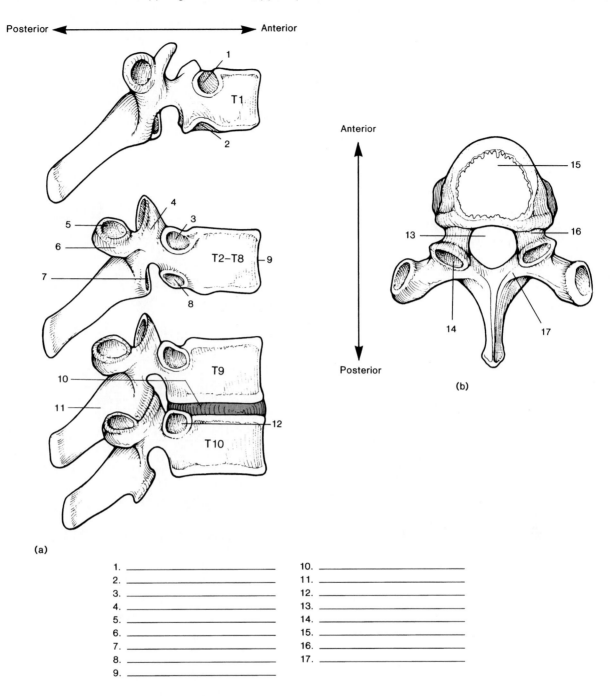

Posterior ← → Anterior

T 1

T2–T8

T9

T10

(a)

Anterior

Posterior

(b)

1. _____   10. _____
2. _____   11. _____
3. _____   12. _____
4. _____   13. _____
5. _____   14. _____
6. _____   15. _____
7. _____   16. _____
8. _____   17. _____
9. _____

Figure 11.13 Lumbar vertebrae: (*a*) a superior view; (*b*) a right lateral view.

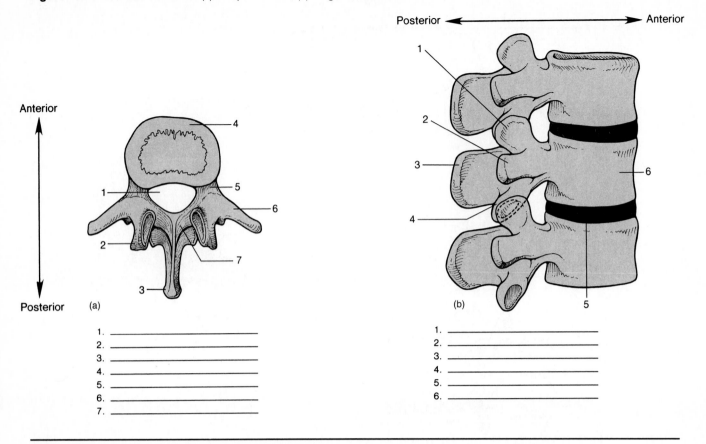

1. _____
2. _____
3. _____
4. _____
5. _____
6. _____
7. _____

1. _____
2. _____
3. _____
4. _____
5. _____
6. _____

Figure 11.14 The sacrum and coccyx: (*a*) an anterior view; (*b*) a posterior view.

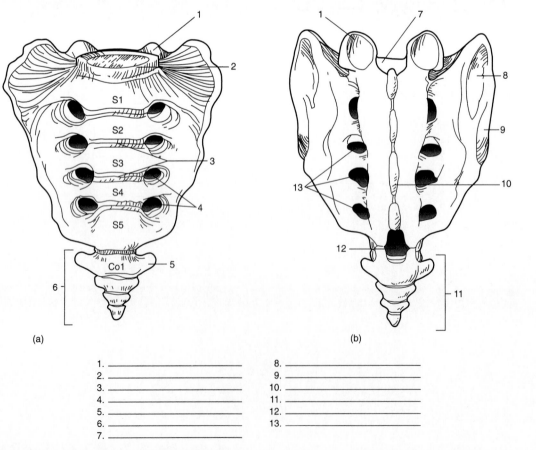

1. _____
2. _____
3. _____
4. _____
5. _____
6. _____
7. _____

8. _____
9. _____
10. _____
11. _____
12. _____
13. _____

Figure 11.15 The rib cage.

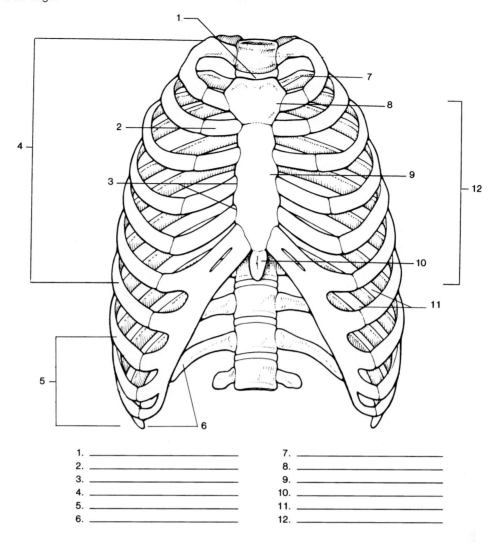

1. _____	7. _____
2. _____	8. _____
3. _____	9. _____
4. _____	10. _____
5. _____	11. _____
6. _____	12. _____

D. Rib Cage

Identify the following bones and structures in your samples and label these bones and structures in figures 11.15 and 11.16:

Ribs
- Angle
- Articular facets
- Costal cartilage
- Costal groove
- Costal notches
- False ribs
- Floating ribs

Head
Intercostal space
Neck
Shaft
True ribs
Tubercle

Sternum
- Body of sternum
- Clavicular notch
- Costal notches
- Manubrium
- Jugular notch
- Xiphoid process

Figure 11.16 The structure of a rib.

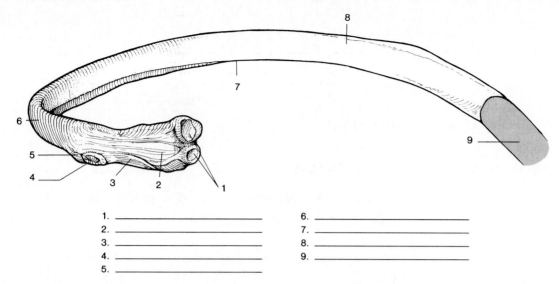

1. _____
2. _____
3. _____
4. _____
5. _____

6. _____
7. _____
8. _____
9. _____

Laboratory Report 11

Name _____

Date _____

Section _____

Skeletal System: The Axial Skeleton

Read the assigned sections in the textbook before completing the laboratory report.

A. Skull

1. List the bones of the skull that form the orbit of the eye.

2. List all the bones of the skull that can be palpated.

3. List the four prominent sutures of the skull. Identify the bones that join to form these sutures.

Suture	Articulating bones

4. Which two bones/processes form the nasal septum?_____
5. Which two bones/processes form the hard palate?_____
6. In which bones are the paranasal sinuses found?_____

B. Hyoid Bone and Auditory Ossicles

1. Describe the location of the hyoid bone. What structures are attached to the hyoid bone?

2. Describe the location of the malleus, incus, and stapes. What is their function?

C. Vertebral Column

List the five regions of the vertebral column, the number of vertebrae in each, and the diagnostic characteristics of each region.

Region	Number of vertebrae	Diagnostic characteristics

D. Rib Cage

1. Distinguish between true ribs, false ribs, and floating ribs, and give the number of each.

2. Describe the articulating structures involved in the connection between a rib and the vertebral column. Does one rib always articulate with only one vertebra? Explain.

Skeletal System:
The Appendicular Skeleton

Before coming to class, review the following sections in chapter 10 of the textbook: "Pectoral Girdle and Upper Extremity" and "Pelvic Girdle and Lower Extremity."

Introduction

The appendicular skeleton includes the bones of the upper and lower extremities and their respective girdles. The upper extremities comprise the bones of the pectoral (shoulder) girdle, the brachii (upper arms), the antebrachii (forearms), and the hands. They are adapted to permit freedom of movement and hand dexterity. The lower extremities comprise the bones of the pelvic girdle, the thighs, the legs, and the feet. They are adapted to support and move the body, and thus contain heavier bones. It is helpful to identify the bones and their processes by palpating them on your own body.

Objectives

Students completing this exercise will be able to:

1. Identify structural differences between the pectoral and pelvic girdles.
2. Identify from an articulated or disarticulated skeleton the individual bones of the appendicular portion, the diagnostic processes of these bones, and the articulating bones.
3. Interpret skeletal elements and processes from radiographs (X rays).
4. Differentiate between an adult male and female skeleton when shown a pelvic diagram or a complete skeleton.

Materials

1. Articulated and disarticulated skeletons
2. Male pelvis and female pelvis
3. Chart of the skeletal system
4. Radiographic viewing box with assorted radiographs
5. Colored pencils
6. Reference text

A. Pectoral Girdle

Identify the following bones and structures in your sample bones and label them in figures 12.1 and 12.2:

Clavicle
    Acromial extremity
    Conoid tubercle
    Costal tuberosity
    Sternal extremity
Scapula
    Acromion
    Coracoid process
    Glenoid cavity
    Inferior angle
    Infraspinous fossa
    Lateral (axillary) border
    Medial (vertebral) border
    Scapular notch
    Spine
    Subscapular fossa
    Superior angle
    Superior border
    Supraspinous fossa

B. Brachium

Identify the following bones and structures in your sample bones and label them in figure 12.3:

Humerus
    Anatomical neck
    Capitulum
    Coronoid fossa
    Deltoid tuberosity
    Greater tubercle
    Head of humerus
    Intertubercular groove

Figure 12.1 The right clavicle: (*a*) a superior view; (*b*) an inferior view.

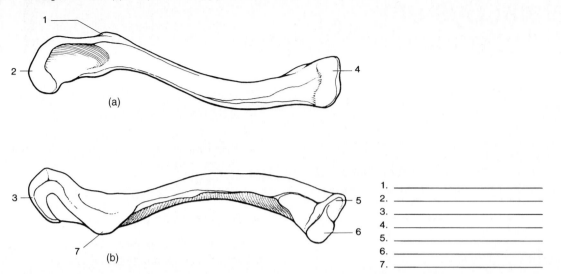

1. _____
2. _____
3. _____
4. _____
5. _____
6. _____
7. _____

Figure 12.2 The right scapula: (*a*) an anterior view; (*b*) a posterior view.

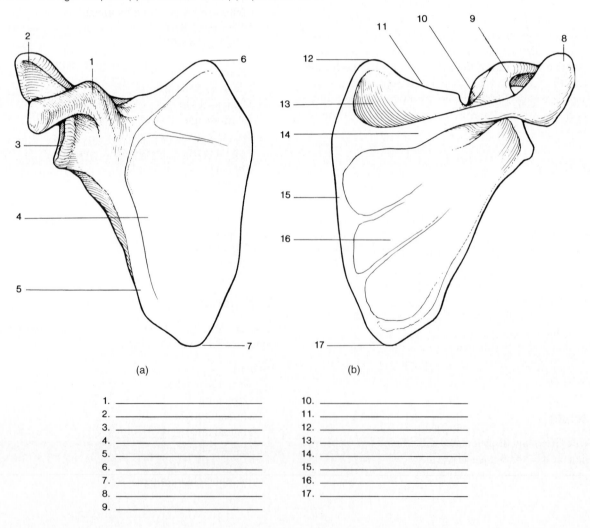

1. _____
2. _____
3. _____
4. _____
5. _____
6. _____
7. _____
8. _____
9. _____

10. _____
11. _____
12. _____
13. _____
14. _____
15. _____
16. _____
17. _____

Figure 12.3 The right humerus: (*a*) an anterior view; (*b*) a posterior view.

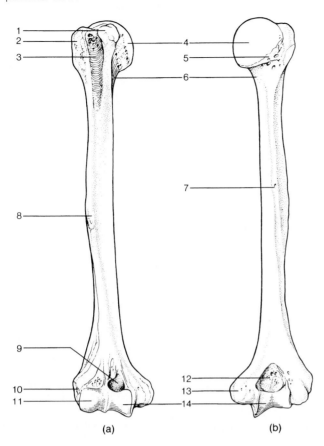

(a)   (b)

Lateral epicondyle
Lesser tubercle
Medial epicondyle
Nutrient foramina
Olecranon fossa
Surgical neck
Trochlea

C. Antebrachium

Identify the following bones and structures in your sample bones and label them in figures 12.4 and 12.5:

Radius
    Head of radius
    Neck of radius
    Radial tuberosity
    Shaft (body) of radius
    Styloid process
Ulna
    Coronoid process
    Head of ulna
    Interosseous crest and ligament
    Olecranon
    Trochlear notch
    Shaft (body) of ulna
    Styloid process

D. Hand

1. Identify the bones and structures listed below in your sample bones and label them in figures 12.6 and 12.7. (Keep in mind the thumb has only a proximal and distal phalanx.)
2. Using colored pencils, color-code the carpal bones, metacarpal bones, and phalanges in figure 12.6.

Carpal bones
    Capitate
    Hamate
    Lunate
    Pisiform
    Scaphoid
    Trapezium (greater multangular)
    Trapezoid (lesser multangular)
    Triquetrum (triangular)
Metacarpal bones
    Base
    Head
    Shaft (body)
Phalanges
    Distal row
    Middle row
    Proximal row

1. _____
2. _____
3. _____
4. _____
5. _____
6. _____
7. _____
8. _____
9. _____
10. _____
11. _____
12. _____
13. _____
14. _____

Figure 12.4 Right radius and ulna (anterior view). A cross section showing the binding interosseous ligament is also shown.

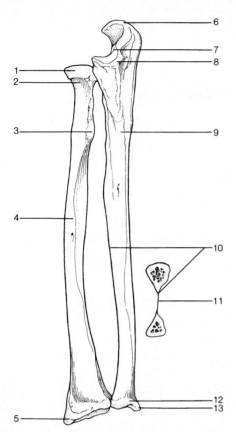

Figure 12.5 The right radius and ulna (a posterior view).

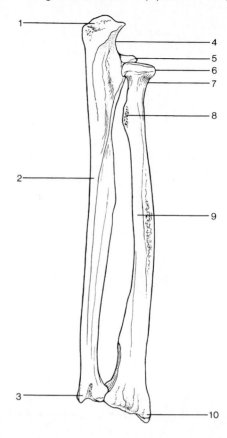

1. _____
2. _____
3. _____
4. _____
5. _____
6. _____
7. _____
8. _____
9. _____
10. _____
11. _____
12. _____
13. _____

1. _____
2. _____
3. _____
4. _____
5. _____
6. _____
7. _____
8. _____
9. _____
10. _____

E. Pelvic Girdle

1. Identify the bones and structures listed below in your sample bones and label them in figures 12.8 and 12.9.
2. Using colored pencils, color-code the regions of the os coxa in figure 12.9 that correspond to the illium, ischium, and pubis.

Os coxa
    Acetabulum
    Acetabular notch
    Symphysis pubis
Ilium
    Anterior inferior iliac spine

    Anterior superior iliac spine
    Greater sciatic notch
    Iliac crest
    Iliac fossa
    Posterior inferior iliac spine
    Posterior superior iliac spine
Ischium
    Body of ischium
    Ischial tuberosity
    Lesser sciatic notch
    Obturator foramen
    Ramus of ischium
    Spine of ischium
Pubis
    Body of pubis
    Inferior ramus
    Pubic tubercle
    Superior ramus
Sacroiliac joint

Figure 12.6 The bones of the right hand (a posterior view).

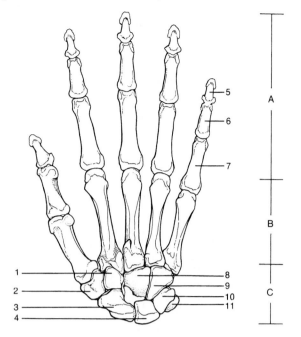

Figure 12.7 The bones of the right hand (an anterior view).

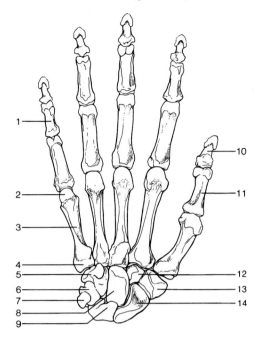

1. _____
2. _____
3. _____
4. _____
5. _____
6. _____
7. _____
8. _____
9. _____
10. _____
11. _____

1. _____
2. _____
3. _____
4. _____
5. _____
6. _____
7. _____
8. _____
9. _____
10. _____
11. _____
12. _____
13. _____
14. _____

F. Thigh

Identify the following bones and structures in your sample bones and label them in figure 12.10:

Femur
    Greater trochanter
    Head of femur
    Intercondylar fossa
    Intertrochanteric crest
    Intertrochanteric line
    Lateral condyle
    Lateral epicondyle
    Lesser trochanter
    Linea aspera
    Medial condyle
    Medial epicondyle
    Neck of femur
    Patellar surface
    Shaft (body) of femur

G. Leg

Identify the following bones and structures in your sample bones and label them in figure 12.11:

Fibula
    Head of fibula
    Lateral malleolus
Patella
    Apex
    Articular facets
    Base
Tibia
    Anterior crest
    Intercondylar eminence
    Lateral condyle

Figure 12.8 An anterior view of the pelvic girdle.

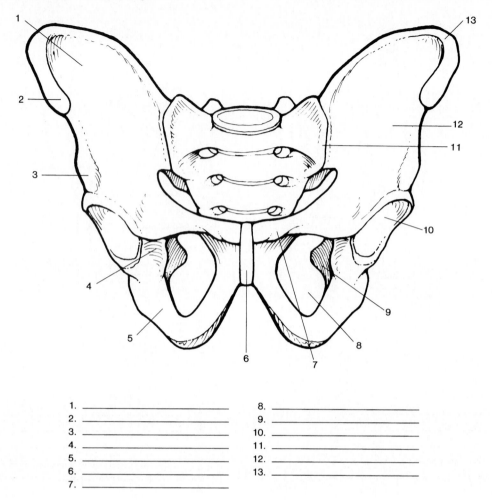

1. _____   8. _____
2. _____   9. _____
3. _____   10. _____
4. _____   11. _____
5. _____   12. _____
6. _____   13. _____
7. _____

Medial condyle
Medial malleolus
Tibial tuberosity

H. Foot

1. Identify the bones and structures listed below in your sample bones and label them in figure 12.12.
2. Using colored pencils, color-code the tarsal bones, metatarsal bones, and phalanges.

Tarsal bones
    Calcaneus
        Tuberosity of calcaneus
    Cuboid bone

Medial cuneiform
Scaphoid
Intermediate cuneiform
Talus
Lateral cuneiform
Metatarsal bones
    Base
    Head
    Shaft (body)
Phalanges
    Distal row
    Middle row
    Proximal row

Figure 12.9 The lateral aspect of the right os coxa.

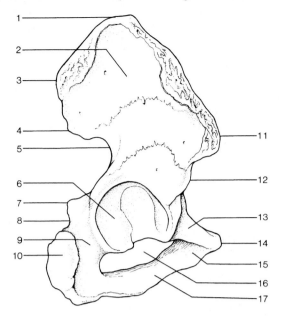

Figure 12.10 The right femur: (*a*) an anterior view; (*b*) a posterior view.

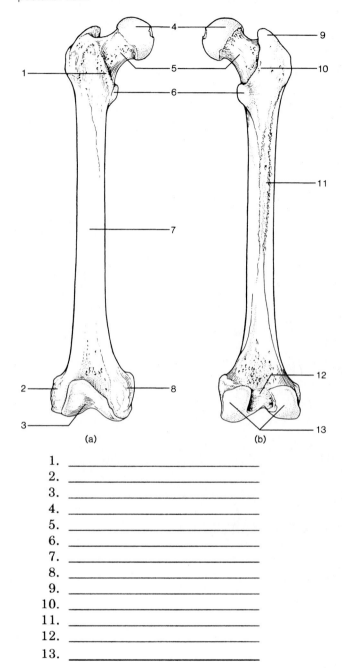

(a)   (b)

1. _____
2. _____
3. _____
4. _____
5. _____
6. _____
7. _____
8. _____
9. _____
10. _____
11. _____
12. _____
13. _____
14. _____
15. _____
16. _____
17. _____

1. _____
2. _____
3. _____
4. _____
5. _____
6. _____
7. _____
8. _____
9. _____
10. _____
11. _____
12. _____
13. _____

Figure 12.11 The right tibia, fibula, and patella: (*a*) an anterior view; (*b*) a posterior view.

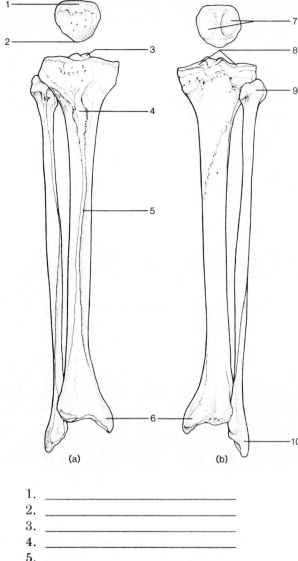

(a) (b)

1. _____
2. _____
3. _____
4. _____
5. _____
6. _____
7. _____
8. _____
9. _____
10. _____

Figure 12.12 The right foot: (*a*) a superior view; (*b*) an inferior view.

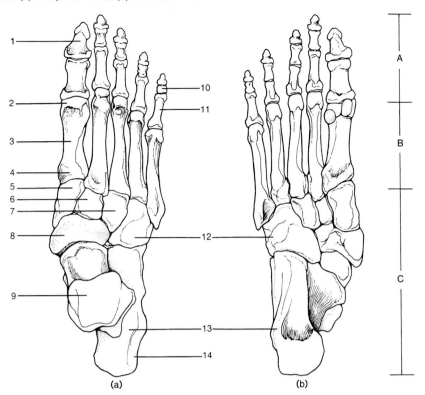

(a) (b)

1. _____
2. _____
3. _____
4. _____
5. _____
6. _____
7. _____

8. _____
9. _____
10. _____
11. _____
12. _____
13. _____
14. _____

Laboratory Report 12

Name _____

Date _____

Section _____

Skeletal System: The Appendicular Skeleton

Read the assigned sections in the textbook before completing the laboratory report.

A–D. Pectoral Girdle and Upper Extremity

1. Is the pectoral girdle firmly attached to the axial skeleton? What precautions should be taken, especially with children, because of this?

2. Using the correct anatomical term, name the bone on which a wedding ring is worn.

3. Explain what the following terms mean or why they are so named:
   a. Infraspinous fossa _____
   b. Medial border _____
   c. Subscapular fossa _____
   d. Sternal extremity _____
   e. Supraspinous fossa _____
4. Describe the position of the ulna relative to the radius when the palm of the hand is up and when it is down.

5. Name the bones and structures that articulate with each of the following:
   a. Olecranon of ulna _____
   b. Head of radius _____
   c. Radial notch of ulna _____
   d. Coronoid fossa of humerus _____

E–H. Pelvic Girdle and Lower Extremity

1. Name the bones and structures that articulate with each of the following:
   a. Acetabulum _____
   b. Condyles of the femur _____
   c. Medial malleolus of the tibia _____
   d. Patella _____

2. What is the scientific name of each of the following?

   a. Knee bone _____

   b. Finger bone _____

   c. Shoulder blade _____

   d. Heel bone _____

   e. Elbow (funny bone) _____

   f. Hipbone _____

   g. Collarbone _____

   h. Shinbone _____

   i. Anklebone _____

   j. Knuckle _____

   k. Instep of foot _____

3. Identify the bones of the appendicular skeleton depicted in the radiographs in examples 1 through 4. Label as many diagnostic processes, surfaces, and fossae as possible. After completing the following chapter on articulations, refer to these radiographs and identify the types of joints between the articulating bones.

Example 1 The upper extremity.

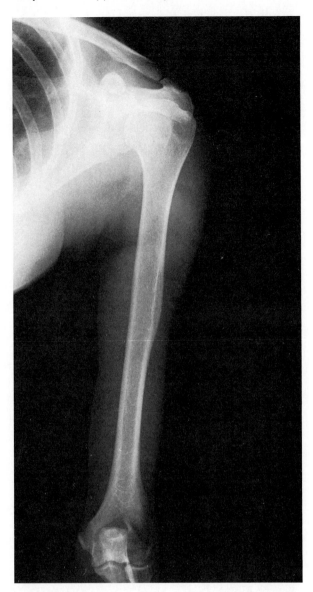

Example 2 The lower extremity.

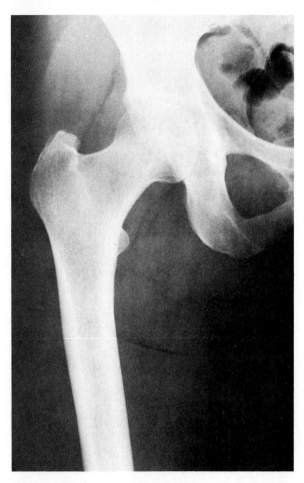

Example 4 The foot.

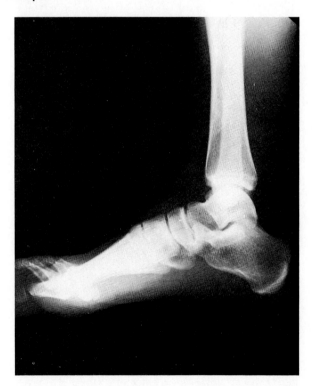

Example 3 The hand.

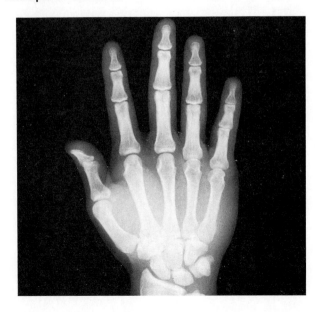

Articulations

Introduction

Articulations, or joints, are sites within the skeleton where adjacent bones meet. The structure of the joint determines the function and range of movement of the bones that come together at the joint. Structurally, joints are classified as fibrous, cartilaginous, and synovial.

Objectives

Students completing this exercise will be able to:

1. Explain the relationship of the skeleton and joints to the overall biomechanics of the body.
2. Classify the joints and understand in detail the parts of a synovial joint.
3. Demonstrate the various types of movements of the body and understand the advantages and limitations of each type of movement.

Materials

1. Articulated skeleton
2. Disarticulated skeleton
3. Skull of an infant and of an adult
4. Radiographic viewing box with assorted radiographs
5. Colored pencils
6. Reference text
7. Fresh beef or pig knee joint and dissection instruments

A. Joints

1. Identify and label the following joint types and structures in figures 13.1 and 13.2.
2. Use colored pencils to highlight the structures depicted in figure 13.1.

Cartilaginous joints
    Symphysis
    Syndesmosis
Synovial joints
    Ball-and-socket
    Condyloid joint
    Gliding joint
    Hinge joint
    Pivot joint
    Saddle joint
Structure of a synovial joint
    Articular cartilage
    Bursa: (a) prepatellar, (b) infrapatellar
    Joint capsule
    Meniscus
    Synovial membrane
Fibrous joints
    Sutures

B. Movements at Synovial Joints

Identify and label the following movements depicted in figure 13.3:

Abduction
Adduction
Circumduction
Dorsiflexion
Eversion
Extension
Flexion
Hyperextension
Inversion
Plantar flexion
Pronation
Protraction
Retraction
Rotation
Supination

Figure 13.1 Various joints of the body.

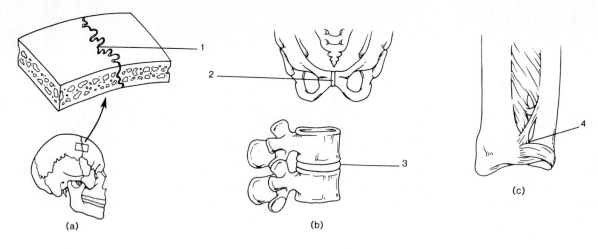

(a) (b) (c)

1. _____
2. _____
3. _____
4. _____
5. _____

6. _____
7. _____
8. _____
9. _____
10. _____

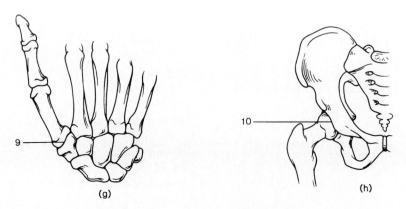

(d) (e) (f)

(g) (h)

Figure 13.2 A synovial joint is represented by this diagrammatic lateral view of the knee joint.

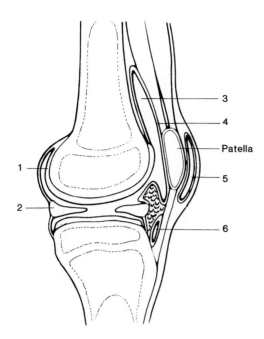

Patella

1. _____
2. _____
3. _____
4. _____
5. _____
6. _____

C. Joint Dissection

A fresh beef or pig knee joint (stifle joint) should have been procured by the laboratory instructor from a livestock processing plant. Place the limb on a dissection tray, and carefully remove the musculature that crosses over the joint and the fat surrounding it. Note the reinforcing collateral ligaments on the medial and lateral sides of the joint. The joint capsule is the tough, shiny connective sheet that completely surrounds the joint. The joint capsule is firmly attached to the periosteum of both the femur and tibia. Cutting into the lateral side of this capsule will expose the smoothly lined synovial membrane, which encloses the synovial cavity and produces the synovial fluid.

Extend your cut anterosuperiorly above the patella, and identify the tendon of the quadriceps femoris muscle. The articular cartilages capping the condyles of the bones can be observed as the patella is reflected back. Flex the knee joint, and observe the gliding action. With the knee flexed, the menisci are exposed. The cruciate ligaments can be best identified from a posterior view of the popliteal fossa.

Figure 13.3 Movements at synovial joints.

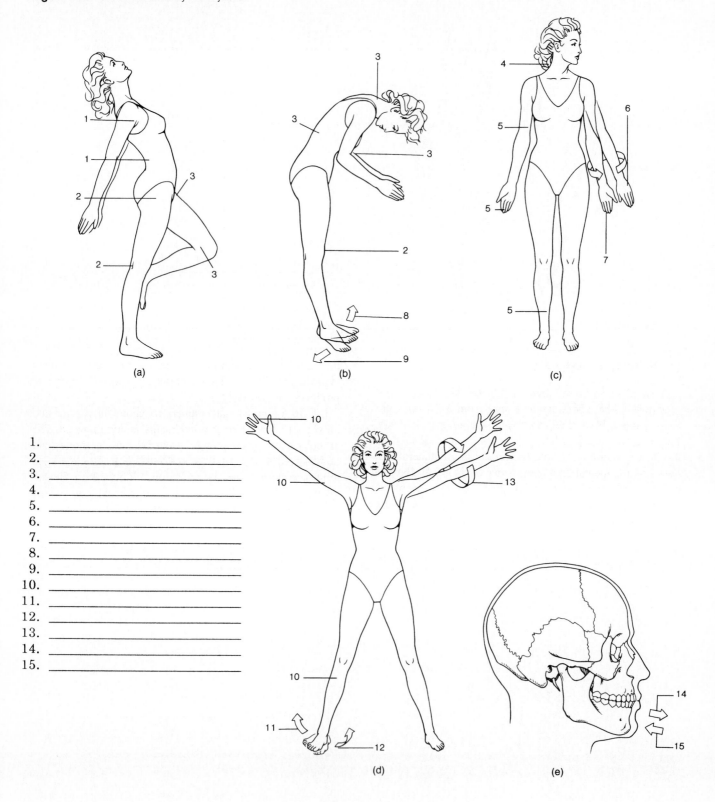

(a)

(b)

(c)

(d)

(e)

1. _____
2. _____
3. _____
4. _____
5. _____
6. _____
7. _____
8. _____
9. _____
10. _____
11. _____
12. _____
13. _____
14. _____
15. _____

Laboratory Report 13

Name _____

Date _____

Section _____

Articulations

Read the assigned sections in the textbook before completing the laboratory report.

A. Joints

1. By definition, all joints are movable. Is this statement true or false? Explain.

2. Give the function or significance of each of the following:
   a. Synovial fluid _____
   b. Joint capsule _____
   c. Meniscus _____
   d. Articular cartilage _____
   e. Ligament _____
   f. Synovial membrane _____
   g. Bursa _____

3. Identify the bones that articulate in each of the following joints:
   a. Knee _____
   b. Elbow _____
   c. Hip _____
   d. Wrist _____
   e. Ankle _____
   f. Shoulder _____

4. Why aren't sutures present in an infant's skull?

5. Define the following terms:
   a. Interosseous ligament _____
   b. Hyaline cartilage _____
   c. Intervertebral disc _____
   d. Symphysis _____

B. Movements at Synovial Joints

1. Compare the shoulder joint of the pectoral girdle with the hip joint of the pelvic girdle; contrast their structure, motion permitted, and general function.

2. Compare the knee joint with the elbow joint; contrast their size, shape, number of bones, and potential for movement.

Neural Control of Skeletal Muscle Contraction

Before coming to class, review the following sections in chapter 12 of the textbook: "Sliding Filament Theory of Contraction," "Motor Units," and "Regulation of Contraction."

Introduction

The physiology of skeletal muscle contraction can be studied using isolated muscles from a pithed frog. Isolated frog muscles can be stimulated directly with an electric current, and indirectly by activating the appropriate motor nerve. Contraction of the muscle exerts tension on a transducer, which causes a pen to move and record the contraction on paper.

Objectives

Students completing this exercise will be able to:

1. Prepare a pithed frog for the study of muscle physiology.
2. Explain how muscle contraction occurs from stimulation by a motor nerve, and directly from application of an electrical shock to the muscle.
3. Describe the sliding filament theory of contraction and the mechanisms by which this process is regulated.

Materials

1. Frog
2. Surgical scissors, forceps, sharp probes and pins, dissecting trays, glass probes
3. Recording equipment (either a kymograph or an electrical recorder, such as a Physiograph), electrical stimulators
4. Straight pins bent into Z shapes, thread
5. Myograph transducer (if Physiograph is used)

6. Frog Ringer's solution: dissolve 6 g NaCl, 0.075 g KCl, 0.10 g $CaCl_2$, and 0.10 g $NaHCO_3$ in 1 liter of water

Recording Procedures

The **Physiograph** is a modern device for recording the mechanical aspects of muscular contraction. Its mechanism is much more sensitive than the kymograph's straightforward coupling mechanism, because muscle movement is *transduced* (changed) into an electrical current, which is amplified before it is recorded. Because it transduces mechanical energy into electrical energy, the Physiograph is more versatile than the kymograph recorder. It can record not only mechanical events with different energies (muscle contraction, sound waves), but also primarily electrical events, such as nerve impulses, and events recorded by the electrocardiograph (ECG), electromyograph (EMG), and the electroencephalograph (EEG). A number of physiological parameters can be recorded simultaneously on different *channels* of the Physiograph, to allow study of the temporal relationships between these events.

The Physiograph consists of four basic parts: the **transducer** changes the original energy of the physiological event into electrical energy; the interchangeable **coupler** takes the input energy from the transducer and makes it compatible with the built-in amplifier; the **amplifier** increases the strength of the electrical current to activate the galvanometer; the **galvanometer** measures this current and moves a pen. The movement of the pen is proportional to the strength of the electrical current generated by the physiological event being measured. Since recording paper moves continuously at a known speed under the pen, both the *frequency* (number per unit time) and the *strength* (amplitude of the pen deflection compared to a baseline) of the physiological event can be continuously recorded (fig. 14.1).

Figure 14.1 The Physiograph Mark III recorder.

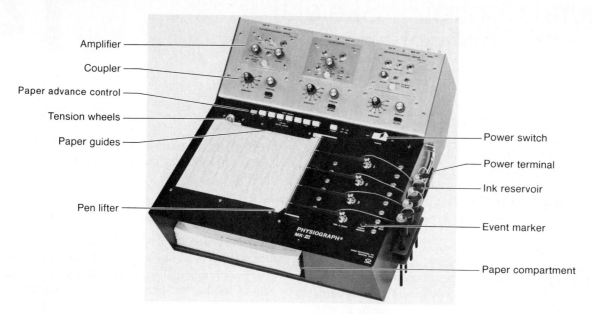

Amplifier

Coupler

Paper advance control

Tension wheels

Paper guides

Pen lifter

Power switch

Power terminal

Ink reservoir

Event marker

Paper compartment

For Physiograph Recording

1. Insert the *transducer coupler* into the Physiograph, and plug the myograph transducer into the coupler (fig. 14.2).[2]
2. Raise the inkwells and lower the pens onto the paper by lowering the *pen lifter* (fig. 14.1). Squeeze the rubber bulbs on the inkwells to force ink into the pens.
3. Turn on the Physiograph with the rocking switch. Depress the *paper-speed* button marked *0.5* centimeters per second (cm/sec) (the *down* position is on). Turn on the paper drive by depressing and releasing the *paper advance* button, allowing it to rise (the *up* position is on).
4. Move the *time* switch to *on.* The bottom pen, labeled *time and event,* will make upward deflections every second. Notice that at a paper speed of 0.5 cm/sec, these deflections will be separated by the width of one small box on the recording paper. If the paper speed is increased to 1.0 cm/sec, the deflections of the time-and-event pen will be two small boxes apart.
5. Turn the outer knob of the *sensitivity* control to its lowest number (this will be its greatest sensitivity) (fig. 14.2). With the *record* button off (in the *up* position), adjust the position of the recording pen for the appropriate channel with the *position* knob so that the pen writes exactly on the heavy horizontal line corresponding to the channel being recorded.

[2] For Physiograph MK-III, Narco Bio-Systems.

Figure 14.2 The Narco Mark III Physiograph with an inserted transducer coupler. The transducer coupler is connected by means of a cable to a myograph transducer.

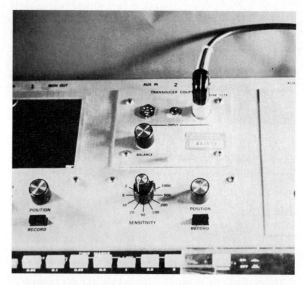

6. Depress the *record* button (the *down* position is on). This will cause the pen to move away from the heavy horizontal line. Bring the pen back to the line by rotating the *balance* knob. The pen should now remain on the heavy line whether or not the *record* button is depressed and regardless of the setting of the *sensitivity* knob.

Figure 14.3 The procedure for pithing a frog: (*a*) a probe is first inserted through the foramen magnum into the skull; (*b*) then it is inserted through the spinal cord (procedure was simulated with a preserved frog).

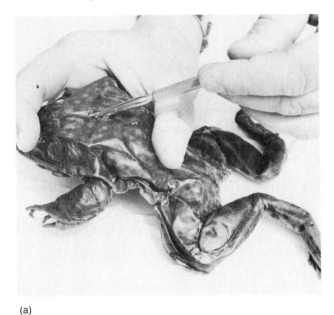

(a)

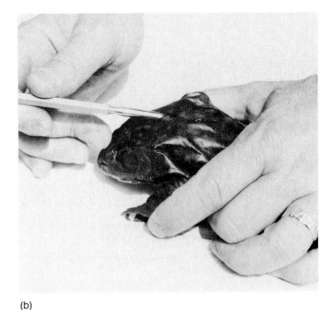

(b)

A. Frog Muscle Preparation

To study the physiology of frog muscle and nerve, the frog must be killed but its tissues kept alive. This can be accomplished by destroying the frog's central nervous system—a procedure called *pithing*. The frog is thus clinically dead (clinical death is defined as the irreversible loss of higher brain function), but its muscles and peripheral nerves are functional as long as their cells remain alive. Under the proper conditions, this can be prolonged for several hours.

There are two techniques for pithing a frog. To use the first procedure, grasp the frog securely in one hand and flex its neck so that the base of the skull can be felt with the fingers of the other hand. Then perform these steps:

1. Insert a sturdy metal probe into the skull through the foramen magnum (the opening in the skull where the spinal cord joins the brain stem). This is shown in figure 14.3*a*.
2. Move the probe around in the skull. This destroys the brain and prevents the frog from feeling any pain (it is now clinically dead).
3. Partially withdraw the probe, turn it so that it points toward the hind end of the frog (keeping the neck flexed), and insert the probe downward into the spinal cord (fig. 14.3*b*). This destroys spinal reflexes. The frog's legs will straighten out as the probe is inserted. When the spinal nerves are destroyed, the frog will become limp.

To use the second procedure, force one blade of a pair of sharp scissors into the frog's mouth as shown in figure 14.4*a*. Quickly decapitate the frog by cutting behind its eyes (the frog is dead as soon as its brain is severed from its spinal cord). Insert a probe down into the exposed spinal cord to destroy its spinal reflexes (fig. 14.4*b*).

After the frog has been pithed, skin one of its legs to expose the underlying muscle. Then run one blade of a pair of scissors under the tendo calcaneus (Achilles tendon) and cut it, leaving part of the tendon attached to the gastrocnemius muscle.

Setup for Recordings

1. Secure the frog to a dissecting tray by inserting sharp probes through its arms and legs.
2. Push a bent pin through its tendo calcaneus (fig. 14.5), and tie one end of a cotton thread to the hook of a myograph transducer. Position the myograph so that it is directly above the muscle, and adjust the height of the myograph so that the muscle is under tension. Insert two stimulating electrodes into the muscle.
3. Establish the **threshold stimulus** (the minimum stimulus that will evoke a particular response). To do this, set the stimulus intensity on 1.0 volts (V), and deliver a single shock. Increase the strength of the stimulus in 0.5" V increments until the muscle responds with a contraction (*twitch*) that is recorded on the kymograph or Physiograph. Record this voltage.

Threshold: _____ V

☞ **Note:** *Rinse the muscle periodically with Ringer's solution (a salt solution balanced to the extracellular fluid of the frog). Do not allow the muscle to dry out.*

Figure 14.4 Alternative pithing procedure: (*a*) the frog is first decapitated; (*b*) then a probe is inserted into the spinal cord (procedure was simulated with a preserved frog).

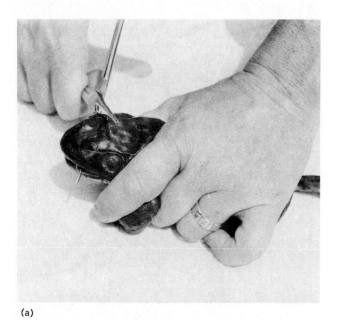

(a)

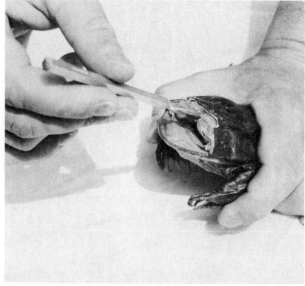

(b)

B. Stimulation of Motor Nerve

In the body, or in vivo, skeletal muscles are stimulated to contract by impulses through **somatic motor neurons.** Nerve impulses in the motor nerve stimulate the release of a chemical neurotransmitter called *acetylcholine* (*ACh*) from the axon terminal (ending). ACh diffuses to the muscle cell membrane, where it evokes electrical impulses in the muscle, which leads to contraction.

Clinical Significance

Some types of muscle degeneration are secondary to nerve damage or to dysfunction of the neuromuscular junctions. Muscle degeneration follows damage to the motor nerve pathway because proper neuromuscular activity and resulting muscle tone seem to be required for muscle health. In the autoimmune disease *myasthenia gravis* (*myasthenia* means that a muscle fatigues too easily), antibodies attack the muscle membrane receptors for acetylcholine, the neurotransmitter of somatic motor neurons. This prevents the muscle from being properly stimulated by somatic motor neurons.

Procedure

1. Starting at the pelvis, skin the leg that is going to be used for kymograph or Physiograph recordings.
2. With the frog on its belly, part the muscles of the thigh around the femur to reveal the sciatic nerve.
3. Using glass probes, free the nerve from its attached connective tissue, and raise it tightly held between two glass probes (fig. 14.5*b*).
4. Place both stimulating electrodes against the nerve. Starting with the stimulator set at 0 V, gradually increase the stimulus voltage by small increments until the minimum stimulus is attained that will produce a muscle twitch. Record this threshold in the space provided, and compare it with the threshold obtained previously by placing the electrodes directly on the muscle.

Threshold: _____ V
5. Turn off the stimulator. Using cotton thread, tie a knot in the nerve, and observe the response of the gastrocnemius muscle.

Figure 14.5 After the frog's leg is skinned (*a*), the tendo calcaneus is cut (*b*) and a bent pin is inserted into it (*c*). A length of thread attaches this pin to the hook of the myograph transducer (not shown). In (*b*) the sciatic nerve is shown between two glass probes.

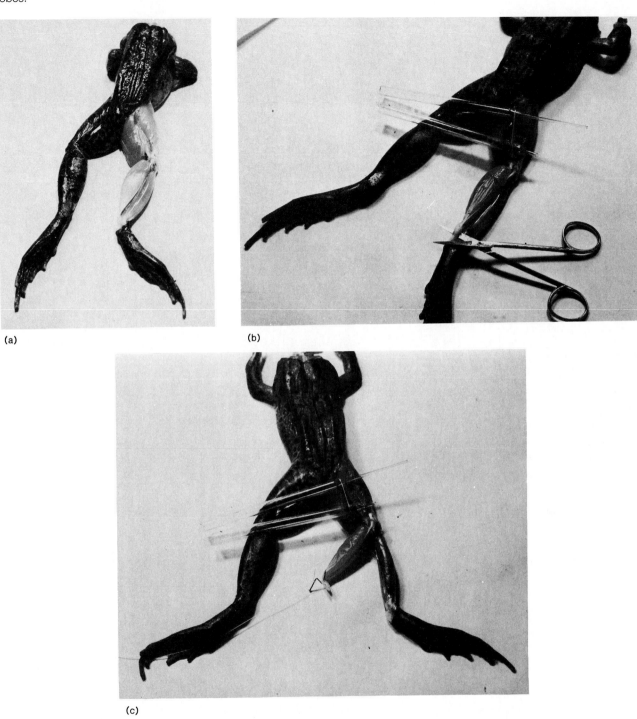

(a)

(b)

(c)

Laboratory Report 14

Name _____

Date _____

Section _____

Neural Control of Skeletal Muscle Contraction

Read the assigned sections in the textbook before completing the laboratory report.

1. The basic subunit of contraction in the muscle fiber is the _____ .

2. Thick myofilaments are composed of _____ ; thin myofilaments are composed primarily of
    _____ .

3. Describe how the lengths of the A, I, and H bands are affected by muscle contraction. What can you conclude from this information about the lengths of the thick and thin myofilaments?

4. What couples electrical excitation of a muscle fiber to contraction? Explain how this coupling occurs.

5. Define the term *threshold*. Explain why the threshold stimulus for producing a muscle twitch is lower when the nerve is stimulated than when the muscle is directly stimulated.

Summation, Tetanus, and Fatigue

Before coming to class, review the following sections in chapter 12 of the textbook: "Types of Muscle Contractions," and "Muscle Fatigue."

Introduction

Muscle twitches can summate so quickly that they produce a smooth, sustained contraction called *complete tetanus.* This occurs in the body in response to neural stimulation, and it can be demonstrated in vitro by means of the isolated frog muscle preparation. Twitch, summation, and tetanus can also be produced in students' muscles by an electrical stimulator.

Objectives

Students completing this exercise will be able to:

1. Describe twitch, summation, tetanus, and fatigue, and demonstrate these processes in an isolated frog gastrocnemius muscle and in human muscles.
2. Explain how the contraction of individual muscle fibers produces twitch, summation, and tetanus.
3. Explain the causes of muscle fatigue.

Materials

1. Frogs
2. Equipment and setup used in exercise 14
3. Electrocardiograph (ECG) plates and electrolyte gel
4. Alternative equipment: *Physiogrip* (Intelitool, Inc.) with IBM or Apple computer and square wave stimulator

Individual skeletal muscle fibers cannot sustain a contraction; they can only twitch. Muscle fibers likewise cannot produce a graded contraction; they can only contract maximally to any stimulus above threshold (they contract in an all-or-none fashion). Smooth, graded skeletal muscle contractions are produced by the **summation** of fiber twitches. This occurs when fibers twitch asynchronously, so that some are in the process of contraction before the muscle

has had time to relax completely from the twitch of the previously stimulated fibers. Maintenance of a sustained muscle contraction is called **tetanus.**

Tetanus can be demonstrated in the laboratory by setting the stimulator to deliver shocks automatically to the muscle at an ever-increasing frequency until the twitches fuse into a smooth contraction. This is similar to what occurs in the body when different motor neurons in the spinal cord are activated at slightly different times.

If the stimulator is left on so that the muscle remains in tetanus, a gradual decrease in contraction strength will be observed. This is due to muscle **fatigue.** True muscle fatigue rarely occurs in the body because the sensations of muscle pain and depletion of the neurotransmitter at the neuromuscular junction usually causes exercise to cease before the muscle's energy stores have been depleted.

A. Summation, Tetanus, and Fatigue in Frog Gastrocnemius Muscle

Summation, tetanus, and fatigue can be demonstrated with the frog gastrocnemius muscle preparation of the previous exercise; the stimulating electrodes should be placed within the muscle.

Procedure

1. Set the stimulus voltage above threshold, and press down rapidly two or three times on the switch that delivers a single pulse to the muscle. If this is done rapidly enough, successive twitches can "ride piggyback" on preceding twitches (fig. 15.1).
2. Set the stimulus switch to deliver shocks automatically to the muscle at a frequency of about one per second. Gradually increase the frequency of stimulation until the twitches fuse into a smooth, sustained contraction.
3. Maintain stimulation until the strength of contraction gradually diminishes because of muscle fatigue.
4. Enter your recordings in the laboratory report.

Figure 15.1 A recording of the summation of two muscle twitches on a Physiograph. Note that contraction to the second stimulus is greater than contraction to the first stimulus (the intensity of the first and second stimulus is the same).

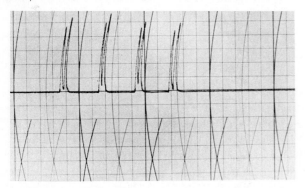

B. Twitch, Summation, and Tetanus in Human Muscle

The properties of frog muscle observed in vitro duplicate the behavior of human muscle in vivo in many ways. A single pulse of electrical stimulation produces a single short contraction (twitch), and many pulses of stimulation delivered in rapid succession produce a summation of twitches resulting in a smooth, graded muscular contraction and eventually in tetanus.

Clinical Significance

Sustained muscular spasm (*tetany*) may be produced by hypocalcemia and alkalosis. (The most common cause of tetany is alkalosis produced by hyperventilation.) Cramps may be caused by a variety of conditions including salt depletion. General muscle weakness may be caused by alterations in plasma potassium levels (which can be due to excessive diarrhea or vomiting).

Muscular dystrophy is a name given to a variety of diseases that involve progressive weakening of skeletal (and sometimes also heart) muscles not associated with inflammation or neural disease. In severe forms of these diseases, great amounts of myofilaments and sarcomeres are lost and replaced by fibrous connective tissue and fat.

Procedure

1. Rub a small amount of electrolyte gel on the skin near the wrist, and attach an ECG electrode plate to this area with an elastic band. Rub electrolyte gel on a second ECG electrode plate, and place it on the anterior, medial area of the arm just below (distal to) the elbow. Do not attach this electrode to the arm, since this is the exploring electrode (fig. 15.2).

Figure 15.2 The placement of electrodes for eliciting finger twitches in response to electrical stimulation.

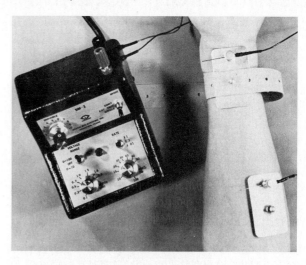

2. Attach the electrode plates to a stimulator. Make sure that the stimulator is *off* at this time.
3. Set the stimulus intensity at 15 volts (V) and deliver a single electrical pulse. If no twitch is observed or felt in the fingers, move the exploring electrode around the medial area of the forearm until an effect is seen or felt. (See fig. 15.2 for the approximate position of the electrode.)

☞ **Note:** *Increased stimulus intensity may be necessary for some people, **but do not exceed 30 V! Most of the time an effect can be obtained at a lower voltage by moving the exploring electrode to a slightly different position. A tingling sensation means that the stimulus intensity is adequate, although the electrode position may have to be changed.***

4. Once a muscle twitch has been observed, set the stimulator so that it automatically delivers one electrical pulse per second. Adjust the exploratory plate so that only one finger twitches.
5. Keep the stimulus intensity constant, and increase the **frequency** of stimulation gradually until a maximum contraction is reached. Gradually decrease the stimulus frequency until the individual twitching responses return.

Procedure for Physiogrip

1. As in the previous procedure, determine the correct points for placement of the electrodes on the forearm so that the *flexor digitorum superficialis* muscle is stimulated. This will result in flexion of the finger that grips the trigger of the Physiogrip.

Figure 15.3 The student is using the Physiogrip to demonstrate muscle twitches, summation, and tetanus.

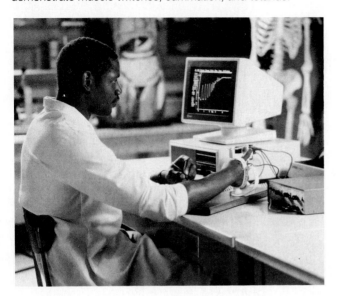

2. Hold the Physiogrip (fig. 15.3) with light pressure while another student delivers electrical shocks using the stimulator. Start with about 15 V at a duration of 1 millisecond, and gradually increase the voltage until threshold is observed.
3. Continue to increase the voltage in small increments, demonstrating the gradation of twitches in response to stronger stimuli.
4. With the voltage constant, gradually increase the frequency of stimulation to demonstrate tetanus.

Laboratory Report 15

Name _____

Date _____

Section _____

Summation, Tetanus, and Fatigue

Read the assigned sections of the textbook before completing the laboratory report.

A. Summation, Tetanus, and Fatigue in Frog Gastrocnemius Muscle

1. Tape your recording to (or draw a facsimile in) the space below.

2. Label twitch, summation, tetanus, and fatigue in your recording.

3. Describe how summation of muscle twitches is produced. Using this information, explain how variations in the strength of muscle contraction are produced.

4. Explain the causes of muscle fatigue. What would happen if a muscle could continue to contract until it ran out of *ATP?*

B. Twitch, Summation, and Tetanus in Human Muscle

1. Describe the results of your procedure in the space below.

2. What can you conclude about the production of normal muscular movements?

3. Explain how a sustained muscle contraction was produced in vitro in this exercise and how sustained muscle contractions are produced in vivo.

Muscles of the Head and Neck

Before coming to class, review the following sections in chapter 13 of the textbook: "Muscles of Facial Expression," "Muscles of Mastication," "Muscles That Move the Tongue," and "Muscles of the Neck."

Introduction

There are more than 600 muscles in the body. Learning the muscles will be more meaningful if their positions are noted under the skin. Many of the muscles to be learned can be felt (palpated) and identified as they are contracted. Knowing the location of a muscle helps when learning its primary action. In general, muscles that cooperate to accomplish a particular movement are said to be *synergistic*. Muscles that oppose each other are said to be *antagonistic*.

Objectives

Students completing this exercise will be able to:

1. Identify the muscles of facial expression and their actions.
2. Identify the muscles of the head and neck, and describe the action(s) of each muscle.

Materials

1. Drawing pencils
2. Laboratory manikin
3. Wall charts of human muscles
4. Articulated human skeleton
5. Reference text and atlas
6. Colored pencils

A. Muscles of Facial Expression

1. Identify, label, and color-code the following muscles in figure 16.1:

   Buccinator
   Depressor labii inferioris
   Frontalis
   Occipitalis
   Levator labii superioris
   Mentalis
   Nasalis
   Orbicularis oculi
   Orbicularis oris
   Platysma
   Risorius
   Depressor anguli oris
   Zygomaticus
   Temporalis

2. Complete table 16.1 by writing in the actions of the muscles of facial expression.

B. Muscles of Mastication

1. Identify, label, and color-code the following muscles on figures 16.1 and 16.2:

   Lateral pterygoid
   Masseter
   Medial pterygoid
   Temporalis

2. Complete table 16.2 by writing in the actions of the muscles of mastication.

Figure 16.1 Superficial facial muscles.

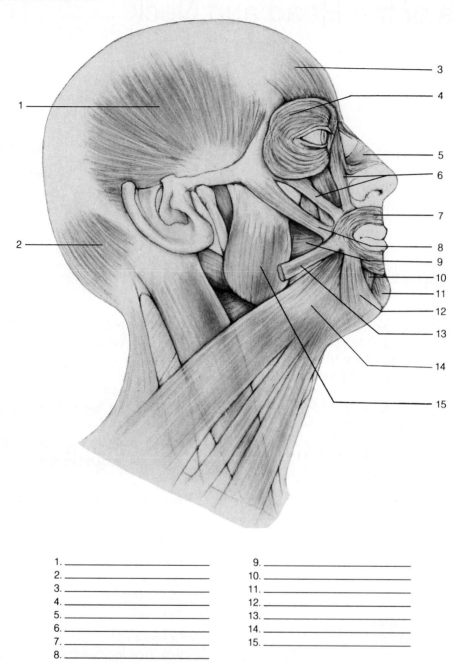

1. _____
2. _____
3. _____
4. _____
5. _____
6. _____
7. _____
8. _____

9. _____
10. _____
11. _____
12. _____
13. _____
14. _____
15. _____

C. Muscles of the Neck

1. Identify, label, and color-code the following muscles in figure 16.3:

Infrahyoid muscles
    Omohyoid
    Sternohyoid
    Thyrohyoid
Middle pharyngeal constrictor

Posterior muscles
    Sternocleidomastoid
    Trapezius
Suprahyoid muscles
    Digastric: (a) anterior belly, (b) posterior belly
    Hyoglossus
    Mylohyoid
    Stylohyoid

2. Complete table 16.3, indicating the action of the muscles of the neck.

Table 16.1 Muscles of Facial Expression

Muscle	Action
Buccinator	
Corrugator	
Depressor anguli oris	
Depressor labii inferioris	
Frontalis	
Levator labii superioris	
Mentalis	
Nasalis	
Occipitalis	
Orbicularis oculi	
Orbicularis oris	
Platysma	
Risorius	
Zygomaticus	

Figure 16.2 Muscles of mastication: (*a*) a superficial view; (*b*) a deep view.

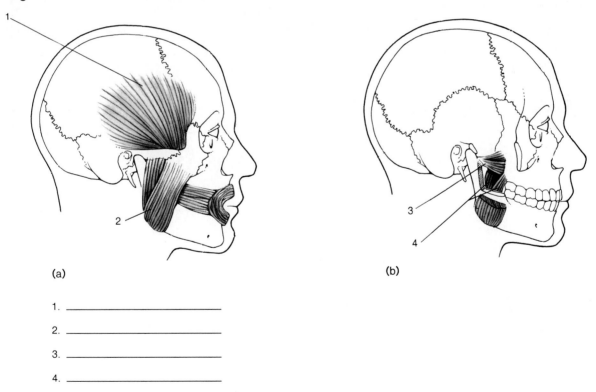

(a)

(b)

1. _____

2. _____

3. _____

4. _____

Figure 16.3 Muscles of the neck.

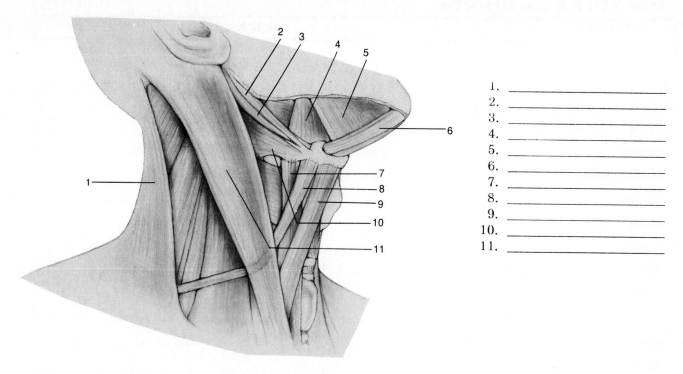

1. _____
2. _____
3. _____
4. _____
5. _____
6. _____
7. _____
8. _____
9. _____
10. _____
11. _____

Table 16.2 Muscles of Mastication	
Muscle	**Action**
Temporalis	
Masseter	
Medial pterygoid	
Lateral pterygoid	

Table 16.3 Muscles of the Neck

Muscle	Action
Sternocleidomastoid	
Trapezius	
Digastric	
Hyoglossus	
Mylohyoid	
Stylohyoid	
Omohyoid	
Sternohyoid	
Thyrohyoid	

Laboratory Report 16

Name _____

Date _____

Section _____

Muscles of the Head and Neck

Read the assigned sections in the textbook before completing the laboratory report.

1. Name the muscle that fits each of the following descriptions:

Description	Muscle
Puckers the mouth	_____
Raises the eyebrows	_____
Closes the eye	_____
Winks the eye	_____
Produces a smile	_____
Produces a frown	_____

2. Name the muscle that is synergistic with the temporalis. _____
3. What muscle opens the mouth? _____
4. What muscle turns the head sideways? _____
5. Name the muscle that retracts the tongue. _____
6. What muscle moves the jaw laterally? _____

Muscles of the Pectoral Girdle and Upper Extremity

17

Before coming to class, review the following sections in chapter 13 of the textbook: "Muscles That Act on the Pectoral Girdle," "Muscles That Move the Humerus (Brachium)," "Muscles of the Forearm That Move the Wrist, Hand, and Fingers," and "Muscles of the Hand."

Introduction

Of the nine muscles that span the shoulder joint to insert on the humerus, only two of them, the pectoralis major and latissimus dorsi, do not originate on the scapula. The powerful muscles of the brachium are responsible for movement of the forearm at the elbow and at the radioulnar joint. The muscles that cause hand movement are positioned along the forearm; they perform four primary actions on the hand: supination, pronation, flexion, and extension.

Objectives

Students completing this exercise will be able to:

1. Identify the muscles of the pectoral girdle and upper extremity.
2. Describe the actions of each muscle, and identify synergists and antagonists.

Materials

1. Drawing pencils
2. Laboratory manikin
3. Wall charts of human muscles
4. Articulated human skeleton
5. Reference text and atlas
6. Colored pencils

A. Muscles That Act on the Pectoral Girdle

1. Identify, label, and color-code the following muscles in figures 17.1, 17.2, and 17.3:

   Anterior group
   > Pectoralis minor
   > Serratus anterior
   > Subclavius

   Posterior group
   > Levator scapulae
   > Rhomboideus major
   > Rhomboideus minor
   > Trapezius

2. Complete table 17.1 by writing in the actions of the muscles that act on the pectoral girdle.

B. Muscles That Move the Brachium

1. Identify, label, and color-code the following muscles in figures 17.1, 17.2, and 17.3:

   Biceps brachii
   Brachialis
   Coracobrachialis
   Deltoid
   Infraspinatus
   Latissimus dorsi
   Long head of biceps brachii
   Pectoralis major
   Subscapularis
   Supraspinatus
   Teres major
   Teres minor

2. Complete table 17.2 providing the actions of the muscles that move the humerus.

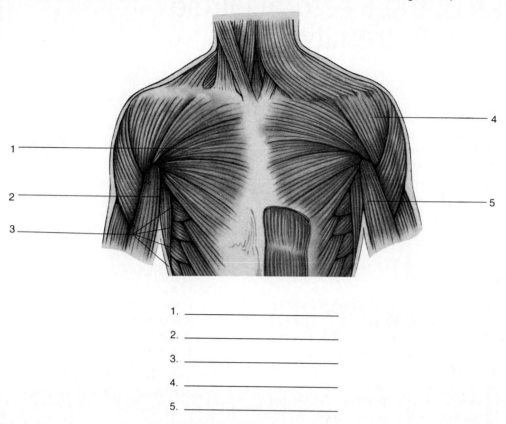

1. _____

2. _____

3. _____

4. _____

5. _____

Table **17.1** Muscles That Act on the Pectoral Girdle	
Muscle	**Action**
Serratus anterior	_____
Pectoralis minor	_____
Trapezius	_____
Levator scapulae	_____
Rhomboideus major	_____
Rhomboideus minor	_____
Subclavius	_____

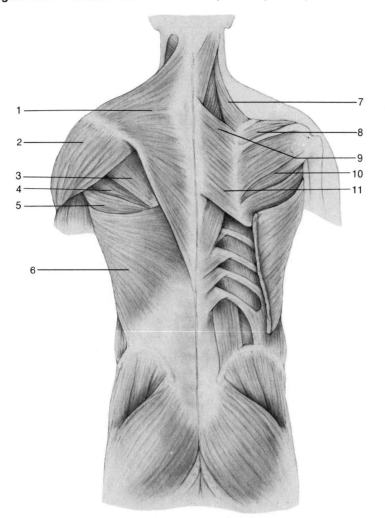

1. _____
2. _____
3. _____
4. _____
5. _____
6. _____
7. _____
8. _____
9. _____
10. _____
11. _____

Table 17.2 Muscles That Move the Humerus

Muscle	Action
Latissimus dorsi	
Pectoralis major	
Coracobrachialis	
Deltoid	
Infraspinatus	
Subscapularis	
Supraspinatus	
Teres major	
Teres minor	

Figure 17.3 Deep muscles of the thorax and brachium.

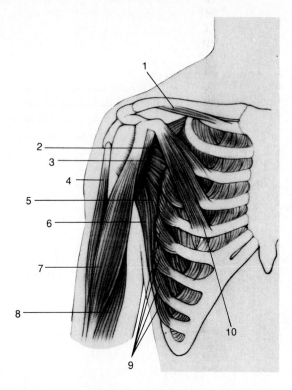

1. _____
2. _____
3. _____
4. _____
5. _____
6. _____
7. _____
8. _____
9. _____
10. _____

C. Muscles That Act on the Antebrachium

1. Identify, label, and color-code the following muscles in figures 17.3 and 17.4:

   Biceps brachii (long and short heads)
   Brachialis
   Brachioradialis
   Triceps brachii (long, medial, and lateral heads)

2. Review the anatomy of the upper extremity by referring to figure 17.5 and identifying similar structures on your own body.
3. Complete table 17.3, providing the action of the muscles that act on the forearm.

Figure 17.4 Right shoulder and brachium: (*a*) an anterior view; (*b*) a posterior view.

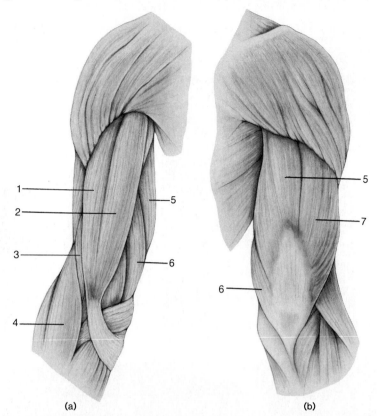

1. _____
2. _____
3. _____
4. _____
5. _____
6. _____
7. _____

(a)   (b)

Table **17.3** Muscles That Act on the Forearm	
Muscle	**Action**
Biceps brachii	
Brachialis	
Brachioradialis	
Triceps brachii	

Figure 17.5 Surface anatomy showing some of the muscles of the thorax and upper extremity: (*a*) an anterolateral view of the trunk; (*b*) flexion of the upper extremities; (*c*) extension of the upper extremities; and (*d*) an anterior view of the upper extremity.

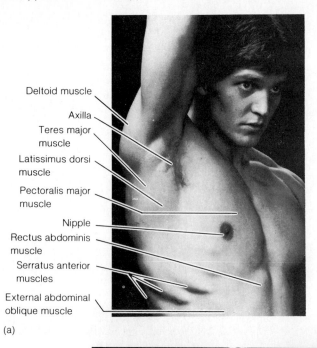

Deltoid muscle

Axilla

Teres major muscle

Latissimus dorsi muscle

Pectoralis major muscle

Nipple

Rectus abdominis muscle

Serratus anterior muscles

External abdominal oblique muscle

(a)

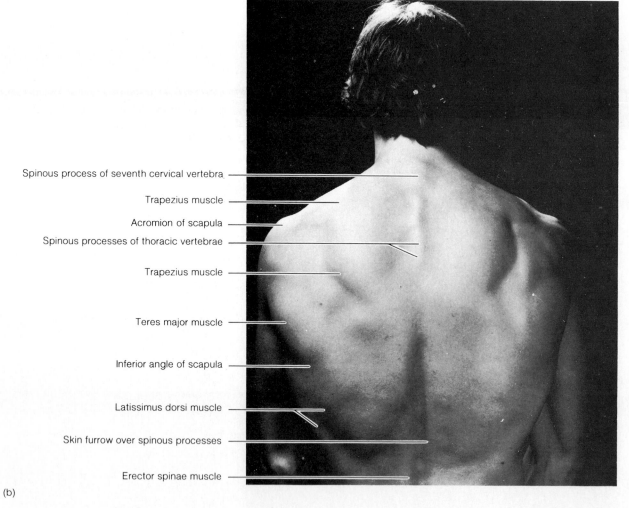

Spinous process of seventh cervical vertebra

Trapezius muscle

Acromion of scapula

Spinous processes of thoracic vertebrae

Trapezius muscle

Teres major muscle

Inferior angle of scapula

Latissimus dorsi muscle

Skin furrow over spinous processes

Erector spinae muscle

(b)

Figure 17.5 Continued

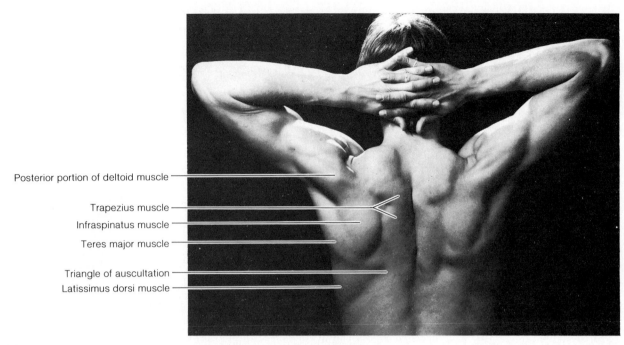

Posterior portion of deltoid muscle

Trapezius muscle
Infraspinatus muscle
Teres major muscle

Triangle of auscultation
Latissimus dorsi muscle

(c)

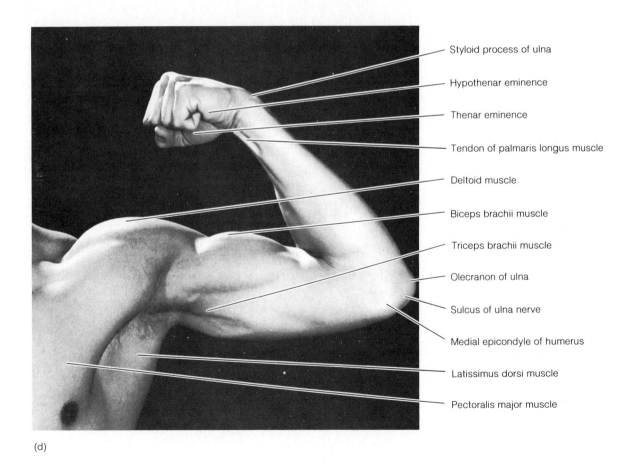

Styloid process of ulna

Hypothenar eminence

Thenar eminence

Tendon of palmaris longus muscle

Deltoid muscle

Biceps brachii muscle

Triceps brachii muscle

Olecranon of ulna

Sulcus of ulna nerve

Medial epicondyle of humerus

Latissimus dorsi muscle

Pectoralis major muscle

(d)

Muscles of the Pectoral Girdle and Upper Extremity   137

Figure 17.6 Deep muscles of the forearm: (a) rotators; (b) flexors; (c) extensors.

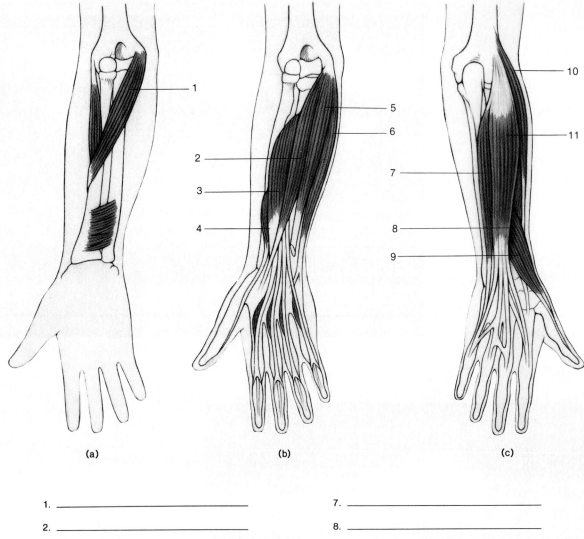

(a) (b) (c)

1. _____   7. _____

2. _____   8. _____

3. _____   9. _____

4. _____   10. _____

5. _____   11. _____

6. _____

D. Muscles That Move the Wrist, Hand, and Fingers

1. Identify, label, and color-code the following muscles in figure 17.6:

   Pronation of the hand
   Pronator teres
   Flexion of the wrist, hand, and fingers
   Flexor carpi radialis
   Flexor carpi ulnaris
   Superficial digital flexor
   Flexor pollicis longus
   Palmaris longus

   Extension of the hand
   Abductor pollicis longus
   Extensor carpi radialis longus
   Extensor carpi ulnaris
   Extensor digitorum communis
   Extensor pollicis longus

2. Complete table 17.4, providing the action of the muscles that move the wrist, hand, and fingers.

Table 17.4 Muscles That Move the Wrist, Hand, and Fingers

Muscle	Action
Pronator teres	
Flexor carpi radialis	
Flexor carpi ulnaris	
Superficial digital flexor	
Flexor pollicis longus	
Palmaris longus	
Abductor pollicis longus	
Extensor carpi radialis brevis	
Extensor carpi radialis longus	
Extensor carpi ulnaris	
Extensor digitorum communis	
Extensor pollicis longus	

Laboratory Report 17

Name _____

Date _____

Section _____

Muscles of the Pectoral Girdle and Upper Extremity

Read the assigned sections of the textbook before completing the laboratory report.

A and B. Muscles That Act on the Pectoral Girdle and Muscles That Move the Humerus

1. The muscles that act on the scapula originate on the _____ skeleton.

2. Except for the pectoralis major and latissimus dorsi, the muscles that insert on the humerus have their origin on the _____.

3. The large, triangular muscle that powerfully adducts the arm is the _____.

4. Describe exercises that would strengthen each of the following:

   a. Pectoralis major _____

   b. Deltoid _____

   c. Triceps brachii _____

   d. Latissimus dorsi _____

5. On the following diagram, sketch the position of the trapezius, levator scapulae, rhomboideus, latissimus dorsi, deltoid, supraspinatus, teres minor, and teres major. Draw the superficial muscles on the right side of the skeleton and the deep muscles on the left.

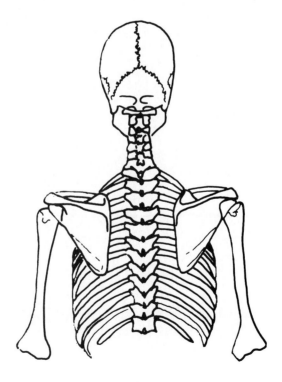

C and D. Muscles That Act on the Forearm and Muscles That Move the Wrist, Hand, and Fingers

1. Three muscles that flex the forearm are the _____ , _____ , and _____ .

2. Which muscles of the forearm have their origin on the medial epicondyle of the humerus?

3. Two primary flexors of the hand are the _____ and _____ .

Muscles of the Trunk, Pelvic Girdle, and Lower Extremity

Before coming to class, review the following sections in chapter 13 of the textbook: "Muscles of the Abdominal Wall," "Muscles of the Pelvic Outlet," "Muscles of the Vertebral Column," "Muscles That Move the Thigh," "Muscles of the Thigh That Move the Leg," "Muscles of the Leg That Move the Ankle, Foot, and Toes," and "Muscles of the Foot."

Introduction

Because humans are bipedal (walk on two appendages), the spinal column, gluteal region, knee joint, ankle, and foot are highly modified to support the weight of the body. The bipedal stance allows the hands to be free to manipulate objects, but some potential structural weaknesses result. The vertebral column and abdominal area are particularly susceptible to strain from improper lifting. The knee joint and ankle are common areas for shearing stress and sprains. In certain sports there is a high incidence of injury involving these regions of the body.

Objectives

Students completing this exercise will be able to:

1. Identify the muscles of the trunk and lower extremities.
2. Identify the muscles that form specified groups.
3. Describe the actions of each muscle, and identify the synergists and antagonists.

Materials

1. Drawing pencils
2. Laboratory manikin
3. Wall charts of human muscles
4. Articulated human skeleton
5. Reference text and atlas
6. Colored pencils

A. Muscles of the Trunk

1. Identify, label, and color-code the following muscles and structures in figures 18.1, 18.2, and 18.3:

   Muscles of the abdominal wall
       External abdominal oblique
       Internal abdominal oblique
       Linea alba
       Rectus abdominis
       Sheath of rectus abdominis
       Transverse abdominis
   Muscles of the vertebral column
       Diaphragm
       Iliacus
       Inguinal ligament
       Psoas major
       Quadratus lumborum
           Iliocostalis thoracis
           Longissimus thoracis
           Spinalis thoracis
       Semispinalis capitis
       Splenius capitis

2. Review the anatomy of the abdomen by referring to figure 18.4.
3. Complete table 18.1 by providing the actions of the abdominal muscles.

Figure 18.1   Muscles of the anterior abdominal wall.

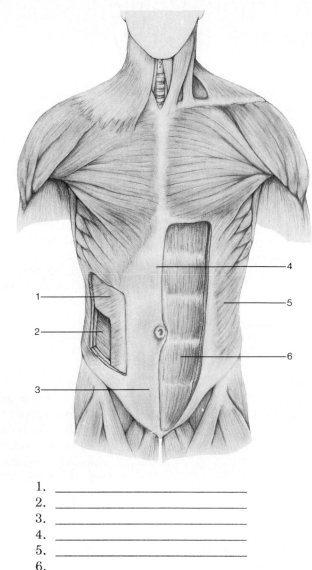

1. _____
2. _____
3. _____
4. _____
5. _____
6. _____

Figure 18.2 Muscles of the vertebral column.

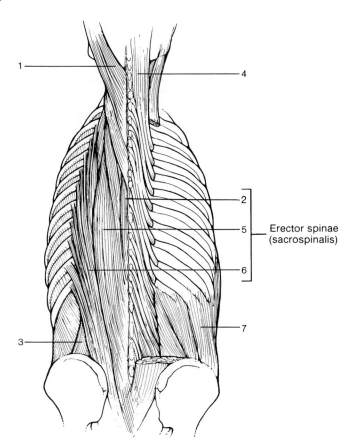

Erector spinae
(sacrospinalis)

1. _____
2. _____
3. _____
4. _____
5. _____
6. _____
7. _____

Figure 18.3 Anterior pelvic muscles.

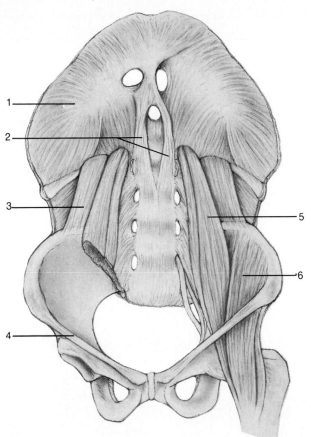

1. _____
2. _____
3. _____
4. _____
5. _____
6. _____

Figure 18.4 The surface anatomy of the anterior abdominal region.

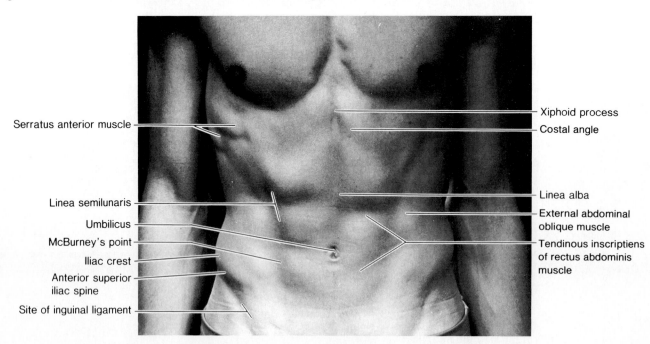

Serratus anterior muscle

Linea semilunaris

Umbilicus

McBurney's point

Iliac crest

Anterior superior
iliac spine

Site of inguinal ligament

Xiphoid process

Costal angle

Linea alba

External abdominal
oblique muscle

Tendinous inscriptiens
of rectus abdominis
muscle

Table 18.1 Muscles of the Abdominal Wall

Muscle	Action
External abdominal oblique	
Internal abdominal oblique	
Transverse abdominis	
Rectus abdominis	

Figure 18.5 The deep gluteal muscles (a posterior view).

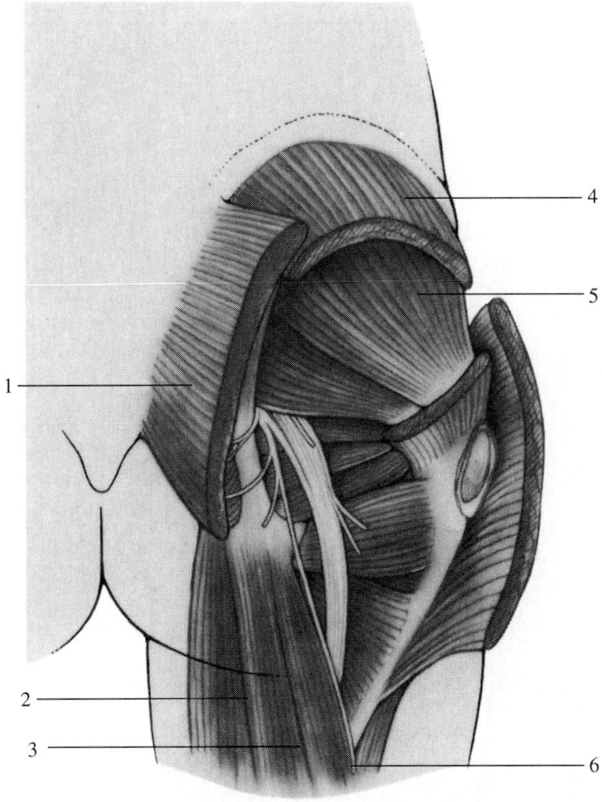

1. _____
2. _____
3. _____
4. _____
5. _____
6. _____

B. Muscles of the Hip

1. Identify, label, and color-code the following muscles and structures in figures 18.3, 18.5, and 18.6:

   Anterior group
       Iliacus
       Psoas major
   Posterior and lateral group
       Gluteus maximus
       Gluteus medius
       Gluteus minimus
       Tensor fasciae latae

2. Complete table 18.2 by providing the actions of the muscles that move the thigh at the hip.

Table 18.2 Anterior and Posterior Muscles That Move the Thigh at the Hip

Muscle	Action
Illiacus	
Psoas major	
Gluteus maximus	
Gluteus medius	
Gluteus minimus	
Tensor fasciae latae	

Figure 18.6 Muscles of the thigh (an anterior view).

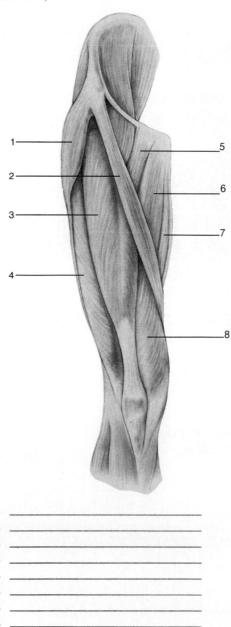

1. _____
2. _____
3. _____
4. _____
5. _____
6. _____
7. _____
8. _____

C. Muscles of the Thigh

1. Identify, label, and color-code the following muscles in figures 18.5, 18.6, and 18.7:

   Anterior muscles
       Quadriceps femoris
           Rectus femoris
           Vastus intermedius
           Vastus lateralis
           Vastus medialis
       Sartorius
   Medial muscles
       Adductor brevis
       Adductor longus
       Adductor magnus
       Gracilis
       Pectineus
   Posterior muscles (hamstrings)
       Biceps femoris
       Semimembranosus
       Semitendinosus

2. Review the anatomy of the thigh by referring to figure 18.8 and by palpating these structures on your own thigh.

3. Complete tables 18.3, 18.4, and 18.5, providing the action of the indicated muscles.

Figure 18.7 Muscles of the thigh (a posterior view).

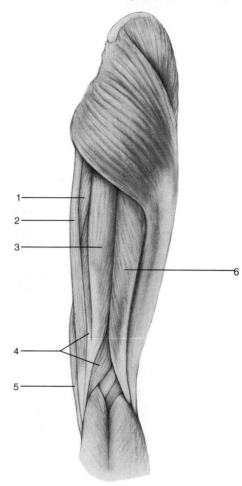

1. _____
2. _____
3. _____
4. _____
5. _____
6. _____

Figure 18.8 The surface anatomy of the right thigh and knee: (a) an anterior view; (b) a lateral view.

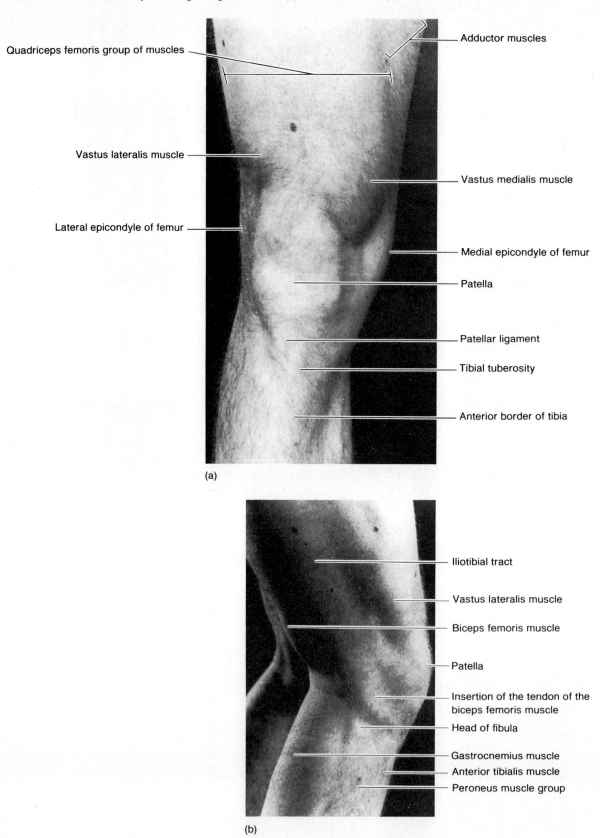

Quadriceps femoris group of muscles

Adductor muscles

Vastus lateralis muscle

Vastus medialis muscle

Lateral epicondyle of femur

Medial epicondyle of femur

Patella

Patellar ligament

Tibial tuberosity

Anterior border of tibia

(a)

Iliotibial tract

Vastus lateralis muscle

Biceps femoris muscle

Patella

Insertion of the tendon of the biceps femoris muscle

Head of fibula

Gastrocnemius muscle

Anterior tibialis muscle

Peroneus muscle group

(b)

Table 18.3 Anterior Thigh Muscles That Move the Leg

Muscle	Action
Sartorius	
Quadriceps femoris	
Rectus femoris	
Vastus lateralis	
Vastus medialis	
Vastus intermedius	

Table 18.4 Medial Thigh Muscles That Move the Leg

Muscle	Action
Gracilis	
Pectineus	
Adductor longus	
Adductor brevis	
Adductor magnus	

Table 18.5 Posterior Thigh Muscles That Move the Leg

Muscle	Action
Biceps femoris	
Semitendinosus	
Semimembranosus	

D. Muscles of the Leg

1. Identify, label, and color-code the following muscles in figures 18.9, 18.10, and 18.11:

   Anterior group
   Extensor digitorum longus
   Extensor hallucis longus
   Tibialis anterior
   Lateral group
   Peroneus brevis
   Peroneus longus
   Peroneus tertius
   Posterior group
   Flexor digitorum longus
   Flexor hallucis longus (not shown)
   Gastrocnemius
   Plantaris
   Soleus
   Tendo calcaneus
   Tibialis posterior (not shown)

2. Review the anatomy of the leg by referring to figure 18.12 and by palpating these structures on your own leg.

3. Complete table 18.6 by providing the actions of the muscles of the leg.

Figure 18.9 Anterior crural muscles.

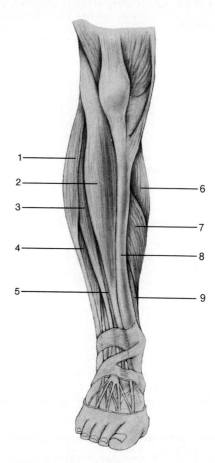

1. _____
2. _____
3. _____
4. _____
5. _____
6. _____
7. _____
8. _____
9. _____

Figure 18.10 Lateral crural muscles.

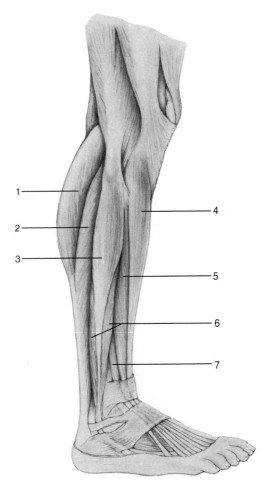

1. _____
2. _____
3. _____
4. _____
5. _____
6. _____
7. _____

Figure 18.11 Posterior crural muscles and the popliteal fossa.

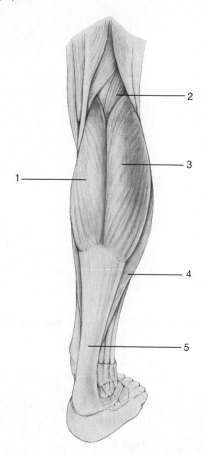

1. _____
2. _____
3. _____

4. _____
5. _____

Figure 18.12 The surface anatomy of the right leg and foot: (*a*) a lateral view; (*b*) a medial view; (*c*) an anterior view; (*d*) a posterior view.

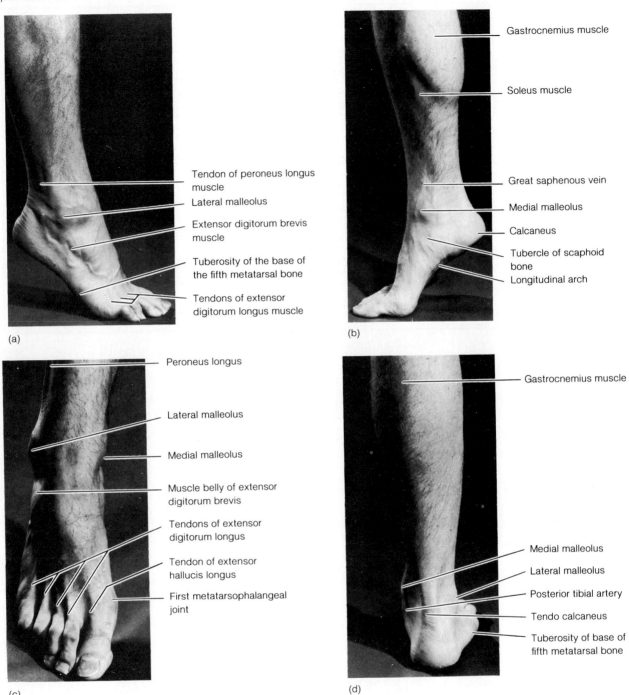

(a)

Tendon of peroneus longus muscle
Lateral malleolus
Extensor digitorum brevis muscle
Tuberosity of the base of the fifth metatarsal bone
Tendons of extensor digitorum longus muscle

(b)

Gastrocnemius muscle
Soleus muscle
Great saphenous vein
Medial malleolus
Calcaneus
Tubercle of scaphoid bone
Longitudinal arch

(c)

Peroneus longus
Lateral malleolus
Medial malleolus
Muscle belly of extensor digitorum brevis
Tendons of extensor digitorum longus
Tendon of extensor hallucis longus
First metatarsophalangeal joint

(d)

Gastrocnemius muscle
Medial malleolus
Lateral malleolus
Posterior tibial artery
Tendo calcaneus
Tuberosity of base of fifth metatarsal bone

Table 18.6 Muscles of the Leg That Move the Ankle, Foot, and Toes

Muscle	Action
Tibialis anterior	
Extensor digitorum longus	
Extensor hallucis longus	
Peroneus tertius	
Peroneus longus	
Peroneus brevis	
Gastrocnemius	
Soleus	
Plantaris	
Flexor hallucis longus	
Flexor digitorum longus	
Tibialis posterior	

Muscles of the Trunk, Pelvic Girdle, and Lower Extremity

Read the assigned sections in the textbook before completing the laboratory report.

A. Muscles of the Trunk

1. What exercises can be done to strengthen specific muscles of the abdomen?

2. Which muscles form the boundary of the femoral triangle? Why is this region clinically significant?

B and C. Muscles of the Hip and Thigh

1. What are the biomechanical advantages of having four distinct muscles composing the quadriceps femoris rather than one large muscle? In the answer, consider that the rectus femoris acts on two joints.

2. Explain how the adductor muscles assist in stabilizing the hip joint.

D. Muscles of the Leg

1. Firmly press on the front, sides, and back of the ankle as the foot is moved. The tendons of which muscles can be palpated anteriorly, laterally, and posteriorly?

2. Which muscles of the leg have their tendons of insertion passing behind the lateral malleolus? Which behind the medial malleolus?

3. Identify the popliteal fossa. What structures are found in this region?

General Questions

Name the muscle or muscles that fit the following descriptions:

Description	Muscle (s)
Hamstrings	
Lower back muscle	
Abdominal "strap" muscle	
Tailor's muscle	
Calf muscle	
Adductor muscles	
Abdominal compression	

Cat Musculature

Introduction

The best way to learn anatomy is to dissect a laboratory specimen. Human cadavers are ideal for human anatomy courses; but they are not always available, nor are they practical for all courses. Often embalmed cats are dissected instead. Despite the definite anatomical differences between cats and humans, there are far greater similarities than differences; the dissection of cats is thus a suitable substitute for human cadaver dissections.

Objectives

Students completing this exercise will be able to:

1. Demonstrate the proper technique of muscle dissection.
2. Identify the muscles of the cat, and correlate them with corresponding muscles in the human.
3. Identify the muscles that are present in the cat but not in the human.
4. Describe the action of each muscle and its synergists and antagonists.

Materials

1. Embalmed cats
2. Dissecting trays
3. Dissection instruments
4. Plastic bags and name tags
5. Muslin cloth
6. Reference text

Student Instructions for Skin Removal

Cats purchased from biological supply houses are specially embalmed for dissection with an alcohol-formaldehyde-glycerol solution, then packed individually in plastic bags for shipment and storage. Embalming solution may have an objectionable smell and may irritate the skin. People with sensitive skin may want to wear rubber surgical gloves while dissecting. Good ventilation is important. Excess embalming fluid should be drained from the specimen before removing its skin.

Students should work in pairs to dissect their specimen. Instructions for skin removal and specimen preparation are as follows:

1. Place the cat on a dissecting tray with its dorsal side up. Using a sharp scalpel, make a short, shallow incision through the skin across the nape of the neck.
2. With the scissors, continue to cut a dorsal midline incision forward over the skull, and then down the back to about two inches onto the tail. Sever the tail with bone shears or a saw and discard.
3. Make an incision around the neck and down each foreleg to the paws. Continuing with scissors, make a circular incision around each wrist.
4. Beginning at the base of the tail, make incisions down each of the hind legs to the ankles. Make a circular cut around each ankle.
5. Carefully remove the skin, using a blunt probe to separate the hypodermis of the skin from the fascia of the superficial muscles. Where it is necessary to use a scalpel, keep the cutting edge directed toward the skin and away from the muscle. As the skin is removed, note that a thin sheet of muscle adheres to the underside of it. This muscle sheet moves the skin and is called the *cutaneous maximus* in the body region and the *platysma* in the neck region. The cutaneous maximus is absent in humans.
6. For a male specimen, make an incision around the genitalia, leaving the skin intact. For a female specimen, remove the mammary glands—longitudinal, glandular masses along the underside of the abdomen and thorax—with the skin.
7. After the specimen is skinned, remove the excess fat and connective tissue to expose the underlying muscles. As you dissect the muscles, be sure to separate them at their natural boundaries. When transection of a muscle is necessary, isolate the muscle from its attached connective tissue, and make a clean cut across the belly of the muscle, leaving the origin and insertion intact.
8. At the end of the laboratory period, discard the skin and removed body tissue in the plastic shipment bag. Wrap the specimen in muslin and store it in a new plastic bag. It might be necessary from time to time to

Figure 19.1 Muscles of the hyoid and larynx regions and extrinsic muscles of the tongue (deep muscles are exposed on the right, superficial muscles on the left).

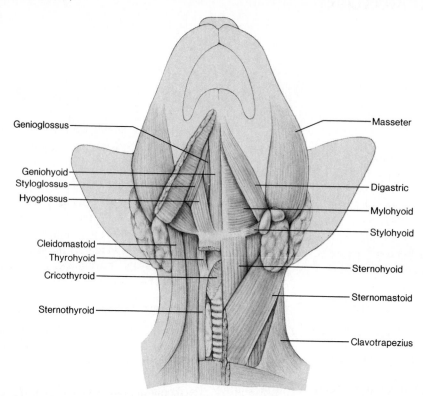

wet the specimen in a preservative solution (usually 2% to 3% phenol). Use caution with a phenol wetting solution; phenol is caustic, poisonous, and may damage the skin and eyes if misused or if used in a concentrated form.

A. Muscles of the Head and Neck

Once the skin has been removed from the cat, place the specimen on a dissection tray, ventral side up. Three large salivary glands on both sides of the neck will partially obscure the underlying muscles. These glands will be examined in more detail at a later date when the digestive system is studied, so tease them away from the supporting connective tissue to expose the neck muscles. Refer to figure 19.1 and plate 1 to identify the muscles in this region.

Superficial Muscles

Identify the *masseter* at the angle of the jaw, the *temporalis* (see fig. 19.2) on the temporal fossa of the skull, the *digastric* on the ventral border of the mandible, and the *mylohyoid,* whose fibers course horizontally deep to the digastric. These are the cat's major muscles of mastication; essentially their origins, insertions, and functions (table 19.1) are the same as those of the human.

The *sternomastoid, sternohyoid,* and *stylohyoid* muscles, named on the basis of their origin and insertion, can be easily located.

Deep Muscles

Remove the mylohyoid, sternohyoid, and sternomastoid muscles from the right side of the specimen (as in fig. 19.1) to expose the deeper neck muscles. Above the hyoid bone are the *geniohyoid,* extending from the hyoid bone to the chin; the *styloglossus,* extending from the skull to the tongue; the *genioglossus,* extending from the chin to the tongue; and the *hyoglossus,* extending from the hyoid to the tongue. Below the hyoid bone and above the thorax are four deep muscles, three of which—*sternothyroid, cricothyroid,* and *thyrohyoid*—are attached to the larynx. The *cleidomastoid* muscle (deep to the sternomastoid) moves the head. In humans there is only one muscle in this region, the sternocleidomastoid.

B. Muscles of the Posterior Shoulder Region

Superficial Muscles

There are eight superficial shoulder muscles on the back of a cat, but only three corresponding muscles in humans.

Figure 19.2 Superficial (right side) and deep (left side) muscles of the back.

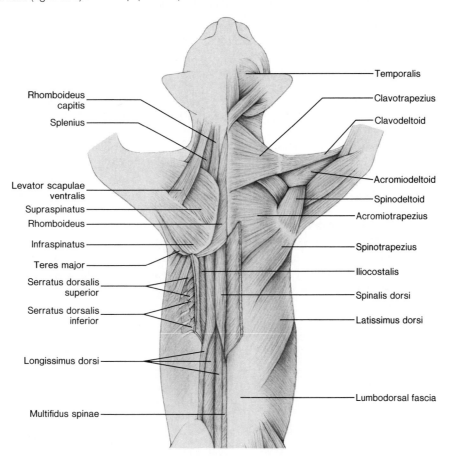

Rhomboideus capitis

Splenius

Levator scapulae ventralis

Supraspinatus

Rhomboideus

Infraspinatus

Teres major

Serratus dorsalis superior

Serratus dorsalis inferior

Longissimus dorsi

Multifidus spinae

Temporalis

Clavotrapezius

Clavodeltoid

Acromiodeltoid

Spinodeltoid

Acromiotrapezius

Spinotrapezius

Iliocostalis

Spinalis dorsi

Latissimus dorsi

Lumbodorsal fascia

Refer to figure 19.2 and table 19.1 as these muscles are identified on the specimen. Also refer to plate 6 to help identify the muscles.

The cat has three separate *trapezius* muscles. The *clavotrapezius* is the upper one, the *acromiotrapezius* is the middle one between the scapulae, and the *spinotrapezius* is the most posterior. The clavotrapezius muscle attaches to the clavicle where the *clavodeltoid* (clavobrachialis) originates. These two muscles are frequently referred to as a single muscle, the *brachiocephalicus*.

Likewise, there are three *deltoid* muscles in the cat. The *acromiodeltoid* is posterior to the clavodeltoid, already identified. The *spinodeltoid* is ventral to the acromiotrapezius and posterior to the acromiodeltoid.

The slender *levator scapulae ventralis* is difficult to identify unless the clavotrapezius is cut and reflected. This muscle has no homologue in humans. The thick *latissimus*

dorsi muscle is immediately posterior to the spinotrapezius. This muscle does not attach to the scapula; its function in the cat is the same as in humans.

Deep Muscles

To identify the deep muscles of the shoulder region, remove the following muscles on the left side: the clavotrapezius, acromiotrapezius, spinotrapezius, spinodeltoid, acromiodeltoid, and latissimus dorsi. The dissection should resemble the left shoulder region depicted in figure 19.2. Also refer to plate 3 to help identify the muscles. Eight muscles can be identified in this dissection, seven of which have their origin or insertion on the scapula (table 19.1).

The *rhomboideus* and *rhomboideus capitis* insert, respectively, on the medial border and on the angle of the scapula. These muscles can best be seen if they are stretched

Table 19.1 Summary of Head, Neck, Shoulder, and Back Musculature

Muscles of the Head and Neck

Muscle	Origin	Insertion	Action
Superficial Group			
Masseter	Zygomatic arch	Mandible	Elevates mandible
Temporal	Temporal fossa of skull	Coronoid process of mandible	Elevates mandible
Digastric	Occipital bone of skull	Ventral border of mandible	Depresses lower jaw
Mylohyoid	Medial surface of mandible	Median raphe	Elevates floor of mouth
Sternomastoid	Manubrium of sternum	Lambdoidal ridge of skull	Turns head
Sternohyoid	First costal cartilage	Body of hyoid	Depresses hyoid
Stylohyoid	Stylohyal bone of hyoid	Body of hyoid	Elevates hyoid
Deep Group			
Geniohyoid	Medial surface of mandible	Body of hyoid	Draws hyoid craniad
Styloglossus	Styloid process of skull	Apex of tongue	Retracts tongue
Genioglossus	Medial surface of mandible	Base of tongue	Draws root of tongue forward
Hyoglossus	Body of hyoid bone	Dorsum of tongue	Retracts tongue
Sternothyroid	First costal cartilage	Thyroid cartilage	Depresses larynx
Cricothyroid	Cricoid cartilage	Thyroid cortilage	Tensor of true vocal cords
Thyrohyoid	Thyroid cartilage	Posterior horn of hyoid	Elevates larynx
Cleidomastoid	Mastoid process of skull	Clavicle	Turns head

Muscles of the Back

Muscle	Origin	Insertion	Action
Serratus dorsalis superior	First nine ribs	Middorsal raphe	Draws ribs craniad
Serratus dorsalis inferior	Last four ribs	Lumbar spinous processes	Draws ribs craniad
Sacrospinalis			
Spinalis dorsi	Last four thoracic vertebrae	Thorac and cervical vertebrae	Extends vertebral column
Longissimus dorsi	Sacral and caudal vertebrae	Trunk and cervical vertebrae	Extends vertebral column
Iliocostalis	As separate muscle bundles from lower thoracic ribs	Three ribs craniad to origin of each bundle	Draws ribs together

Shoulder Muscles

Muscle	Origin	Insertion	Action
Superficial Group			
Clavotrapezius	Lambdoidal crest of skull	Clavicle	Draws scapula dorsocraniad
Acromiotrapezius	Spines of cervical vertebrae	Spine of scapula	Draws scapula dorsad
Spinotrapezius	Spines of thoracic vertebrae	Fascia of scapular muscles	Draws scapula dorsad
Clavodeltoid	Clavicle	Ulna near semilunar notch	Flexes forearm
Acromiodeltoid	Acromion of scapula	Outer surface of spinodeltoid	Flexes and rotates humerus
Spinodeltoid	Spine of scapula	Deltoid ridge of humerus	Flexes and rotates humerus
Levator scapulae ventralis	Atlas and occipital bone	Metacromion process	Draws scapula craniad
Latissimus dorsi	Thoracic and lumbar vertebrae	Shaft of humerus	Pulls arm caudodorsad
Deep Group			
Rhomboideus	Upper thoracic vertebrae	Medial border of scapula	Draws scapula dorsad
Rhomboideus capitis	Lambdoidal ridge of skull	Angle of scapula	Draws scapula craniad
Supraspinatus	Supraspinous fossa	Greater tubercle of humerus	Extends arm
Infraspinatus	Infraspinous fossa	Greater tubercle of humerus	Rotates humerus laterally
Teres major	Axillary border of scapula	Medial surface of humerus	Flexes and rotates humerus
Teres minor	Infraspinous fossa	Greater tubercle of humerus	Rotates humerus
Subscapularis	Subscapular fossa	Lesser tubercle of humerus	Draws humerus medially
Splenius	Middorsal fascial line	Lambdoidal ridge	Turns and elevates head

Figure 19.3 A ventral view of the superficial (left side) and deep (right side) muscles of the thorax and abdomen.

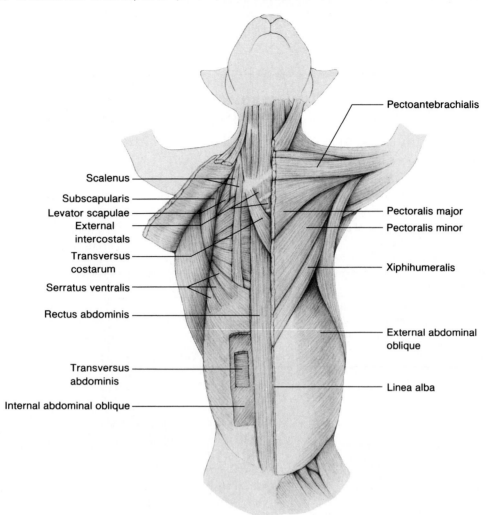

Scalenus

Subscapularis

Levator scapulae

External intercostals

Transversus costarum

Serratus ventralis

Rectus abdominis

Transversus abdominis

Internal abdominal oblique

Pectoantebrachialis

Pectoralis major

Pectoralis minor

Xiphihumeralis

External abdominal oblique

Linea alba

by drawing the front appendages together. Five of the eight muscles of this region originate from the scapula and insert upon the humerus. The *supraspinatus* is positioned on the supraspinous fossa above the scapular spine, while the *infraspinatus* originates on the infraspinous fossa below the spine. The *teres major* and *teres minor* are ventral to the infraspinatus. The teres minor is positioned deep to the spinodeltoid muscle (see fig. 19.5) and may be difficult to find because of its small size. The *subscapularis* muscle is positioned on the subscapular fossa (because it is located on the opposite side of the scapula, one can identify it more easily from a ventral view in a later dissection). The *splenius* is a deep, dorsally positioned neck muscle that turns and raises the head.

C. Muscles of the Back

The muscles of the back (table 19.1) permit movement of the vertebral column and are frequently referred to as *epaxial* muscles. These muscles will be visible after the lumbodorsal fascia is removed on the left side of the specimen. Refer to figure 19.2 to identify these muscles.

The *serratus dorsalis superior* and the *serratus dorsalis inferior* extend at right angles to the spine and attach to the ribs.

Several muscle masses extend over the back from the sacral region to the midthoracic region. These muscles are referred to collectively as the *sacrospinalis,* but they can be divided into three distinct muscle groups: a medial *spinalis dorsi,* an intermediate *longissimus dorsi*, and a lateral *iliocostalis.* The serratus dorsalis superior and inferior partially cover the iliocostalis. The sacrospinalis acts as an extensor of the spinal column to maintain back posture.

D. Thoracic Muscles

The thoracic muscles are illustrated in figure 19.3. Place the specimen on its back and remove the excess fat and fascia on the chest to expose these muscles. The thoracic muscles fall into two groups: the superficial pectoral group and the deep thoracic group.

Table **19.2** Summary of Thoracic and Abdominal Musculature

Thoracic Muscles

Muscle	Origin	Insertion	Action
Pectoral Group			
Pectoantebrachialis	Manubrium	Superficial fascia of forearm	Adducts forelimb
Pectoralis major	Sternum	Pectoral ridge of humerus	Adducts forelimb
Pectoralis minor	Sternum	Pectoral ridge of humerus	Adducts forelimb
Xiphihumeralis	Xiphoid process	Proximal end of humerus	Adducts forelimb
Deep Thoracic Group			
Serratus ventralis	First ten ribs	Medial surface of scapula	Draws scapula to thorax
Levator scapulae	Last five cervical vertebrae	Medial surface of scapula	Draws scapula cranioventrad
Scalenus	Ribs	Cervical transverse processes	Flexes the neck
Transversus costarum	Lateral border of sternum	First rib	Draws sternum craniad
External intercostal	Border of rib	Border of adjacent rib	Protracts the ribs
Internal intercostal	Border of rib	Border of adjacent rib	Retracts the ribs

Abdominal Muscles

Muscle	Origin	Insertion	Action
External abdominal oblique	Lumbodorsal fascia and ribs	Linea alba and pubis	Constricts abdomen
Internal abdominal oblique	Lumbodorsal fascia	Linea alba	Compresses abdomen
Transversus abdominis	Costal cartilages of lower ribs	Linea alba	Constricts abdomen
Rectus abdominis	Pubis	Sternum and costal cartilages	Compresses abdomen

Pectoral Group

Four synergistic muscles that adduct and rotate the brachium are included in the pectoral group (plates 2 and 5). The *pectoantebrachialis* is the most anterior and is superficial to the *pectoralis major*. The straplike pectoantebrachialis should be teased away from the underlying pectoralis major with a sharp dissecting probe. A fan-shaped *pectoralis minor* lies posterior to the pectoralis major. The borders of these two muscles should be isolated. The thin, straplike *xiphihumeralis* is the fourth and most posterior of the group.

Deep Thoracic Group

The pectoral muscles on the right side of the specimen must be severed to expose the deeper muscles of the thorax (fig. 19.3 and plate 2). Clean away the fat, vessels, and connective tissue to provide a clear view of these muscles. Six deep thoracic muscles will be described.

The fan-shaped *serratus ventralis* muscle arises from the ribs and extends obliquely upward to the scapula. The anterior continuation of this muscle is known as the *levator scapulae*. These two muscles cannot be readily separated at their adjunct border.

The *scalenus* extends along the ventrolateral surface of the thorax and can be divided into three distinct muscles: the *scalenus anterior, medius,* and *posterior*. The scalenus medius is the largest of the three. Acting synergistically with the scalenus is the *transversus costarum*

(rectus thoracis), which extends diagonally from the sternum to the first rib. The transversus costarum has no homologue in humans.

Two series of intercostal muscles occupy the intercostal spaces and are associated with respiration. The *external intercostals* are superficial and have caudoventrally directed fibers. The fibers of the deeper *internal intercostals* are directed caudodorsally.

Notice the position of the previously discussed subscapularis muscle on the subscapular fossa of the scapula.

E. Abdominal Muscles

The abdominal muscles of the cat are similar to those of humans. Three of the four paired abdominal muscles—*external abdominal oblique, internal abdominal oblique,* and *transversus abdominis*—appear as layered sheets. The fourth—*rectus abdominis*—is a straplike, paired muscle on both sides of the *linea alba*. These muscles are shown in figure 19.3 and in plate 3, and they are described in table 19.2.

F. Muscles of the Brachium

Six muscles are associated with the brachial region of the cat. These muscles can be grouped as those of the medial and lateral surfaces. Refer to figures 19.4 and 19.5, and table 19.3, for illustrations and descriptions of these muscles. Many of these muscles can also be seen in plates 4 and 5.

Figure 19.4 Muscles of the brachium (the medial surface).

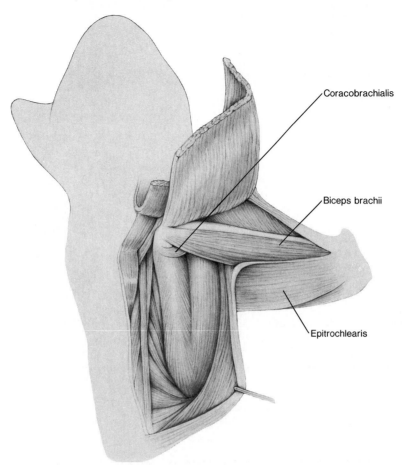

Coracobrachialis

Biceps brachii

Epitrochlearis

Three muscles can be identified on the medial surface: the *coracobrachialis, epitrochlearis,* and *biceps brachii*. In the cat, the coracobrachialis is a small muscle that stabilizes the joint capsule. The epitrochlearis (known also as the *extensor antebrachii*) is a broad, flat muscle on the medial side of the front leg. It is not present in the human. The biceps brachii has two heads and is positioned, for the most part, deep to the insertion of the pectoralis.

Three muscles can be identified on the lateral surface: the *triceps brachii* (three heads), *anconeus,* and *brachialis*. To find the latter two muscles, locate, cut, and reflect the lateral head of the triceps brachii, as illustrated in figure 19.5.

G. Muscles of the Hip and Thigh Regions

The muscles of the hip and thigh region are illustrated in figures 19.6 and 19.7, and described in table 19.4. They can also be seen in the photographs on plates 6 through 8. The superficial fat and connective tissue must be removed to identify these muscles properly. Take care not to remove

the *fascia lata,* an aponeurosis on the anterior and lateral side of the thigh. If the specimen is male, be cautious in the genital area and do not sever the spermatic cord.

Medial Surface

The *sartorius* on the anterior inner surface of the thigh and the *gracilis* on the posterior inner surface (fig. 19.6) are easy to identify. When these muscles have been clearly exposed, transect them across their bellies and reflect their cut edges to reveal the deeper muscles.

The ventral femoral group of muscles called the *quadriceps femoris* comprises the great extensor muscles of the knee. This group includes four muscles: the *vastus medialis, rectus femoris, vastus intermedius* (not shown), and *vastus lateralis*. The vastus lateralis is positioned more laterally than medially on the thigh. The vastus intermedius can be seen when the vastus lateralis and rectus femoris have been separated.

The adductor group of thigh muscles includes the *pectineus, adductor longus,* and *adductor femoris,* as well as the gracilis, already identified. The small pectineus muscle is medial to the vastus medialis. The *adductor magnus* muscle is absent in the cat.

Figure 19.5 Muscles of the brachium (the lateral surface).

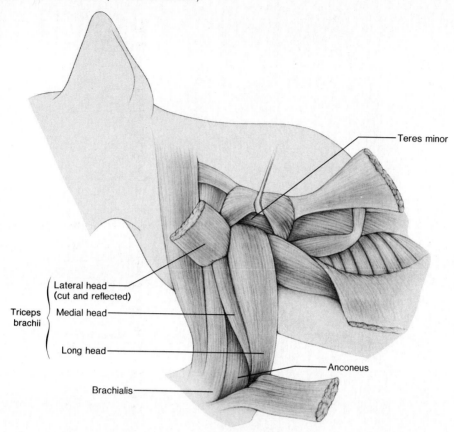

Teres minor

Triceps brachii
- Lateral head (cut and reflected)
- Medial head
- Long head

Anconeus

Brachialis

Table 19.3 Muscles of the Brachium

Muscle	Origin	Insertion	Action
Medial Surface			
Coracobrachialis	Coracoid process	Proximal end of humerus	Adducts humerus
Epitrochlearis	Ventral border of lateral dorsi	Olecranon of ulna	Extends forearm
Biceps brachii	Scapula near glenoid cavity	Radial tuberosity	Flexes forearm
Lateral Surface			
Triceps brachii	Scapula and shaft of humerus	Olecranon of ulna	Extends forearm
Anconeus	Distal end of humerus	Olecranon of ulna	Tensor of elbow joint capsule
Brachialis	Lateral surface of humerus	Proximal end of ulna	Flexes forearm

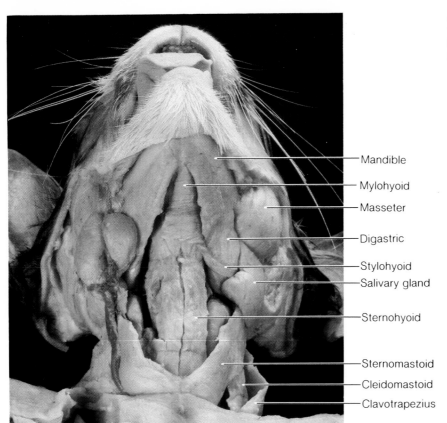

Mandible
Mylohyoid
Masseter
Digastric
Stylohyoid
Salivary gland
Sternohyoid
Sternomastoid
Cleidomastoid
Clavotrapezius

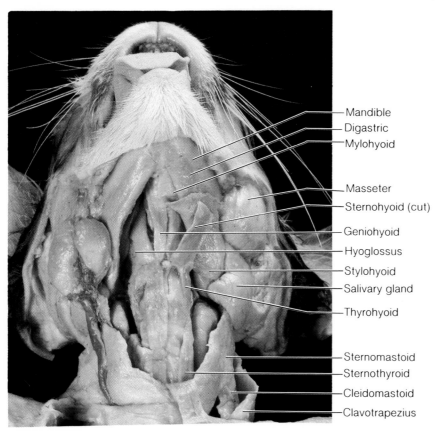

◀ **Exhibit 2**

Deep muscles of the cat's head and neck.

Mandible
Digastric
Mylohyoid
Masseter
Sternohyoid (cut)
Geniohyoid
Hyoglossus
Stylohyoid
Salivary gland
Thyrohyoid
Sternomastoid
Sternothyroid
Cleidomastoid
Clavotrapezius

▶ **Exhibit 3**

Superficial muscles of the cat's thorax.

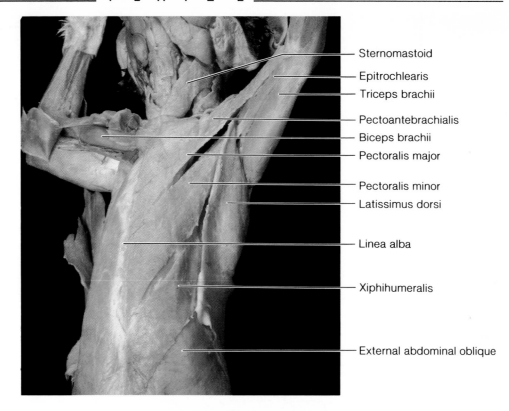

Sternomastoid
Epitrochlearis
Triceps brachii
Pectoantebrachialis
Biceps brachii
Pectoralis major

Pectoralis minor
Latissimus dorsi

Linea alba

Xiphihumeralis

External abdominal oblique

▶ **Exhibit 4**

Deep muscles of the cat's thorax.

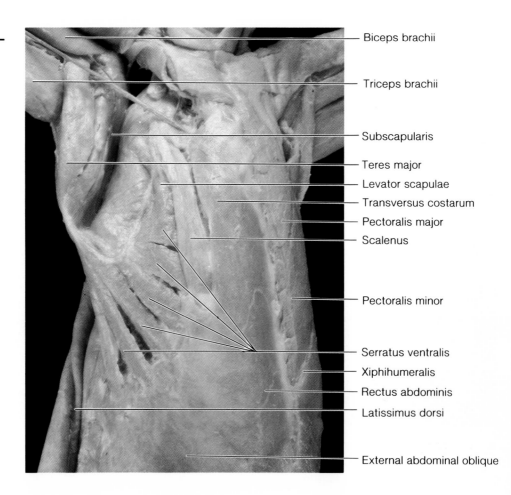

Biceps brachii

Triceps brachii

Subscapularis

Teres major
Levator scapulae
Transversus costarum
Pectoralis major
Scalenus

Pectoralis minor

Serratus ventralis
Xiphihumeralis
Rectus abdominis
Latissimus dorsi

External abdominal oblique

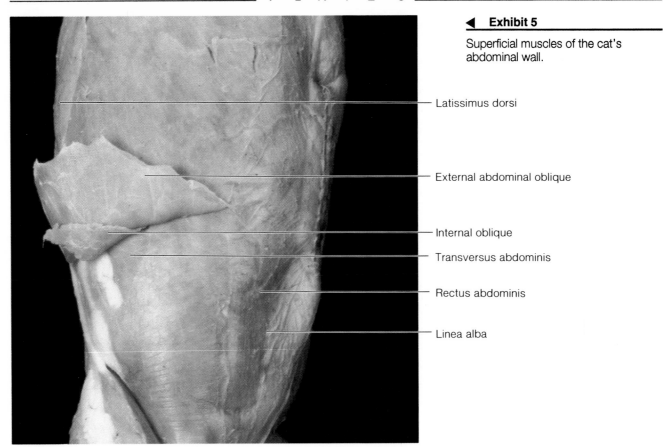

Superficial muscles of the cat's abdominal wall.

— Latissimus dorsi

— External abdominal oblique

— Internal oblique

— Transversus abdominis

— Rectus abdominis

— Linea alba

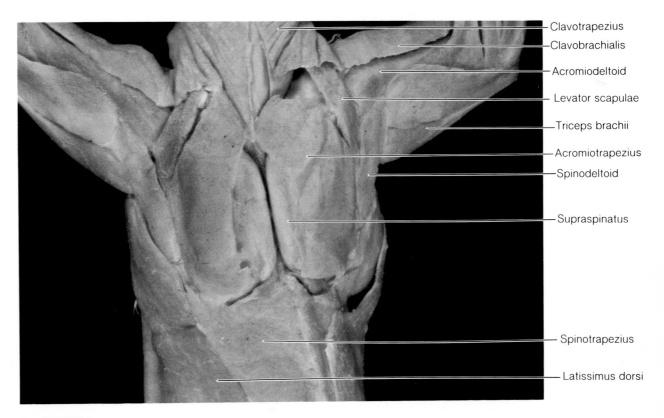

Clavotrapezius
Clavobrachialis
Acromiodeltoid
Levator scapulae
Triceps brachii
Acromiotrapezius
Spinodeltoid
Supraspinatus
Spinotrapezius
Latissimus dorsi

▲ **Exhibit 6**

Superficial muscles of the cat's shoulder and back.

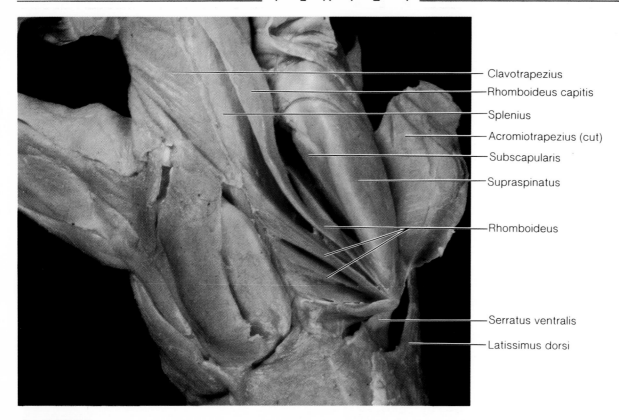

Clavotrapezius
Rhomboideus capitis
Splenius
Acromiotrapezius (cut)
Subscapularis
Supraspinatus

Rhomboideus

Serratus ventralis
Latissimus dorsi

▲ **Exhibit 7**

Deep muscles of the cat's shoulder and back.

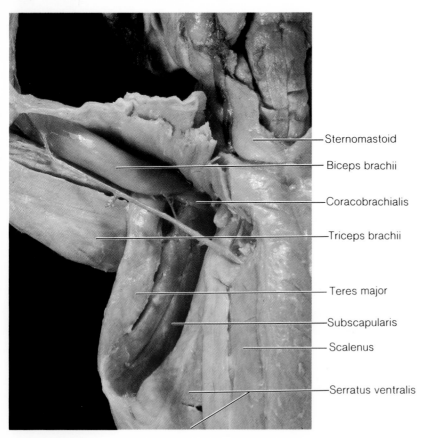

Sternomastoid

Biceps brachii

Coracobrachialis

Triceps brachii

Teres major

Subscapularis

Scalenus

Serratus ventralis

▲ **Exhibit 8**

Deep muscles of the cat's shoulder and back with the scapula reflected.

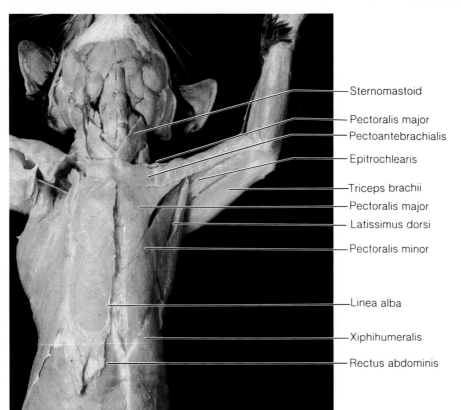

Superficial muscles of the cat's left forelimb, medial view.

— Sternomastoid

— Pectoralis major
— Pectoantebrachialis

— Epitrochlearis

— Triceps brachii
— Pectoralis major
— Latissimus dorsi

— Pectoralis minor

— Linea alba

— Xiphihumeralis

— Rectus abdominis

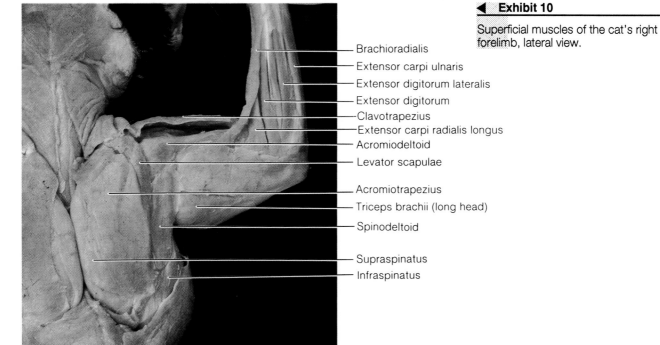

Superficial muscles of the cat's right forelimb, lateral view.

— Brachioradialis
— Extensor carpi ulnaris
— Extensor digitorum lateralis
— Extensor digitorum
— Clavotrapezius
— Extensor carpi radialis longus
— Acromiodeltoid
— Levator scapulae

— Acromiotrapezius
— Triceps brachii (long head)
— Spinodeltoid

— Supraspinatus
— Infraspinatus

— Spinotrapezius

▶ **Exhibit 11**

Superficial muscles of the cat's left thigh, medial view.

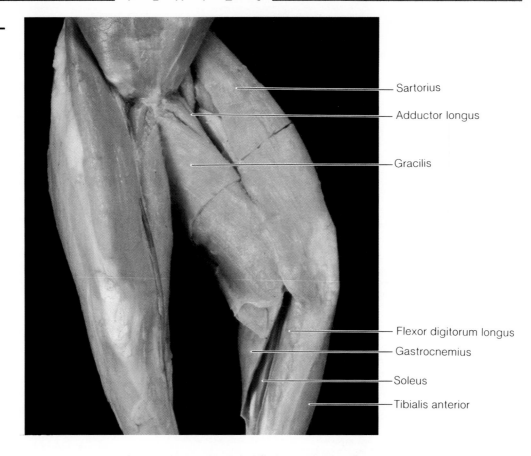

Sartorius

Adductor longus

Gracilis

Flexor digitorum longus

Gastrocnemius

Soleus

Tibialis anterior

▶ **Exhibit 12**

Superficial muscles of the cat's left shank, medial view.

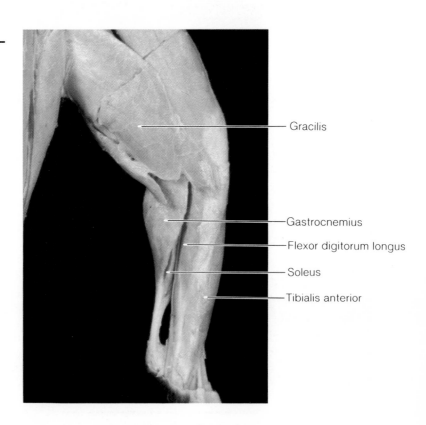

Gracilis

Gastrocnemius

Flexor digitorum longus

Soleus

Tibialis anterior

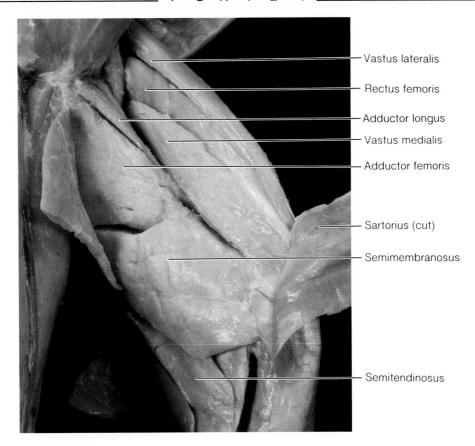

Vastus lateralis
Rectus femoris
Adductor longus
Vastus medialis
Adductor femoris

Sartorius (cut)
Semimembranosus

Semitendinosus

▲ **Exhibit 13**

Deep muscles of the cat's left thigh, medial view.

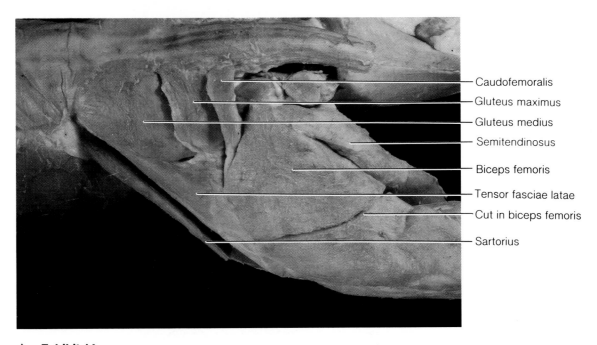

Caudofemoralis
Gluteus maximus
Gluteus medius
Semitendinosus
Biceps femoris
Tensor fasciae latae
Cut in biceps femoris
Sartorius

▲ **Exhibit 14**

Superficial muscles of the cat's left thigh, lateral view.

▶ **Exhibit 15**

Superficial muscles of the cat's left
hindlimb, lateral view.

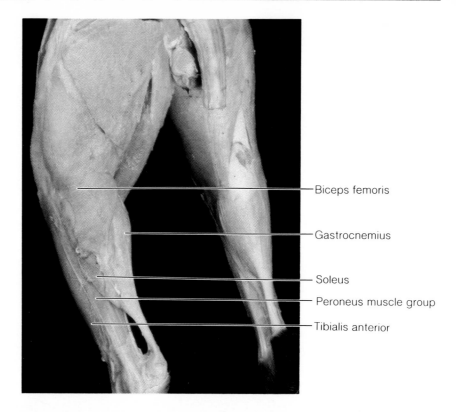

— Biceps femoris

— Gastrocnemius

— Soleus

— Peroneus muscle group

— Tibialis anterior

▶ **Exhibit 16**

Deep muscles of the cat's left
hindlimb, lateral view.

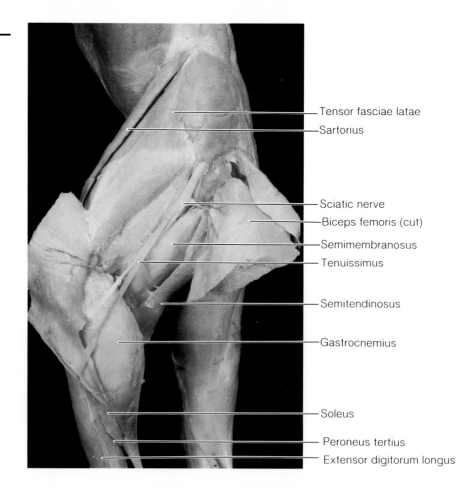

— Tensor fasciae latae
— Sartorius

— Sciatic nerve
— Biceps femoris (cut)
— Semimembranosus
— Tenuissimus

— Semitendinosus

— Gastrocnemius

— Soleus

— Peroneus tertius
— Extensor digitorum longus

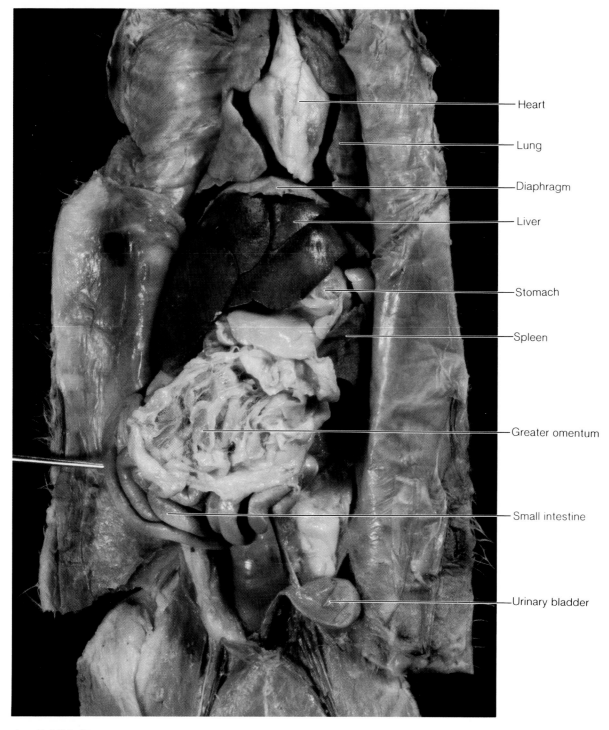

Heart

Lung

Diaphragm

Liver

Stomach

Spleen

Greater omentum

Small intestine

Urinary bladder

▲ **Exhibit 17**

Organs of the cat's thoracic and abdominal cavities.

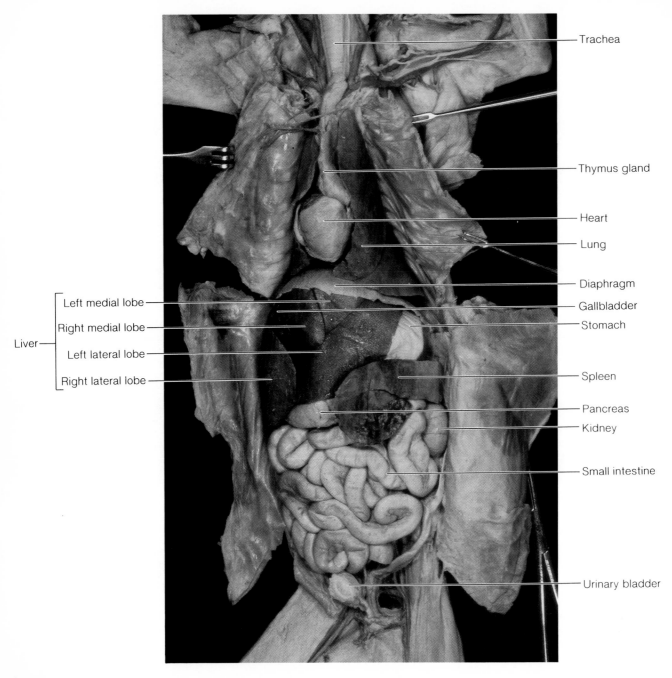

Trachea

Thymus gland

Heart

Lung

Diaphragm

Gallbladder

Stomach

Spleen

Pancreas

Kidney

Small intestine

Urinary bladder

Liver —
- Left medial lobe
- Right medial lobe
- Left lateral lobe
- Right lateral lobe

▲ **Exhibit 18**

Organs of the cat's thoracic and abdominal cavities with the greater omentum removed.

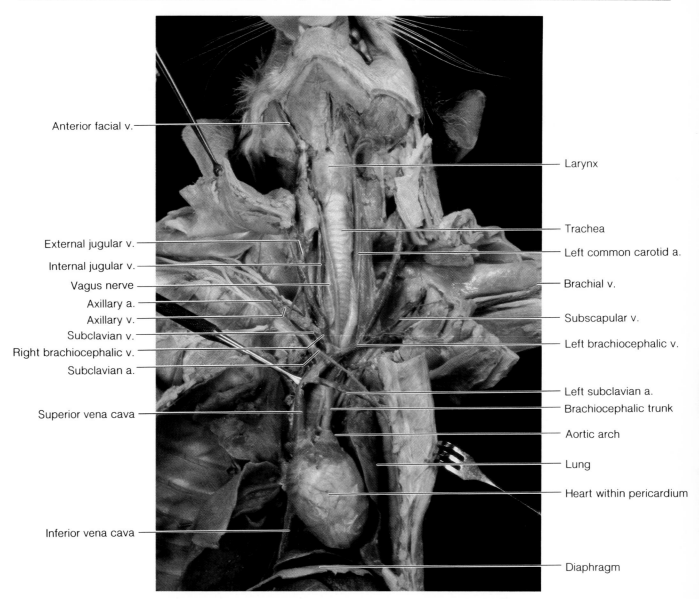

Anterior facial v.

External jugular v.

Internal jugular v.

Vagus nerve

Axillary a.

Axillary v.

Subclavian v.

Right brachiocephalic v.

Subclavian a.

Superior vena cava

Inferior vena cava

Larynx

Trachea

Left common carotid a.

Brachial v.

Subscapular v.

Left brachiocephalic v.

Left subclavian a.

Brachiocephalic trunk

Aortic arch

Lung

Heart within pericardium

Diaphragm

▲ **Exhibit 19**

Blood vessels of the cat's neck and thoracic cavity.

Brachial a.

Nerves of the
brachial plexus

Axillary a.

Right
subclavian v.

Right
subclavian a.

Right
brachiocephalic v.

Superior
vena cava

Lung

Trachea

Left
external jugular v.

Right
internal jugular v.

Right
common carotid a.

Brachial v.

Subscapular v.

Left
brachiocephalic v.

▲ **Exhibit 20**

Brachial plexus and blood vessels of the cat's neck.

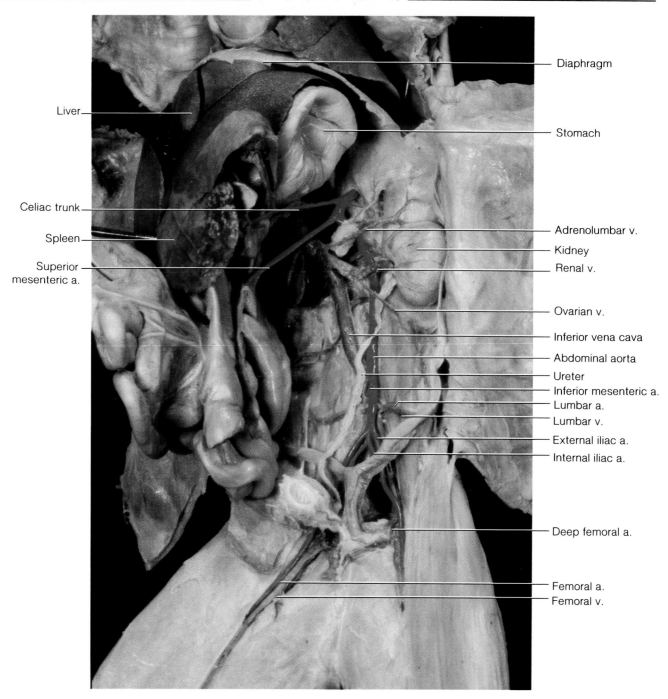

Diaphragm

Liver

Stomach

Celiac trunk

Adrenolumbar v.

Spleen

Kidney

Superior
mesenteric a.

Renal v.

Ovarian v.

Inferior vena cava

Abdominal aorta

Ureter

Inferior mesenteric a.

Lumbar a.

Lumbar v.

External iliac a.

Internal iliac a.

Deep femoral a.

Femoral a.

Femoral v.

▲ **Exhibit 21**

Blood vessels of the cat's abdominal cavity.

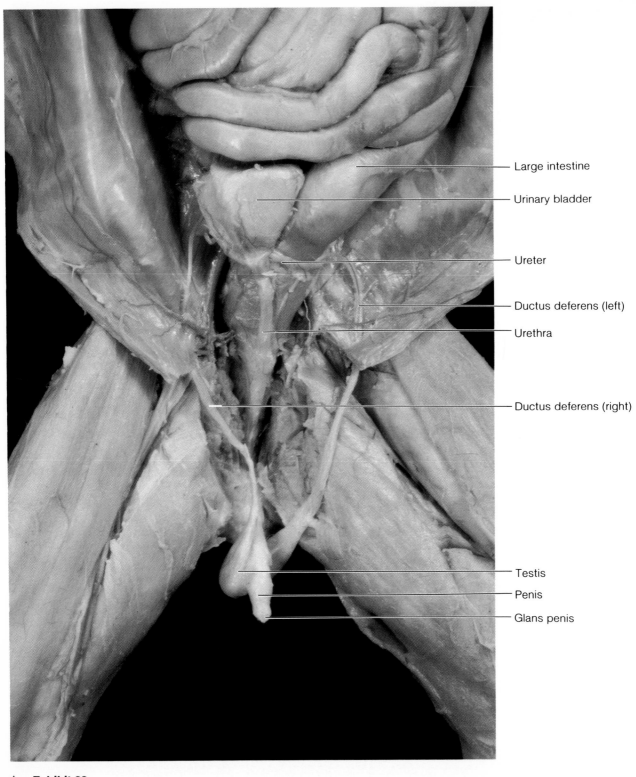

Large intestine

Urinary bladder

Ureter

Ductus deferens (left)

Urethra

Ductus deferens (right)

Testis

Penis

Glans penis

▲   **Exhibit 22**

Male reproductive organs of the cat.

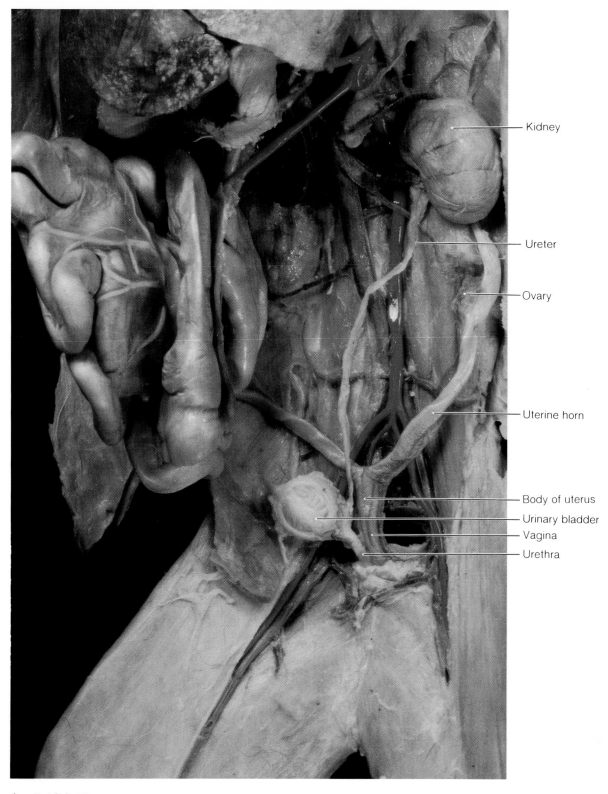

Kidney

Ureter

Ovary

Uterine horn

Body of uterus
Urinary bladder
Vagina
Urethra

▲ **Exhibit 23**

Female reproductive organs of the cat.

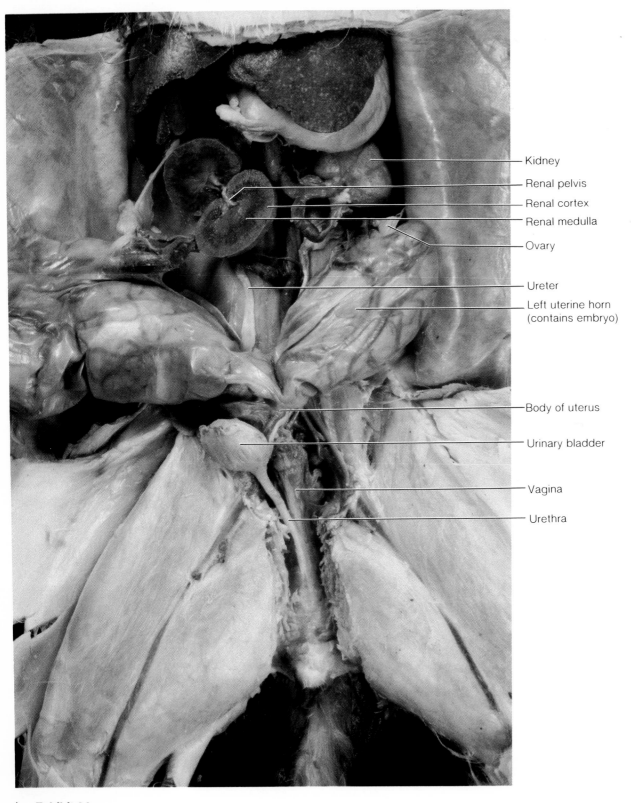

Kidney

Renal pelvis

Renal cortex

Renal medulla

Ovary

Ureter

Left uterine horn
(contains embryo)

Body of uterus

Urinary bladder

Vagina

Urethra

▲ **Exhibit 24**

Female reproductive organs of a pregnant cat.

Figure 19.6 Muscles of the thigh (a ventral view): (a) Superficial muscles; (b) deep muscles.

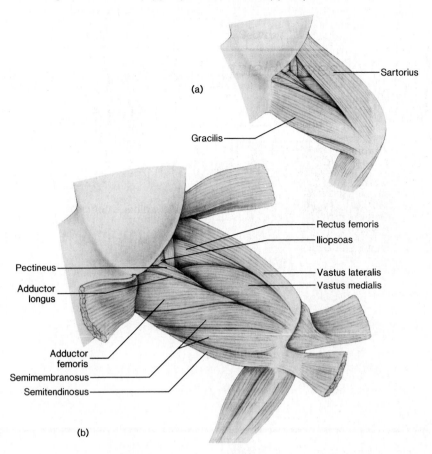

(a)

Sartorius

Gracilis

Rectus femoris

Iliopsoas

Pectineus

Adductor longus

Vastus lateralis

Vastus medialis

Adductor femoris

Semimembranosus

Semitendinosus

(b)

Two additional muscles can be seen on the medial surface: the *iliopsoas* emerges from the abdominal cavity; the thick *semimembranosus* lies deep to the gracilis.

Lateral Surface

Two large, superficial thigh muscles can be identified in a lateral view (fig. 19.7*a*). The fan-shaped *tensor fasciae latae* is on the anteroproximal surface of the thigh. The large *biceps femoris* is posterior to the tensor fasciae latae. Transect both of these muscles across their bellies and reflect them back to expose the deeper muscles (fig. 19.7*b*). When the biceps femoris has been pulled back, the long, thin *tenuissimus* muscle can be observed. Do not confuse

this muscle with the *sciatic nerve,* which is about the same size and immediately anterior to the tenuissimus. The sciatic nerve can be seen in fig. 19.7 and in plate 8. The tenuissimus has no homologue in humans.

The hamstring muscles of the cat consist of the thick *semitendinosus,* the stringlike tenuissimus, the semimembranosus, and the biceps femoris.

The *caudofemoralis,* which is not present in humans, extends from the base of the tail to its tendinous attachment on the patella. It is immediately posterior to the vastus lateralis. The short *gluteus maximus* muscles are just anterior to the origin of the caudofemoralis. The *gluteus medius* is anterior to and actually larger than the gluteus maximus. It is entirely covered by the tensor fasciae latae muscle.

Figure 19.7 Muscles of the hip and thigh (a dorsal view): (a) Superficial muscles; (b) deep muscles.

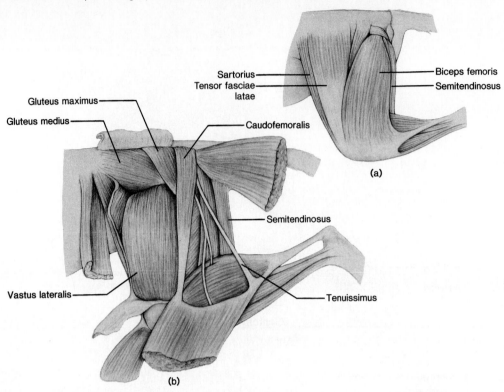

Table 19.4 Muscles of the Hip and Thigh Regions

Muscle	Origin	Insertion	Action
Medical Surface			
Sartorius	Ilium	Proximal end of tibia	Extends and rotates leg
Gracilis	Ischium and symphysis pubis	Proximal end of tibia	Adducts leg
Quadriceps femoris			
Vastus medialis	Shaft of femur	Patella	Extends leg
Rectus femoris	Ilium	Patella	Extends leg
Vastus intermedius	Shaft of femur	Patella	Extends leg
Vastus lateralis	Proximal part of femur	Patella	Extends leg
Pectineus	Pubis	Shaft of femur	Adducts thigh
Adductor longus	Pubis	Shaft of femur	Adducts thigh
Adductor femoris	Ischium and pubis	Shaft of femur	Adducts thigh
Iliopsoas	Lumbar vertebrae and ilium	Lesser trochanter of femur	Rotates and flexes thigh
Semimembranosus	Ischium	Distal end of femur	Extends thigh
Lateral Surface			
Tensor fasciae latae	Ilium	Patella	Extends thigh
Biceps femoris	Ischium	Patella and tibia	Abducts thigh and flexes leg
Semitendinosus	Ischium	Proximal end of tibia	Flexes leg
Tenuissimus	Second caudal vertebrae	Fascia of biceps femoris	Abducts thigh
Caudofemoralis	Caudal vertebrae	Patella	Abducts thigh and extends leg
Gluteus maximus	Sacral and caudal vertebrae	Greater trochanter of femur	Abducts thigh
Gluteus medius	Sacral and caudal vertebrae	Greater trochanter of femur	Abducts thigh

Laboratory Report 19

Name _____

Date _____

Section _____

Cat Musculature

Read the assigned sections in the textbook before completing the laboratory report.

A–C. Muscles of the Head, Neck, Shoulder, and Back Regions

1. What is the function of the cutaneous maximus muscle? Why don't humans need this muscle?

2. List a muscle synergistic to each of the following muscles:

   Sternomastoid _____   Hyoglossus _____

   Masseter _____   Geniohyoid _____

3. List five muscles (three superficial and two deep) that insert on the scapula.

4. Name at least seven muscles that originate on the scapula. Indicate which are superficial and which are deep.

D and E. Muscles of the Thoracic and Abdominal Regions

1. List three thoracic muscles of the cat with no human homologues. Discuss the reasons these muscles are necessary in the cat but not in humans.

2. Contrast the origin, insertion, and action of the levator scapulae with those of the levator scapulae ventralis.

3. Compare the problems of structural support in the abdominal region of a cat (quadruped) with those of a human (biped).

F. Muscles of the Brachium

1. List all the muscles that insert on the olecranon of the ulna.

2. Why is the triceps brachii muscle larger than the biceps brachii muscle?

3. Why does the forearm have more individual muscles but less muscle mass than the brachium?

G. Muscles of the Hip and Thigh Regions

1. Compare the function and structure of the cat sartorius muscle with those of the human sartorius muscle.

2. Which of the cat's hind limb muscles are not present in humans? Are there any muscles in humans that are not present in cats?

Integration and Control Systems of the Human Body

The following exercises are included in this unit:

20. Brain and Cranial Nerves
21. Spinal Cord, Spinal Nerves, and Spinal Reflexes
22. Cutaneous Receptors
23. Eyes and Vision
24. Ears and Hearing
25. Taste
26. Endocrine System

These exercises are based on information presented in the following chapters of *Concepts of Human Anatomy and Physiology,* 4th ed.:

14. Functional Organization of Neural Tissue
15. Central Nervous System
16. Peripheral Nervous System
17. Autonomic Nervous System
18. Sensory Organs
19. Endocrine System

In this unit, the structure of the human brain can be studied by using figures and models, and first-hand experience with brain structure can be gained by dissecting a sheep or cat brain. The spinal sensory and motor neuron activity will be evident through various tests of spinal reflexes. In the simple reflex arc, stimulation of muscle stretch receptors activates sensory neurons, which in turn synapse directly with somatic motor neurons in the spinal cord. The structure and function of other sensory receptors—the cutaneous senses and the special senses of sight, hearing, equilibrium, and taste—are explored in these exercises. Since the endocrine system and the nervous system work together to regulate body function, the structure of endocrine glands and the functions of hormones are covered in the last laboratory exercise.

Brain and Cranial Nerves

Introduction

The human brain is composed of more than 100 billion neurons and weighs nearly 1.5 kg (3.0 lbs). These individual neurons and their synapses are responsible for perceptions, emotions, thought, and complex motor activity. The enormous complexity of structure and function required for these activities makes brain physiology one of the great frontiers in science. In this exercise, students will develop a working knowledge of the anatomy of the brain by examining figures and laboratory specimens.

Objectives

Students completing this exercise will be able to:

1. Identify the developmental regions of the brain.
2. Describe the structure and position of the cranial meninges.
3. Identify the principal structures of the brain in figures and laboratory specimens.
4. Describe the blood supply to the brain.
5. Identify the 12 pairs of cranial nerves.
6. Describe the formation, function, and circulation of cerebrospinal fluid.

Materials

1. Dissecting instruments and trays
2. Preserved mammalian brains (human, if available; otherwise, sheep or cat)
3. Colored pencils
4. (Optional) Preserved cat or fresh sheep head, autopsy saw, large screwdriver

A. Embryonic Development of the Brain

1. Identify and label the following structures in figure 20.1:

   Diencephalon
   Mesencephalon
   Metencephalon
   Myelencephalon
   Neural tube
   Prosencephalon
   Rhombencephalon
   Telencephalon

2. Color-code the five advanced brain regions shown in figure 20.1*c*.

Figure 20.1   The developmental sequence of the brain: (*a*) undifferentiated; (*b*) the three-region brain; (*c*) the five-region brain.

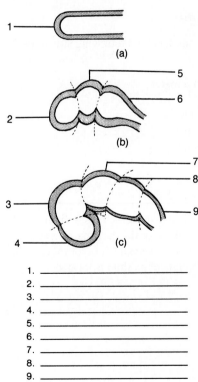

1. _____
2. _____
3. _____
4. _____
5. _____
6. _____
7. _____
8. _____
9. _____

Figure 20.2 The cranial meninges and the layers of the cerebrum as shown in a coronal section through the cranium and brain.

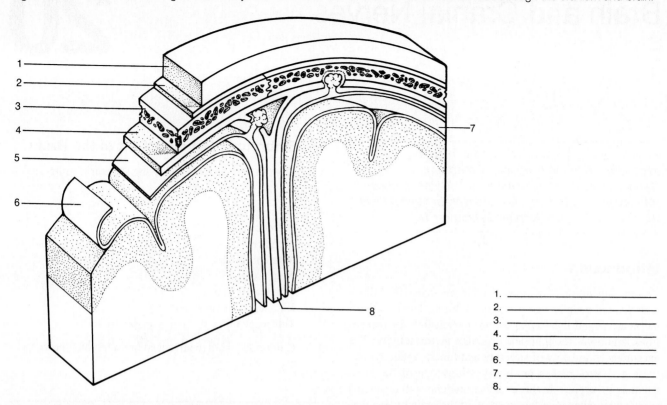

1. _____
2. _____
3. _____
4. _____
5. _____
6. _____
7. _____
8. _____

B. Meninges

1. Identify and label the following structures in figure 20.2:

   Arachnoid mater
   Bone
   Dura mater
   Falx cerebri
   Periosteum
   Pia mater
   Scalp
   Subarachnoid space

2. Color the identified structures in figure 20.2. Use different colors for the dura mater, arachnoid mater, and pia mater. Use another color to shade the space where cerebrospinal fluid is located.

C. External Observation of the Brain

1. Identify and label the following structures in figures 20.3 and 20.4:

   Cerebellum
      Cerebellar hemisphere
      Vermis
   Cerebrum
      Central fissure
      Central selcus
      Convolutions
      Frontal lobes
      Gyri
      Lateral fissure
      Left hemisphere
      Longitudinal fissure
      Occipital lobes
      Parietal lobes
      Parietooccipital fissure
      Right hemisphere
      Sulci
      Temporal lobes
   Mammillary body
   Medulla oblongata
   Pons

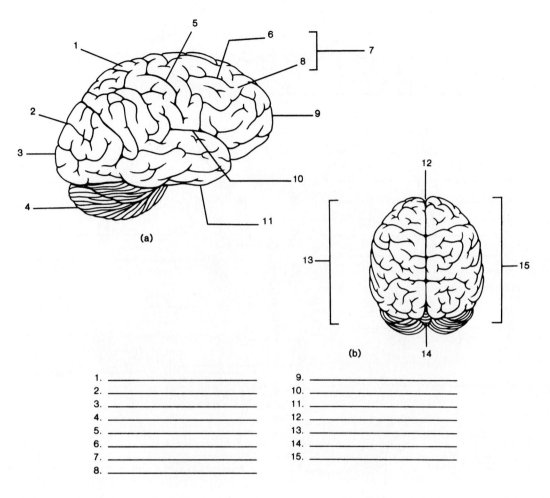

(a)

(b)

1. _____	9. _____
2. _____	10. _____
3. _____	11. _____
4. _____	12. _____
5. _____	13. _____
6. _____	14. _____
7. _____	15. _____
8. _____	

2. Using colored pencils, color-code the different lobes of the cerebrum that can be seen in figure 20.3.
3. Identify as many of these structures as possible on the brain specimens available in the laboratory.

D. Cranial Nerves

1. Identify and label the following nerves and structures in figure 20.4:

Abducens nerve
Accessory nerve
Facial nerve
Glossopharyngeal nerve
Hypoglossal nerve
Oculomotor nerve
Olfactory bulb
Olfactory tract
Optic chiasma
Optic nerve
Trigeminal nerve
    Ophthalmic nerve
    Maxillary nerve

    Mandibular nerve
Trochlear nerve
Vagus nerve
Vestibulocochlear nerve

2. Color-code the cranial nerves in figure 20.4.
3. Identify as many of these structures as possible on the specimens available in the laboratory.

E. Blood Supply to the Brain

1. Identify and label the following arteries and structures in figure 20.5:

Anterior cerebral artery
Basilar artery
Cerebral arterial circle (circle of Willis)
Internal carotid artery
Middle cerebral artery
Posterior cerebral artery
Superior cerebellar artery
Vertebral artery

Figure 20.4 An inferior view of the brain and cranial nerves.

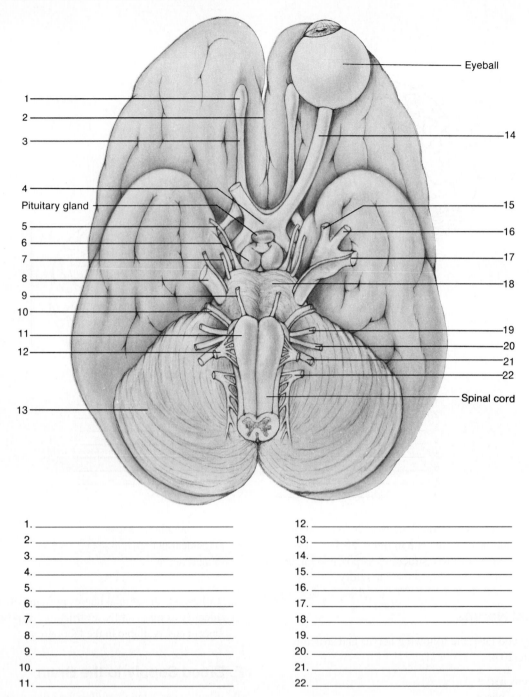

1. _____
2. _____
3. _____
4. _____
5. _____
6. _____
7. _____
8. _____
9. _____
10. _____
11. _____

12. _____
13. _____
14. _____
15. _____
16. _____
17. _____
18. _____
19. _____
20. _____
21. _____
22. _____

F. Midsagittal Section of the Brain

1. Identify and label the following structures in figure 20.6:

Cerebrum
    Corpus callosum
        Genu
        Splenium
    Septum pellucidum
Diencephalon
    Hypothalamus
        Infundibulum
        Mammillary body
        Optic chiasma
    Pituitary gland
    Thalamus
Mesencephalon (midbrain)
    Corpora quadrigemina
    Pineal body

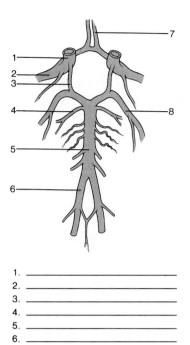

1. _____
2. _____
3. _____
4. _____
5. _____
6. _____
7. _____
8. _____

Figure 20.6 A midsagittal section of the brain.

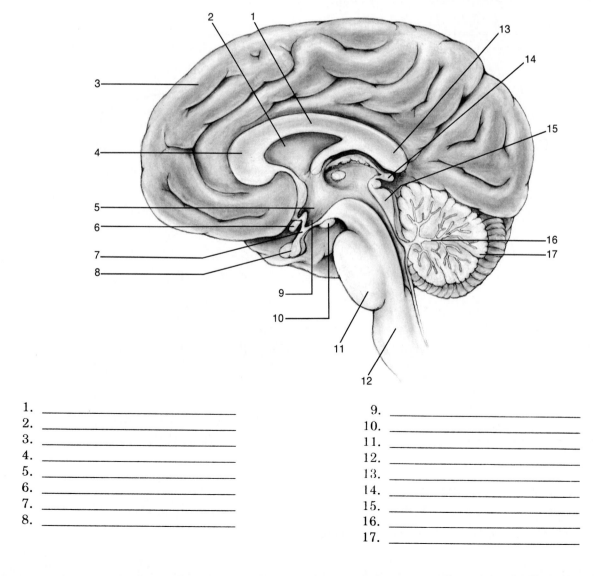

1. _____
2. _____
3. _____
4. _____
5. _____
6. _____
7. _____
8. _____

9. _____
10. _____
11. _____
12. _____
13. _____
14. _____
15. _____
16. _____
17. _____

Metencephalon
    Arbor vitae
    Cerebellum
    Pons
Myelencephalon (hindbrain)
    Medulla oblongata

2. Color-code the corpus callosum, the arbor vitae, and the optic chiasma in one color; color the pituitary gland and pineal body in another color.
3. Identify as many of these structures as possible on a midsagittal section of a brain specimen in the laboratory.

G. Ventricles of the Brain

1. Identify and label the following structures in figure 20.7:

Arachnoid villi
Mesencephalic aqueduct
Choroid plexuses of third and fourth ventricle
Fourth ventricle
Interventricular foramen
Lateral ventricle
Subarachnoid space
Third ventricle

☞ **Note:** *The depiction of subdural space is shown diagrammatically. Septum pellucidum has been removed.*

2. Using colored pencils, color-code the cerebrospinal fluid in one color and the choroid plexus in another color in figure 20.7

H. Removal of the Brain from a Sheep or Preserved Cat Head

1. Remove the skin from the skull if this was not previously done. Strip the muscle and periosteum away from the top and sides of the cranium. With a felt-tipped pen, draw a line on the top of the cranium across the frontal bone between the eyes, diagonally across the temporal bones, and over the occipital condyles (fig. 20.8).
2. Your laboratory partner should hold the skull securely, making certain that his or her *hands are away from the area to be cut.* An autopsy saw has a vibrating, reciprocating action and is, therefore, relatively safe to use. However, to avoid accidents caution must always be exercised. Starting at the line on the frontal bone, carefully make a shallow cut through the skull. Do not extend the cut through the dura mater into the brain.

Figure 20.7 A midsagittal section of the brain showing the ventricles of the brain and the direction of flow of the cerebrospinal fluid.

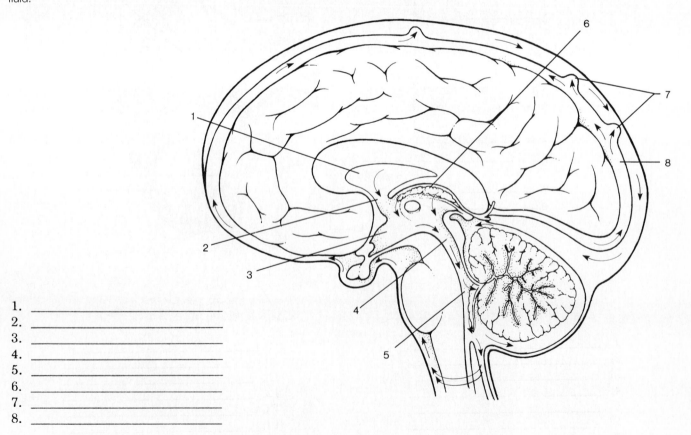

1. _____
2. _____
3. _____
4. _____
5. _____
6. _____
7. _____
8. _____

Figure 20.8 After the skin, muscle, and periosteum are removed, the skull is marked with a felt-tipped pen to indicate the line of cut, which is a rough circle.

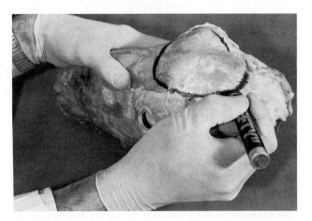

Figure 20.9 The skull is held securely by one person while another cuts with an autopsy saw.

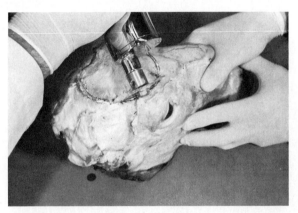

Figure 20.10 The skull is pried up with a screwdriver to determine if the cut is complete.

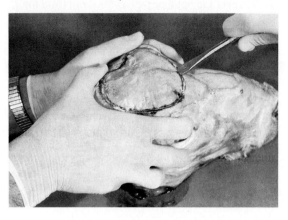

Figure 20.11 The surface of the brain is exposed by lifting off the skull cap.

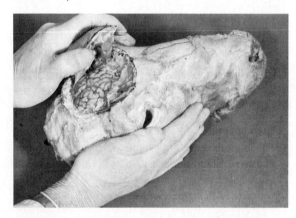

Continue the cut along the sides of the skull, stopping every inch or so to rotate the skull and rest your hands (fig. 20.9).

3. After making the cut around the top of the cranium, place the tip of a screwdriver into the cut, and gently pry upward to determine if the cut is complete (fig. 20.10). If there is any bony resistance, cut the remaining areas free.

4. Lift the cut section from the top of the skull (fig. 20.11). The *dura mater* will adhere to the bone and should be carefully cut loose as the skull cap is removed.

5. Examine the inside surface of the skull cap, and notice the impressions of the brain on the bone. The dura mater is extremely tough and thick because it has an abundance of collagenous fibers. Secure the dura mater with forceps, and make an elongated incision through it with a scalpel or scissors. Identify the *dural sinus,* and notice that the dura mater continues between the

right and *left cerebral hemispheres* as the *falx cerebri.* Try to identify any *arachnoid villi* located within the dural sinus.

6. Observe the surface of the brain, noting the high degree of vascularity and the convolutions. The elevated ridges are called *gyri* (*gyrus,* singular), and the depressions or grooves are the *sulci* (*sulcus,* singular). Examine the sulci carefully, and observe the thin, netlike, porous membrane that appears to adhere to the brain. This is the *arachnoid mater.* Below it is the *arachnoid space.* Both the arachnoid mater and arachnoid space are bathed in *cerebrospinal fluid.* The *pia mater* follows the contour of the convolutions of the brain; it adheres to the *cerebral cortex.*

7. The brain is now ready to be removed from the cranium. Probe anteriorly with your fingers to the forepart of the skull and locate the *olfactory bulbs.* Gently lift the bulbs, and observe how the *olfactory nerves* traverse through the cribriform plate into the olfactory bulbs. Sever the nerves with the scalpel to free the olfactory bulbs from the skull.

8. Continue to elevate the brain to expose the *optic nerves* where they cross and form the *optic chiasma* (see fig. 20.4). Sever the optic nerves as far away from the brain as possible, and gently elevate the brain. The next structures observed will be the two *internal carotid arteries* that branch from the common carotid arteries in the neck. Cut these vessels close to the foramina through which they enter the cranial cavity.

9. Gently raise the brain a bit farther to expose the *infundibulum* (see fig. 20.6) that supports the *pituitary gland*. The pituitary is secured within the sella turcica and is difficult to remove intact. If dissected carefully, the gland can be teased free without breaking the infundibulum.

10. As the brain is lifted, the remaining cranial nerves will be exposed. Identify as many as possible before severing them (see fig. 20.4). As you cut these nerves, do so as far away from the brain as possible.

11. To complete the removal of the brain, sever the spinal cord. It is best if the cord is cut between the occipital condyles of the skull and the atlas vertebra. Identify the stumps of the cranial nerves and the foramina they traverse once more before discarding the remains of the skull.

I. Sheep or Cat Brain

Using brain, identify the structures labeled in figures 20.12 through 20.15.

Figure 20.12   A lateral view of a sheep brain.

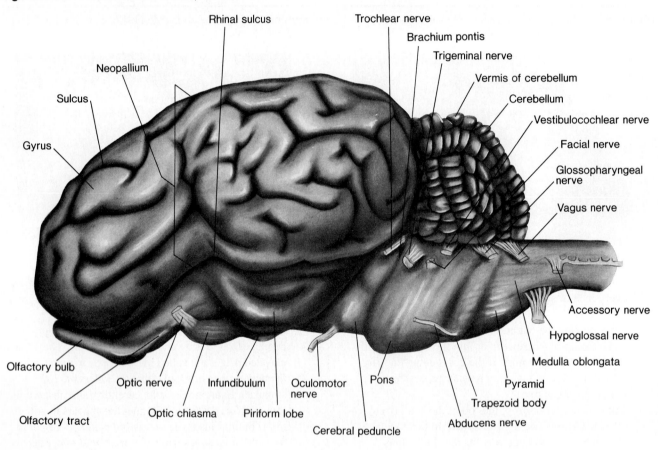

Figure 20.13 A sagittal view of a sheep brain.

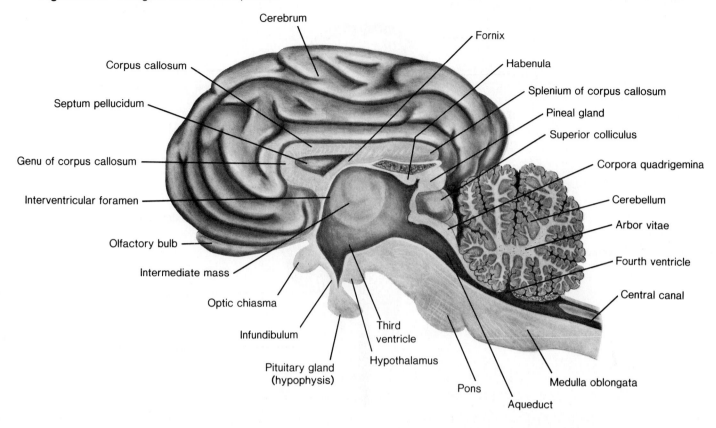

Cerebrum

Fornix

Habenula

Corpus callosum

Septum pellucidum

Splenium of corpus callosum

Pineal gland

Superior colliculus

Genu of corpus callosum

Corpora quadrigemina

Interventricular foramen

Cerebellum

Arbor vitae

Olfactory bulb

Fourth ventricle

Intermediate mass

Central canal

Optic chiasma

Third ventricle

Infundibulum

Hypothalamus

Pituitary gland (hypophysis)

Medulla oblongata

Pons

Aqueduct

Figure 20.14 A ventral view of a sheep brain.

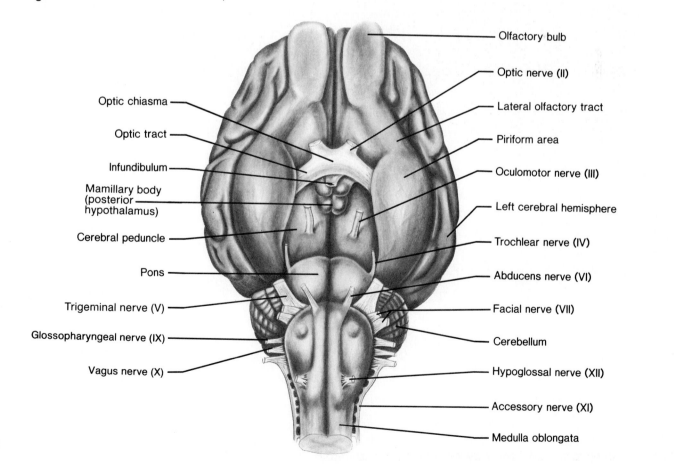

Olfactory bulb

Optic nerve (II)

Optic chiasma

Lateral olfactory tract

Optic tract

Piriform area

Infundibulum

Oculomotor nerve (III)

Mamillary body (posterior hypothalamus)

Left cerebral hemisphere

Cerebral peduncle

Trochlear nerve (IV)

Pons

Abducens nerve (VI)

Trigeminal nerve (V)

Facial nerve (VII)

Glossopharyngeal nerve (IX)

Cerebellum

Vagus nerve (X)

Hypoglossal nerve (XII)

Accessory nerve (XI)

Medulla oblongata

Figure 20.15 A dorsal view of a sheep brain.

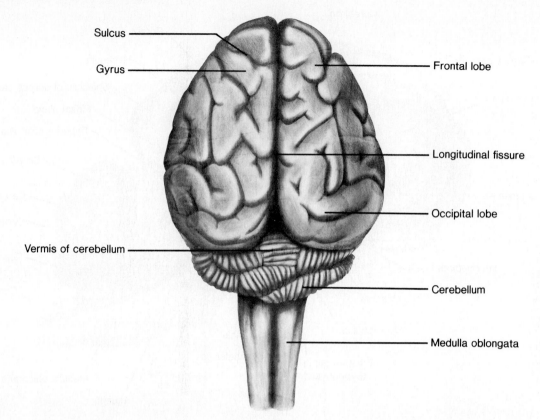

Sulcus

Gyrus

Frontal lobe

Longitudinal fissure

Occipital lobe

Vermis of cerebellum

Cerebellum

Medulla oblongata

Integration and Control Systems of the Human Body

Laboratory Report 20

Name _____

Date _____

Section _____

Brain and Cranial Nerves

Read the assigned sections of the textbook before completing the laboratory report.

1. Draw a frontal section through the skull and cerebrum to illustrate the structural relationship of the cerebral cortex to the meninges and the bone of the skull.

2. Identify the brain region that contains each of the following structures:
   a. Pons _____
   b. Medulla oblongata _____
   c. Cerebral cortex_____
   d. Hypothalamus _____

3. List the cranial nerves that arise from each of the following:
   a. Pons _____
   b. Medulla oblongata _____
   c. Midbrain _____

4. Name the arteries that contribute to the cerebral arterial circle (circle of Willis). What is the functional importance of this structure?

5. If the following symptoms were observed in a patient undergoing a neurological examination, which cranial nerves might be damaged?

a. Inability to smile with the left side of face _____

b. Occasional dizziness _____

c. Persistently dilated pupil in the left eye _____

d. Difficulty swallowing _____

e. Inability to shrug the shoulders and turn the head _____

f. Numbness on the right forehead area _____

g. Inability to turn the eyes upward _____

h. No sensation of sweet tastes on the tip of the tongue _____

i. Cannot look "cross-eyed" when following the movements of a finger _____

6. Provide the location (region of the brain) and general function of the following structures:

Structure	Location	Function
Thalamus		
Optic chiasma		
Pituitary gland		
Pons		
Medulla oblongata		
Arbor vitae		
Pineal gland		
Pia mater		
Inferior colliculus		

7. Describe the location of the ventricles within the brain. Where is cerebrospinal fluid produced, and where does it drain?

Spinal Cord, Spinal Nerves, and Spinal Reflexes

Before coming to class, read the following sections in chapter 16 of the textbook: "Introduction to the Peripheral Nervous System," "Spinal Nerves," "Nerve Plexuses," and "Reflex Arcs and Reflexes." Also, read the section entitled "Proprioceptors" in chapter 18.

Introduction

Spinal nerves are mixed nerves; that is, they contain sensory dendrites and motor axons. Sensory neurons carry impulses into the central nervous system (CNS) from sensory receptors; motor neurons carry impulses out of the CNS and stimulate effector organs. In a reflex, specific sensory stimuli evoke characteristic motor responses very rapidly because of the small number of synapses involved. In the simplest reflex—the muscle stretch reflex—the sensory neurons synapse directly with motor neurons. Thus only one synapse must be crossed. Because a specific stretch reflex involves only a specific spinal cord segment, tests of muscle stretch reflexes are very useful in neurological diagnosis.

Objectives

Students completing this exercise will be able to:

1. Describe the structure of the spinal cord and the composition of spinal nerves.
2. Define a plexus, and identify the plexuses that arise from spinal nerves.
3. Describe the neurological pathways involved in a muscle stretch reflex and in a Babinski reflex.
4. Define the term *referred pain,* and explain its clinical significance.

Materials

1. Microscope
2. Histological slides of a spinal cord and spinal nerve
3. Rubber mallets
4. Reference text
5. Optional: Flexicomp (Intelitool, Inc.) with IBM or Apple computer

A. Spinal Cord

1. Identify and label the following structures in figures 21.1 and 21.2:

   Anterior, posterior, and lateral horns
   Anterior median fissure
   Brachial plexus
       Ulnar nerve
       Radial nerve
       Median nerve
   Cauda equina
   Central canal
   Cervical plexus
   Posterior (dorsal) and anterior (ventral) root
   Posterior root ganglion
   Gray matter
   Lumbosacral plexus
       Sciatic nerve
   Meninges
       Dura mater
       Arachnoid mater (not shown)
       Pia mater (not shown)
   Posterior median sulcus
   White matter

2. Using colored pencils, color-code the posterior root in one color and the anterior root in another color in figure 21.2.
3. Examine histological slides of the spinal cord and spinal nerves.

B. Structure of the Reflex Arc

1. Identify and label the following structures in figure 21.3:

       Alpha motor neuron
       Cell body of sensory neuron
       Posterior (dorsal) root ganglion
       Sensory neuron
       Skeletal muscle fiber
       Somatic motor neuron
       Spindle apparatus

2. Color-code the sensory neuron in one color and the motor neuron in another color in figure 21.3.

Figure 21.1 The spinal cord and plexuses.

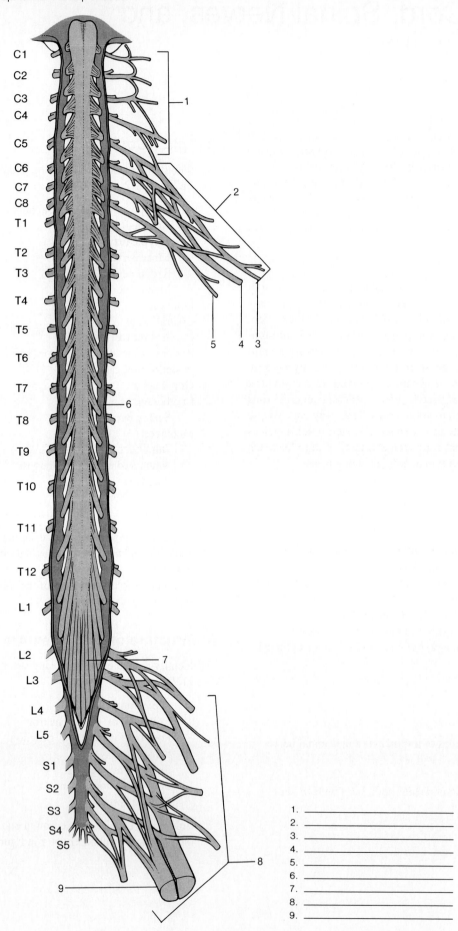

1. _____
2. _____
3. _____
4. _____
5. _____
6. _____
7. _____
8. _____
9. _____

Figure 21.2 A cross section of the spinal cord.

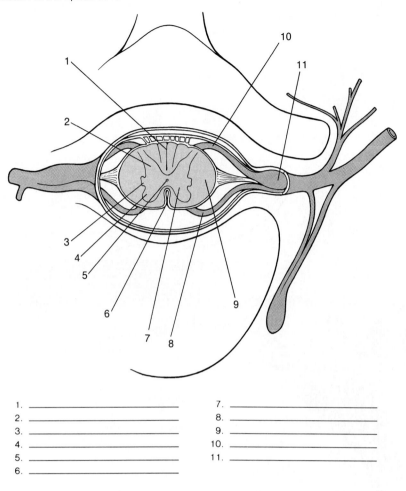

1. _____   7. _____
2. _____   8. _____
3. _____   9. _____
4. _____   10. _____
5. _____   11. _____
6. _____

C. Tests for Spinal Nerve Stretch Reflexes

This exercise illustrates the function of many reflex arcs. These reflexes are initiated by stretch receptors called **muscle spindles,** which are embedded in the muscle's connective tissue. The muscle spindle consists of specialized thin muscle fibers (*intrafusal fibers*), which are connected to a sensory (afferent) nerve. The intrafusal fibers are parallel to the fibers of ordinary muscle cells (*extrafusal fibers*), so that when the muscle is stretched, the resulting tension on the intrafusal fibers stimulates the sensory neuron (fig. 21.3). In a typical clinical examination, these receptors are stimulated by the momentary stretch that occurs from striking the tendon with a rubber mallet. The sensory neuron arising from the intrafusal fiber synapses with a motor neuron that innervates the extrafusal fibers. Contraction of the muscle (extrafusal fibers) releases tension of the intrafusal fibers and decreases stimulation of the stretch receptors.

Clinical Significance

Tests for simple muscle reflexes are done first in any physical examination when motor nerve or spinal cord damage is suspected. If there is spinal cord damage, they can help determine the level of the spinal cord that is damaged: motor nerves that exit the spinal cord above the damaged level may not be affected, whereas nerves that originate at or below the damaged level may not elicit their normal responses. In this regard, testing the plantar reflex is particularly useful, because damage to the pyramidal motor tract at any level may result in a positive Babinski sign.

Figure 21.3 The knee-jerk reflex, an example of a monosynaptic stretch reflex.

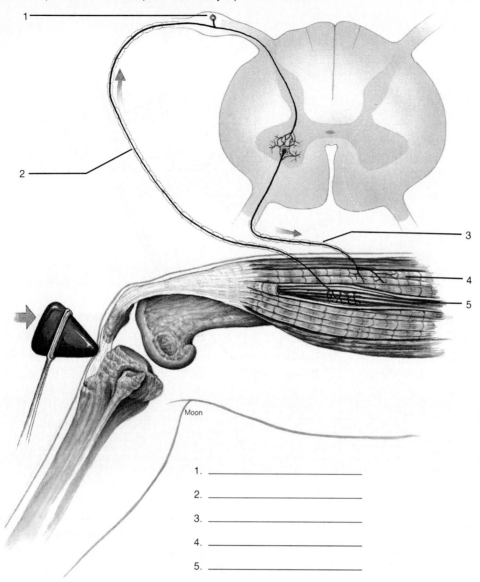

Moon

1. _____
2. _____
3. _____
4. _____
5. _____

Procedure for Knee Jerk Reflex (fig. 21.4)
Tests Femoral Nerve

1. The subject sits comfortably with his or her legs free.
2. Strike the patellar ligament, just below the patella (kneecap), and observe the resulting contraction of the quadriceps muscles and extension of the leg.

Procedure for Ankle Jerk Reflex (fig. 21.4)
Tests Popliteal Nerve

1. The subject kneels on a chair, with his or her back to the examiner, and feet (shoes and socks off) projecting over the chair edge.
2. Strike the tendo calcaneus at the level of the ankle, and observe the resulting plantar flexion of the foot.

Figure 21.4 Some reflexes of clinical importance.

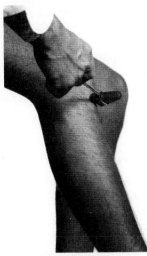

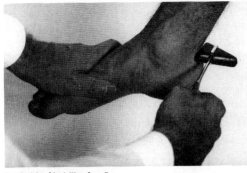

2. Ankle (Achilles) reflex

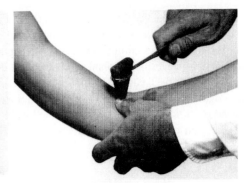

3. Biceps brachii reflex

1. Knee (patellar) reflex

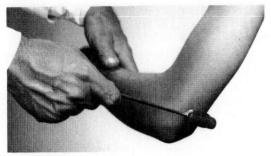

4. Triceps brachii reflex

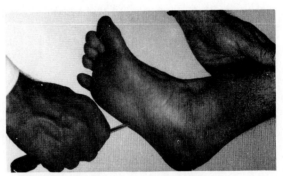

5. Abnormal plantar reflex (Babinski sign)

Procedure for Biceps Brachii Jerk Reflex (fig. 21.4)
Tests Musculocutaneous Nerve

1. With the subject's arm relaxed on the desk, gently press the tendon of the biceps brachii muscle in the cubital fossa with the thumb or forefinger, and strike this finger with the mallet.
2. If this procedure is performed correctly, the biceps brachii muscle will twitch (the contraction is not usually strong enough to cause arm movement).

Procedure for Triceps Brachii Jerk Reflex (fig. 21.4)
Tests Radial Nerve

1. The subject lies supine with the elbow flexed so that the arm lies loosely across the abdomen.
2. Strike the tendon of the triceps brachii muscle about 2 inches above the elbow. If there is no response, repeat this procedure, but strike either side of the original point.
3. If this procedure is correctly performed, the triceps brachii muscle will twitch but will not usually contract strongly enough to produce arm movement.

D. A Cutaneous Reflex: The Plantar Reflex (Babinski Sign)

The *plantar reflex* is one of the most important neurological tests. It is initiated by stimulating cutaneous receptors in the sole of the foot. Proper stimulation of these receptors in the normal individual results in plantar flexion (i.e., downward movement) of the great toe and the other toes, which also come together. Damage to the pyramidal motor tracts results in the response to this stimulation called the *Babinski sign,* in which the great toe extends (i.e., moves upward).

Procedure

1. The subject lies supine, knees slightly bent, with the thigh rotated so that the lateral (outer) side of the foot rests on the couch.
2. Applying firm (but not painful) pressure, draw the tip of a blunt probe along the lateral border of the sole, starting at the heel and ending at the base of the great toe (fig. 21.4). Observe the response of the toes to this procedure.

E. Referred Pain

Damage to one area of the body may cause pain to be perceived in a different area. In the *phantom limb phenomenon* and when the ulnar nerve ("funny bone") is struck, (as in the following exercise), pain is perceived because sensory neurons are stimulated proximal to their receptors. Pains that are referred from the viscera to a somatic location are extremely important in medical diagnosis. The most famous of these referred pains is **angina pectoris,** in which ischemic damage to the heart often causes pain to be felt in the left pectoral region and shoulder areas.

Clinical Significance

Amputees frequently report that they feel pain in their missing limbs (known as the **phantom limb phenomenon**). The source of nerve stimulation is trauma to the cut nerve fibers, but the pain is perceived as coming from the amputated region of the body that normally produces action potentials along these nerves. This is a referred pain because the *actual* location of nerve stimulation is different from the *perceived* location.

Procedure

1. Gently tap the ulnar nerve where it passes lateral to the medial epicondyle of the elbow.
2. Describe, in the laboratory report, the location where a tingling sensation is felt.

Laboratory Report 21

Name _____

Date _____

Section _____

Spinal Cord, Spinal Nerves, and Spinal Reflexes

Read the assigned sections in the textbook before completing the laboratory report.

A. Spinal Cord

1. Distinguish between the central nervous system and the peripheral nervous system.

2. Make a cross-sectional diagram of the spinal cord, and label the major structural features.

3. All spinal nerves are "mixed" nerves. What does this mean?

4. Define or describe the structure of each of the following:
   a. Nerve _____
   b. Ganglion _____
   c. Cauda equina _____
   d. Filum terminale _____

B. Structure of the Reflex Arc

1. Match each of the following tests with the nerve involved:
   ___(1) Biceps brachii jerk reflex
   ___(2) Triceps brachii jerk reflex
   ___(3) Knee jerk reflex
   ___(4) Ankle jerk reflex

   (a) femoral nerve
   (b) musculocutaneous nerve
   (c) popliteal nerve
   (d) radial nerve

C and D. Spinal Nerve Stretch Reflexes and Cutaneous Reflex

1. Describe the sequence of events that occurs from the moment the patellar ligament is struck to the extension of the leg (knee-jerk reflex).

2. Failure to elicit a given stretch reflex may indicate damage to a specific motor nerve or region of the spinal cord, whereas a positive Babinski sign may be produced by damage to the spinal cord at any level. Explain why this statement is true.

E. Referred Pain

1. Describe the locations where pain was felt, and explain why pain was perceived in these locations when this exercise was performed.

2. Describe the importance of referred pain in the diagnosis of deep visceral pain, and give examples.

3. "Our perceptions of the external world are created by our brains." Discuss this concept, using the phantom limb phenomenon to support your argument.

Cutaneous Receptors

Before coming to class, review the following sections in chapter 18 of the textbook: "Characteristics of Sensory Receptors," and "Somatic Senses."

Introduction

Specialized sensory organs and free nerve endings in the skin initiate cutaneous sensation—warmth, cold, touch, and pain. The modality and its perceived location is determined by the specific sensory pathway in the brain. The acuteness of sensation depends on the density of the cutaneous receptors. The intensity of the sensation depends on the intensity of the stimulus, on the degree of sensory adaptation that occurs (some senses adapt quickly, others slowly or not at all), and on individual differences.

Objectives

Students completing this exercise will be able to:

1. Describe the structures of the cutaneous receptors and the modality of sensations that they mediate.
2. Describe what is meant by "the punctate distribution of cutaneous receptors."
3. Determine the two-point touch threshold in different areas of the skin, and explain the physiological significance of the differences obtained.
4. Define and demonstrate sensory adaptation, and explain its significance.

Materials

1. Thin bristles
2. Cold and warm metal rods
3. Calipers
4. Cold and warm water baths
5. Colored pencils

A. Structure of the Cutaneous Receptors

Identify, label, and color-code the following structures in figure 22.1:

> Bulbs of Krause
> Corpuscle of touch
> Free nerve endings
> Hair follicle
> Opening of sweat gland
> Organ of Ruffini
> Lamellated (pacinian) corpuscles
> Root hair plexus
> Sebaceous gland

B. Mapping the Temperature and Touch Receptors of the Skin

Four independent cutaneous modalities have traditionally been recognized—warmth, cold, touch, and pain. (Pressure is excluded because it is mediated by receptors deep in the dermis, and the sensations of itch and tickle are usually excluded because of their mysterious origin.) Mapping these sensations on the skin surface reveals that the receptors are not uniformly distributed over the skin but are clustered at different points (have a *punctate distribution*).

Procedure

1. With a ball-point pen, draw a box with sides 2 centimeters long on the anterior surface of the subject's forearm.
2. With the subject's eyes closed, gently touch an ice-cold metal rod to different points in the square. Mark the points of cold sensation with a dark dot.
3. Heat the metal rod in a water bath to about 45°C, wipe the rod, and repeat the mapping procedure, drawing open circles at the points where heat sensation is felt.
4. Gently touch a thin bristle to different areas of the square, and indicate the points of touch sensation with small x's.
5. Reproduce this map in your laboratory report.

Figure 22.1 Diagrammatic section of skin showing cutaneous receptors.

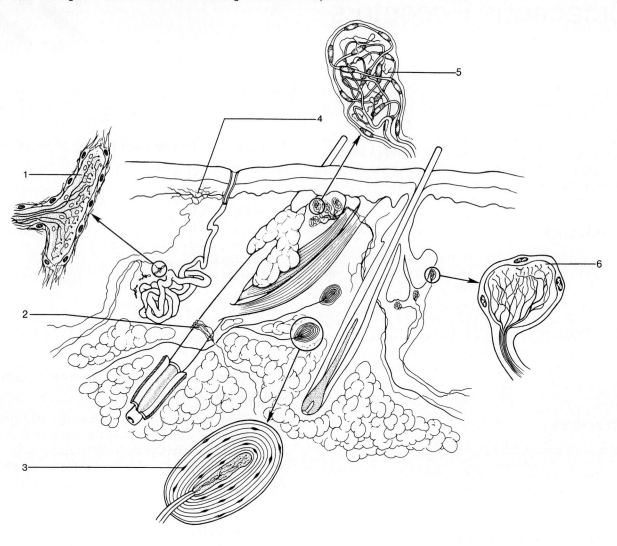

1. _____
2. _____
3. _____

4. _____
5. _____
6. _____

C. Two-Point Touch Threshold

The density of touch receptors is measured by the two-point threshold test. The two points of a pair of adjustable calipers are simultaneously placed on the subject's skin with equal pressure, and the subject is asked if two separate contacts are felt. If so, the points of the divider are brought closer together, and the test is repeated until only one point of contact is felt. The two-point threshold is the minimum distance at which two points of contact can be discriminated.

The density of touch receptors in some parts of the body is greater than in other parts; therefore, the areas of the **sensory cerebral cortex** (*postcentral gyrus*) that correspond to different regions of the body are of different sizes (fig. 22.2). Those areas of the body that have the largest density of touch receptors also receive the greatest motor innervation; the areas of the **motor cerebral cortex** (*precentral gyrus*) that serve these regions are correspondingly larger than other areas. Therefore, a map of the sensory and motor areas of the brain reveals that large areas are devoted to the touch perception and motor activity of the face (particularly the tongue and lips) and hands, whereas relatively small areas are devoted to the trunk, hips, and legs.

Procedure

1. Starting with the calipers wide apart and the subject's eyes closed, determine the two-point threshold on the back of the hand. (Randomly alternate the two-point touch with one-point touches to prevent the subject's second-guessing of the examiner.)
2. Repeat this procedure on the palm of the hand, the fingertip, and the back of the neck.
3. Write the results in the space provided in your laboratory report.

D. Adaptation of Temperature Receptors

When one hand is placed in warm water and another in cold water, the strength of stimulation gradually diminishes until both types of temperature receptors have adapted to their new environment. If both hands are then placed in water at an intermediate temperature, the hand that was in the cold water will feel warm, and the hand that was in the warm water will feel cold. This is due to sensory adaptation of the separate hot and cold receptors.

Procedure

1. Place one hand in warm water (about 40°C) and the other in cold water, and leave them in the water for about 3 minutes.
2. Now place both hands in lukewarm water (about 22°C), and record your sensations in your laboratory report.

Figure 22.2 Motor and sensory areas of the cerebral cortex: (*a*) motor areas that control skeletal muscles; (*b*) sensory areas that receive somatesthetic sensations.

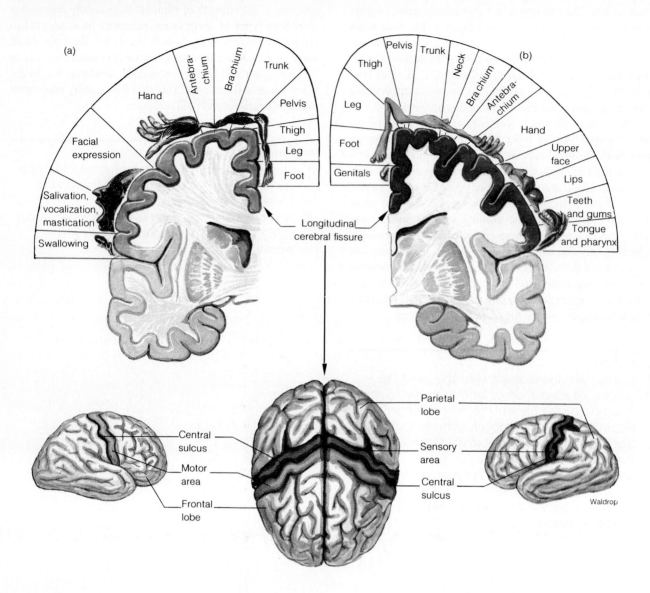

Laboratory Report 22

Name _____

Date _____

Section _____

Cutaneous Receptors

Read the assigned sections of the textbook before completing the laboratory report.

A and B. Structure of the Cutaneous Receptors, and Mapping the Temperature and Touch Receptors of the Skin

1. Reproduce the map of the temperature and touch receptors in the square on the subject's forearm.

2. What can you conclude from your map? Which receptors are believed to be most responsible for the sensations that you mapped?

C. Two-Point Touch Threshold

1. Record the results of the two-point touch procedure in the space below.

2. What is meant by "the punctate distribution of cutaneous receptors"? Describe the distribution of touch receptors in the skin as an example.

3. Most anatomy and physiology textbooks show a picture of an upside-down, odd-looking person in the brain. What does this picture represent, and how was it obtained?

D. Adaptation of Temperature Receptors

1. Did both hands feel the same when placed in the lukewarm water bath? Explain the reasons for the sensations obtained.

2. What does the term *sensory adaptation* mean? Which senses adapt quickly? Which senses adapt slowly, if at all?

Eyes and Vision

Before coming to class, review the sections on vision in chapter 18 of the textbook.

Introduction

The cornea and lens focus light onto the retina, where photoreceptors called *rods* and *cones* are located. A photochemical reaction then occurs in the photoreceptors. This leads to the production of action potentials that are conducted out of the eye in the *optic nerve*.

The *lens* has elastic properties that enable it to change its degree of curvature and refractive power, which helps the eye maintain a clear focus at different distances from the objects being viewed. The amount of light that enters the *pupil* is regulated reflexively by the muscular *iris*. Examinations of the refractive abilities of the eye and of the interior of the eyeball with an ophthalmoscope are common medical procedures.

Objectives

Students completing this exercise will be able to:

1. Describe the structure of the eye and the functions of its components.
2. Define visual acuity and accommodation, and explain how they are tested.
3. Describe the optic disc and fovea centralis, and explain their significance.
4. Describe the three types of cones, and explain how negative and positive afterimages are produced.

Materials

1. Fresh or preserved beef or sheep eyes
2. Dissecting instruments and trays
3. Snellen eye chart and astigmatism chart
4. Wire screen and meter stick
5. Ophthalmoscope
6. Lamp
7. Red, blue, and yellow squares on larger sheets of black paper or cardboard
8. Colored pencils

A. Anatomy of the Eye

1. Identify the following structures of the eye in figures 23.1 and 23.2:
   Associated structures
   - Conjunctiva
   - Levator palpabrae
   - Extrinsic ocular muscles
     - Lateral rectus muscle
     - Medial rectus muscle
     - Superior rectus muscle
     - Inferior rectus muscle
     - Superior oblique muscle
     - Inferior oblique muscle
     - Trochlea
   Structures of the eyeball
   - Chambers of the eye
     - Anterior chamber
     - Posterior chamber
     - Vitreous chamber
   - Choroid layer
   - Ciliary body
   - Cornea
   - Fovea centralis
   - Iris
   - Lens
   - Optic disc
   - Optic nerve
   - Pupil
   - Retina
   - Sclera
   - Suspensory ligament
2. Using colored pencils, color-code the extrinsic ocular muscles shown in figure 23.1.

Dissection of the Eye

1. Examine the external features of the eye. Identify as many of the external structures as possible. Can you determine whether it is a right or left eye?

Figure 23.1 Extrinsic ocular muscles of the left eyeball: *(a)* a lateral view; *(b)* a superior view.

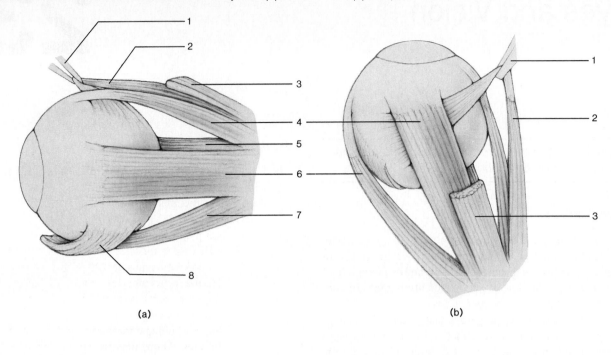

(a)

(b)

1. _____
2. _____
3. _____
4. _____
5. _____
6. _____
7. _____
8. _____

1. _____
2. _____
3. _____

2. With a scalpel, make an incision through the sclera about 1/4 inch behind the cornea. Continue the cut with scissors around the circumference, dividing the eyeball into anterior and posterior halves.
3. Examine the anterior half, noting the lens, suspensory ligament, and ciliary muscle. Carefully remove the lens, and examine the muscle sheets of the iris. Identify the pupil and the chambers of the eye.
4. Study the posterior half, separating and identifying the three layers of the eye. Notice that at the back of the eye the retina separates easily into a clear inner coat and a pigmented *choroid layer.* A portion of the choroid layer is iridescent and is known as the *tapetum lucidum;* this structure, which reflects light and improves night vision, is absent in humans. The *optic disc* (blind spot) is where the retina is attached to the optic nerve. Compare the consistency of the vitreous humor with that of the aqueous humor.

B. Refraction: Tests for Visual Acuity

When a distant object (20 feet or more) is brought to a focus in front of the retina, the individual is said to have **myopia** (nearsightedness). Myopia is usually due to an elongated eyeball (excessive distance from lens to retina) and is corrected by a *concave lens.* In the opposite condition, **hyperopia** (hypermetropia, or farsightedness), the image is brought to a focus behind the retina; this condition is usually due to an eyeball that is too short. In this case, an increase in refractive power is needed, and a convex lens is used. Normal visual acuity is called *emmetropia.*

Visual acuity is frequently tested by means of the Snellen eye chart. A person with normal visual acuity can read the line marked 20/20 from a distance of 20 feet. An individual with 20/40 visual acuity must stand 20 feet away from a line that a normal person can read at 40 feet. An individual with 20/15 visual acuity can read a line at a distance of 20 feet that the average, normal young adult could

Figure 23.2 The internal anatomy of the eyeball.

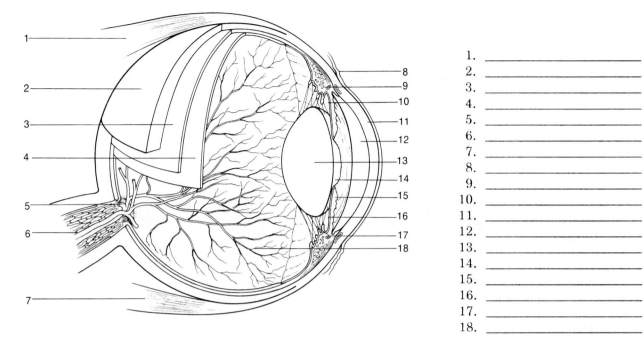

1. _____
2. _____
3. _____
4. _____
5. _____
6. _____
7. _____
8. _____
9. _____
10. _____
11. _____
12. _____
13. _____
14. _____
15. _____
16. _____
17. _____
18. _____

not read at a distance greater than 15 feet. The person with 20/40 vision has myopia, but the person with 20/15 vision does not necessarily have hyperopia. The farsighted person has a decreased ability to see near objects but cannot see distant objects any better than a person with normal vision.

An **astigmatism** is a visual defect produced by an abnormal curvature of the cornea or lens or an irregularity in their surface. Because of this abnormality, the refraction of light rays in the horizontal plane is different from that in the vertical plane. At a given distance, therefore, lines in the visual field oriented in one plane will be clear while lines oriented in the other plane will be blurred. Astigmatism is corrected by means of a *cylindrical lens.*

Procedure

1. Stand 20 feet (6 meters) from the Snellen eye chart. Covering one eye, read the line with the smallest letters you can see (with glasses off, if applicable). Walk up to the chart, and determine the visual acuity of that eye.
2. Repeat this procedure using the other eye (with glasses off, if applicable).
3. Repeat this procedure for each eye with glasses on (if applicable).
4. Stand about 20 feet away from an astigmatism chart, and cover one eye (with glasses off). This chart consists of a number of dark lines radiating from a central point; these lines look like spokes of a wheel. If astigmatism is present, some of the spokes will appear sharp and dark, and others

will appear blurred and lighter because they come to a focus either in front of or behind the retina. Still covering the same eye, slowly walk up to the chart while observing the spokes.
5. Repeat this procedure covering the other eye.
6. Repeat the test for astigmatism for both eyes with glasses on (if applicable).
7. If astigmatism has been corrected with glasses, this can be verified by holding the glasses in front of the face while standing 10 feet from the chart and rotating the glasses 90°. The shape of the wheel will change when the glasses are rotated.

C. Tests for Accommodation

Since the human lens is elastic, its degree of convexity (and, therefore, its refractive power) can be altered by changing the tension placed on it by the suspensory ligament. This tension is regulated by the degree of contraction of the ciliary muscle. When the ciliary muscle is relaxed, the suspensory ligament pulls on the lens, thereby decreasing its convexity and power; distant objects (more than 20 feet away) are thus focused on the retina. Closer objects are brought to a focus on the retina by contraction of the ciliary muscle. The contraction slackens the suspensory ligament, allowing the lens to assume a more convex shape. The ability of the eye to bring into focus objects at different distances from the lens is called **accommodation.**

The elasticity of the lens—and therefore the degree of convexity it can assume for near vision—decreases with age. (This condition is called **presbyopia,** or "old eyes.") Lens elasticity can be tested by measuring the *near point of vision*—the closest an object can be brought to the eyes while still maintaining visual acuity. The near point of vision changes dramatically with age, averaging about 8 centimeters (cm) at age 10 and 100 cm at age 70. Presbyopia is corrected with *bifocals,* in which each lens has two different refractive strengths.

Clinical Significance

In addition to tests of refraction, measurements of intraocular pressure are frequently performed with a device known as a *tonometer.* About 6 milliliters of aqueous humor is formed per day by the *ciliary body.* This fluid is drained by the *scleral venous sinus* (*canal of Schlemm*). If the drainage of aqueous humor is blocked, the intraocular pressure may rise (a condition known as **glaucoma**), resulting in damage to the optic nerve and blindness. Glaucoma may also damage the cornea, resulting in replacement of the normally transparent tissue with opaque scar tissue. When this happens the cornea can be surgically removed and replaced with either a contact lens or a grafted cornea. Because the cornea is avascular, corneal grafts can be performed with less fear that the transplanted tissue will be immunologically rejected.

Procedure

1. Place a square of wire screen about 10 inches in front of the eyes, and observe a distant object through the screen.
2. After closing your eyes momentarily, open them and notice whether the screen or the distant object is in focus.
3. Repeat this procedure, this time focusing the eyes on the screen before opening them.
4. To measure the near point of vision, place a meter stick just under one eye and, holding a pin at arm's length, gradually bring the pin toward the eye.
5. Record the distance at which the pin first appears blurred or doubled.

   _____ cm
6. Repeat this procedure, determining the near point of vision of the other eye.

D. Extrinsic Eye Muscles: Tests for Convergence and Nystagmus

Convergence refers to the medial movement of the eyes, which helps to maintain partial overlap of the visual field of each eye. **Nystagmus** is an oscillation of one or both eyes, produced by weak extrinsic eye muscles or damage to a a cranial nerve that innervates an extrinsic eye muscle.

In a typical examination of the extrinsic ocular muscles, the subject is asked to follow an object (such as a pencil) with his or her eyes as it is moved up and down, right and left. Continued oscillations of the eye (slow phase in one direction, fast phase in the opposite direction) indicate the presence of nystagmus. Inability to move the eye outward indicates damage either to the abducens nerve (VI) or the lateral rectus muscle. Inability to move the eye downward when it is moved inward indicates damage to the trochlear nerve (IV) or the superior oblique muscle. All other defects in eye movement may be due to damage of either the oculomotor nerve (III) or the specific muscles involved.

Procedure

1. Observe retinal disparity by holding a pencil in front of your face with one eye closed, then quickly changing eyes, and noting the apparent position of the pencil.
2. Observe convergence in a subject by asking him or her to focus on the tip of a pencil 2 feet in front of the face. Move it slowly to the bridge of the nose, and notice the change in the diameter of the pupil during this procedure.
3. Hold a pencil about 2 feet away from the bridge of the subject's nose, and then move it to the left, to the right, and up and down, leaving the pencil at least 10 seconds in each position. Observe the movement of the eyes, and note the presence or absence of nystagmus.

E. Test for the Pupillary Reflex

The correct amount of light is admitted into the eye through an adjustable aperture, the *pupil.* The diameter of the pupil is determined by the contraction of the iris muscles surrounding the pupil. Sympathetic nerve stimulation causes contraction of the *radial muscle* of the iris, which dilates the pupil. Parasympathetic nerve stimulation evokes contraction of the *circular muscle* of the iris, which constricts the pupil. Activity of these nerves is regulated by a reflex response to light.

Procedure

1. The examiner and the subject remain in a darkened room for at least 1 minute, allowing their eyes to adjust to the dim light.
2. The examiner shines a narrow beam of light (e.g., from an ophthalmoscope or a pen flashlight) from the right side of the subject into the right eye, and observes the pupillary reflex in the right eye and in the left eye (this is the *consensual reaction*).
3. Repeat this procedure (first dark-adapting the eyes again) from the left side with the left eye.

F. Examination of the Eye with an Ophthalmoscope

In this exercise, the arteries and veins within the eye and two regions of the retina—the **optic disc** and the **macula lutea**—will be observed. The optic disc is the region where the optic nerve exits the eye, otherwise known as the *blind spot.* The macula lutea is a yellowish region containing a central pit, the **fovea centralis,** where the highest concentration of the photoreceptors responsible for visual acuity (the cones) is located. This is the region on which an image is focused when the eyes look directly at an object.

Clinical Significance

Clinical examination of the eye with an ophthalmoscope can aid the diagnosis of many ocular and systemic diseases. The features noted in these examinations include the condition of the blood vessels; the color and shape of the disc; the presence of particles, exudates, or hemorrhage; the presence of edema or inflammation of the optic nerve (*papilledema*); and myopia or hyperopia.

Procedure
1. The subject is seated in a darkened room and is asked to look at a distant object (blinking is allowed).
2. The examiner and subject position themselves face to face. The examiner holds the ophthalmoscope in the right hand and uses the right eye when observing the subject's right eye. (This is reversed when viewing the left eye.)
3. With the examiner's forefinger on the lens adjustment wheel and the examiner's eye as close as possible to the small hole in the ophthalmoscope (glasses off if applicable), the instrument is brought as close as possible to the subject's eye. The examiner's hand can be steadied by resting it on the subject's cheek.
4. The eye will be examined from the front to the back. Looking from the side of the eye (not in front), the iris and lens can be examined by means of a +20 to +15 lens (this will be different if the examiner wears glasses).
5. Rotate the wheel counterclockwise to examine the fundus. If both the examiner's and the subject's eyes are normal, the fundus (back of the eye) can be clearly seen without the need of a lens (set on 0; the refractive strength of the subject's eye is sufficient to focus the light on the fundus).

a. If a positive (convex) lens is necessary to focus on the fundus and the examiner's eyes are normal, the subject has hyperopia (hypermetropia).
b. If a negative (concave) lens is necessary to focus on the fundus and the examiner's eyes are normal, the subject has myopia.
6. Observe the arteries and veins of the fundus, and follow them to their point of convergence at the optic disc.
7. Finally, at the end of the examination, observe the macula lutea by asking the subject to look directly into the light of the ophthalmoscope.

G. Test for the Blind Spot

The axons of all of the ganglion cells gather to form the optic nerve, which leaves the eye at the place in the retina called the optic disc (blind spot). There are no rods or cones in this spot, and an object whose image is focused here will not be seen.

Procedure
1. Hold the drawing of the circle and the cross (fig. 23.3) about 20 inches from your face with the left eye covered or closed. Focus on the circle; this is most easily done if the circle is positioned in line with the right eye.

Figure 23.3 A diagram for demonstrating the blind spot.

2. Keeping your right eye focused on the circle, slowly bring the drawing closer to your face until the cross disappears. Continue moving the drawing slowly toward your face until the cross reappears.
3. Repeat this procedure with the right eye closed or covered and the left eye focused on the cross. Observe the disappearance of the circle as the drawing is brought closer to the face.

H. Production of an Afterimage

When an eye that has adapted to a bright light is closed or quickly turned toward a wall, the bright image of the light will still be seen. This is a *positive afterimage,* caused by the continued "firing" of the photoreceptors. After a short period of time, the dark image of the light—the *negative afterimage*—will appear against a lighter background due to the "bleaching" of the visual pigment of the affected photoreceptors. If one type of cone has been bleached by one's viewing of an appropriately colored object, the positive afterimage of the object will appear in the complementary color.

Procedure

1. Stare at a light bulb, then suddenly shift your gaze to a blank wall. Observe the appearance of the negative afterimage.
2. For 1 minute, stare at a small red square that has been pasted on a larger sheet of black paper.
3. Suddenly shift your gaze to a sheet of white paper and note the color of the positive afterimage.
4. Repeat this procedure using blue squares and yellow squares.

Laboratory Report 23

Name _____

Date _____

Section _____

Eyes and Vision

Read the assigned sections of the textbook before completing the laboratory report.

1. Complete the following sentences:

    a. The _____ is the colored portion of the eye.

    b. The area of the retina that has the highest concentration of cones is the _____ .

    c. The _____ is the first structure that refracts light in the eye.

    d. The optic disc is also known as the _____ .

2. Complete the following table:

Name	Innervation	Function
Levator palpebrae muscle		
Iris		
Lateral rectus muscle		
Medial rectus muscle		
Superior rectus muscle		
Inferior rectus muscle		
Superior oblique muscle		
Inferior oblique muscle		

3. Match the following:

    ___ (1) Myopia
    ___ (2) Hyperopia
    ___ (3) Presbyopia
    ___ (4) Astigmatism
    ___ (5) Glaucoma

    (a) abnormal curvature of cornea or lens
    (b) eye too long
    (c) abnormally high intraocular pressure
    (d) eye too short
    (e) loss of lens elasticity

4. Define the following terms:

    a. Visual acuity _____

    b. Accommodation _____

    c. Convergence of eyes _____

    d. Nystagmus _____

5. What is the function of the iris? What muscles are responsible for pupillary constriction and dilation?

6. List in proper sequence the structures through which light passes as it enters the eye and travels to the photoreceptors. Which of these structures refracts light?

7. A person with myopia does not have to accommodate for near vision as much as a person with normal vision does. Explain why.

8. Explain why the rods provide greater visual sensitivity than the cones, and why the cones provide greater visual acuity than the rods.

9. "You see with your brain, not with your eyes." Explain this statement, using the Young-Helmholtz theory of color vision and the optic disc as examples.

10. Carrots contain large amounts of the compound *carotene,* a precursor of vitamin A. How can eating carrots improve eyesight?

Ears and Hearing

Before coming to class, review "Equilibrium" and "Hearing" in chapter 18 of the textbook.

Introduction

Vibrations of the tympanic membrane caused by sound waves result in movements of the auditory (ear) ossicles. Movements of the stapes against the oval window of the cochlea, in turn, produce pressure waves within the fluid of the cochlea. The resulting displacement of the basilar membrane bends the processes of the hair cells within the spiral organ (organ of Corti); this bending produces action potentials. Damage to the middle ear results in **conduction deafness,** whereas damage to the cochlea or vestibulocochlear nerve results in **sensory deafness.** Simple tests with tuning forks can be used to diagnose deafness and can also demonstrate that both ears must function to accurately determine the position of a sound (binaural localization).

Objectives

Students completing this exercise will be able to:

1. Describe the anatomy of the outer, middle, and inner ear.
2. Describe the function of the auditory ossicles and auditory canal.
3. Describe the detailed structure of the cochlea, and explain how the spiral organ functions.
4. Explain how the ear responds to sounds of different pitches.
5. Distinguish between conduction deafness and sensory deafness, and explain how these conditions can be diagnosed.
6. Explain how the location of a sound can be determined by means of information obtained from both ears.

Materials

1. Tuning forks
2. Rubber mallets
3. Reference text
4. Colored pencils

A. Anatomy of the Ear

1. Identify and label the following structures in figures 24.1 and 24.2:

   Auditory canal
   Auricle (pinna)
   Cochlea
       Basilar membrane
       Cochlear duct
       Cochlear nerve
       Hair cells
       Scala vestibuli
       Scala tympani
       Tectorial membrane
       Vestibular duct
       Vestibular membrane
       Vestibular nerve
   Auditory ossicles
       Incus
       Malleus
       Stapes
   Endolymph (not shown)
   External auditory canal
   Labyrinth: (a) bony; (b) membranous
   Oval window
   Perilymph (not shown)
   Round window
   Semicircular canals
       Hair cells
       Tympanic cavity
   Spiral organ (organ of Corti)
   Tympanic membrane
   Vestibule

2. Using colored pencils, color the cochlea, the semicircular canals, the hair cells, and the tectorial membrane in different colors in figure 24.2.

B. Conduction of Sound Waves through Bone: Rinne and Weber's Tests

Although hearing is normally produced by the vibration of the oval window in response to sound waves conducted through the movements of the auditory ossicles, the endolymph can also be made to vibrate in response to sound waves conducted through the skull bones, which bypass

Figure 24.1 The ear.

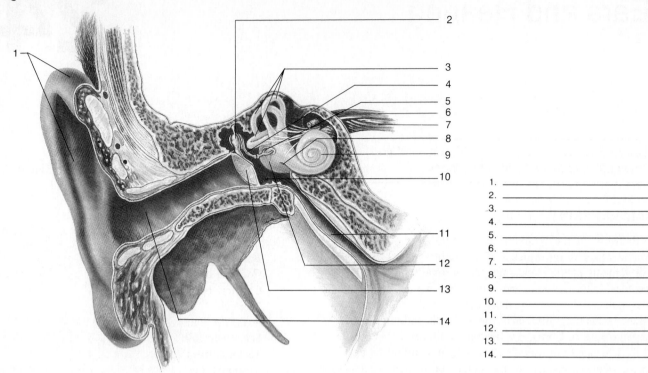

1. _____
2. _____
3. _____
4. _____
5. _____
6. _____
7. _____
8. _____
9. _____
10. _____
11. _____
12. _____
13. _____
14. _____

the middle ear. This is used clinically to differentiate between deafness due to middle ear damage (conduction deafness—e.g., due to damage to the auditory ossicles in *otitis media* or immobilization of the stapes in *otosclerosis*) and deafness caused by damage to the cochlea or vestibulocochlear nerve (sensory deafness—e.g., due to infections, streptomycin toxicity, or prolonged exposure to loud sounds).

Clinical Significance

Conduction deafness may be caused by infections of the middle ear (otitis media), infections of the tympanic membrane (tympanitis), or excessive accumulation of *cerumen* (ear wax). A person with conduction deafness may wear a hearing aid over the mastoid process of the temporal bone, which amplifiessounds and transmits them by bone conduction to the cochlea. Hearing aids do not help if there is complete sensory (nerve) deafness.

Procedure

1. Produce vibrations in a tuning fork by striking it with a rubber mallet.
2. Perform the Rinne test by placing the *handle* of a vibrating tuning fork against the mastoid process of the temporal bone (the bony prominence behind the ear), with the tuning fork pointed down and behind the ear. When the sound almost dies away, move the tuning fork, (by the handle) near the external auditory canal. If there is no damage to the middle ear, the sound will reappear.
3. Simulate conduction deafness by repeating the Rinne test with a plug of cotton in the ear.
4. Perform Weber's test by placing the handle of a vibrating tuning fork on the midsagittal line of the head.
5. Repeat Weber's test with one ear plugged with your finger. Notice that the sound will appear louder in the plugged ear because external room noise is excluded.

Figure 24.2 Structures of the inner ear (*a*); the spiral organ (organ of Corti) (*b*).

1. _____
2. _____
3. _____
4. _____
5. _____
6. _____
7. _____
8. _____
9. _____
10. _____
11. _____
12. _____
13. _____
14. _____
15. _____
16. _____

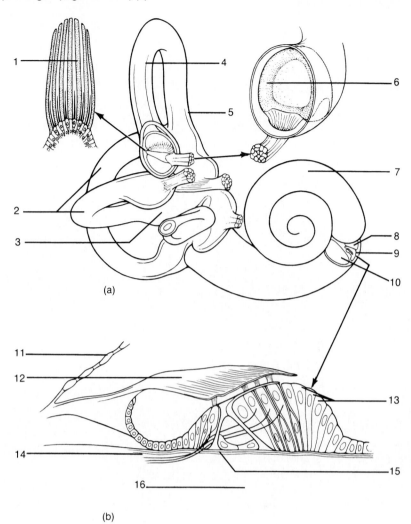

(a)

(b)

Laboratory Report 24

Name _____

Date _____

Section _____

Ears and Hearing

Read the assigned section in the textbook before completing the laboratory report.

A. Anatomy of the Ear

1. Number in correct sequence the sound pathway from the auricle to the cochlear nerve.

    ___ Cochlear duct   ___ Oval window
    ___ Cochlear nerve   ___ Auricle
    ___ External auditory canal   ___ Scala tympani
    ___ Hair cells   ___ Scala vestibuli
    ___ Incus   ___ Stapes
    ___ Malleus   ___ Tympanic membrane
    ___ Spiral organ (organ of Corti)

2. Label the following structures on the schematic diagram of a cochlea: the scala vestibuli, cochlear duct, scala tympani, vestibular membrane, basilar membrane, tectorial membrane, hair cells, cochlear nerve. Which chambers contain perilymph and which contain endolymph?

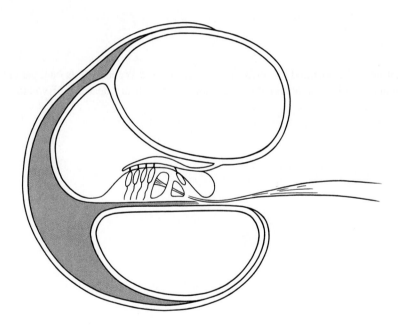

B. Conduction of Sound Waves through Bone: Rinne and Weber's Tests

1. How does the ear transduce the sound waves in air into electrical nerve impulses?

2. How are the pitch and loudness of a sound coded in the nervous system?

3. Explain the results that might be obtained by performing the Rinne and Weber's tests on a patient with otosclerosis. How might these results compare with those obtained from a patient with sensory deafness? Explain.

Taste

Before coming to class, review "Taste and Olfaction" in chapter 18 of the textbook.

Introduction

Taste buds that are sensitive to one of the four tastes—sweet, sour, bitter, and salty—are distributed in a characteristic pattern on the tongue. The sour taste of solutions is due to their acid (H^+) content, and the salty taste is produced by the presence of Cl^- (but this is modified by the cation—NaCl tastes saltier than KCl, for example). The chemical basis for bitter and sweet taste is largely unknown, since these can be produced by a variety of seemingly unrelated compounds. Fructose tastes the sweetest, followed by sucrose, and then by glucose.

Objectives

Students completing this exercise will be able to:

1. Describe the structure of the tongue papillae and associated taste buds.
2. Describe the distribution of different taste buds on the tongue.
3. Describe the innervation of the tongue.

Materials

1. Cotton–tipped applicator sticks
2. Solutions of 5% sucrose, 1% acetic acid, 5% sodium chloride, and 0.5% quinine sulfate

A. Anatomy of Taste

Identify and label the following structures in figure 25.1:
Vallate papillae
Gustatory cell
Gustatory hair
Sensory nerve fiber
Supporting cell
Taste buds
Taste pore

B. Mapping the Modalities of Taste

Each taste quality is perceived most acutely in a particular region of the tongue. Taste buds function as chemoreceptors and activate sensory neurons in two cranial nerves. The facial nerve serves the anterior two–thirds of the tongue, and the glossopharyngeal nerve serves the posterior third.

Procedure

1. Dry the tongue with a paper towel, and use an applicator stick to apply a dab of 5% sucrose solution to the tip, sides, and back of the tongue.
2. Repeat this procedure, using 1% acetic acid, 5% NaCl, and 0.5% quinine sulfate, being sure to rinse the mouth and dry the tongue between solutions.
3. On the tongue diagram in the laboratory report, record the location where each solution could be tasted. Use the symbols *sw* for sweet, *sl* for salty, *sr* for sour, and *b* for bitter.

Figure 25.1 (a) The papillae of the tongue and (b) associated taste.

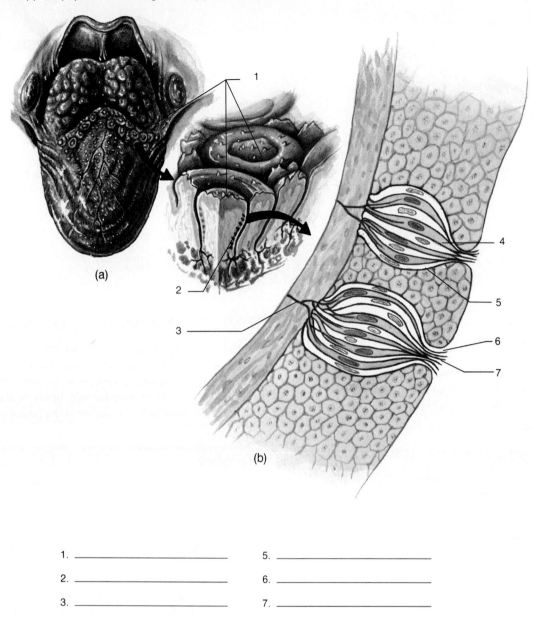

(a)

(b)

1. _____   5. _____

2. _____   6. _____

3. _____   7. _____

4. _____

Laboratory Report 25

Name _____

Date _____

Section _____

Taste

Read the assigned section in the textbook before completing the laboratory report.

1. As described in the last procedure, map the areas of the tongue in the diagram below.

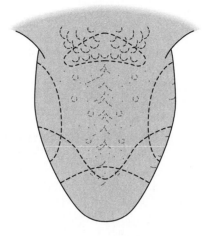

2. Describe the areas of the tongue where the following tastes are best perceived:
   a. Sweet _____
   b. Salty _____
   c. Bitter _____
   d. Sour _____
3. Which taste modalities would be most affected by the destruction of the glossopharyngeal nerve? Which would be most affected by the destruction of the facial nerve? Explain why.

Endocrine System

Before coming to class, review the following sections in chapter 19 of the textbook: "Pituitary Gland," "Adrenal Glands," "Thyroid and Parathyroids," and "Pancreas and Other Endocrine Glands."

Introduction

There are two major categories of glands: exocrine and endocrine. *Exocrine glands,* such as sweat, salivary, and mucous glands, produce secretions that are transported through ducts to the outside of epithelial tissues or membranes. *Endocrine glands,* in contrast, do not have ducts; they secrete their products—known as *hormones*—into the blood. Hormones are biologically active molecules that are transported by the blood to their sites of action in *target organs.* Because hormones from one endocrine gland may influence another endocrine gland, and because the secretions of hormones from different glands are coordinated to promote homeostasis or to regulate reproduction, the endocrine glands constitute a *body system.* The endocrine system is one of the two major control systems of the body, and it functions together with the nervous system to regulate the other body systems.

Objectives

Students completing this exercise will be able to:

1. Define the terms *endocrine gland, hormone,* and *target organ.*
2. Identify the endocrine glands and the major hormones they secrete.
3. Describe the actions of the major hormones of the body.
4. Identify the locations and gross structures of the endocrine glands.
5. Describe the histological structure of the major endocrine glands.

Materials

1. Microscopes
2. Prepared histological slides
3. Human torso model, human brain model, skull
4. Reference text and atlases
5. Colored pencils

A. Pituitary Gland

1. Identify, label, and color-code the following regions of the pituitary gland (hypophysis) in figure 26.1:

   Adenohypophysis (anterior lobe)
       Pars intermedia (intermediate lobe)
   Neurohypophysis (posterior lobe)
       Infundibulum
       Pars nervosa

2. Locate the pituitary gland in a human brain model and where it is supported in the skull.
3. Obtain a histological slide of the pituitary gland and identify as many of the following structures as possible:

Adenohypophysis
    Chromophils (cells with red- or blue-staining cytoplasm)
    Chromophobes (cells with clear cytoplasm)
    Sinusoids (blood capillaries)
Neurohypophysis
    Nerve fibers
    Pituicytes

B. Thyroid and Parathyroid Glands

1. Label the thyroid in figure 26.6.
2. Identify and label the following regions and structures in figures 26.2 and 26.3:

Parathyroid gland
Thyroid gland
    Colloid
    Isthmus
    Simple cuboidal epithelial cells
    Thyroid follicles

3. Obtain a histological slide of the thyroid gland and identify as many of these structures as possible.

Figure 26.1 The pituitary gland.

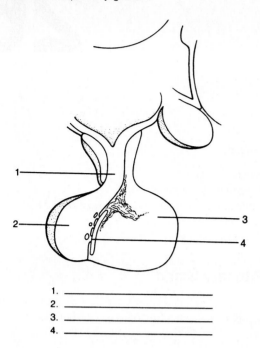

1. _____
2. _____
3. _____
4. _____

Figure 26.2 The thyroid gland.

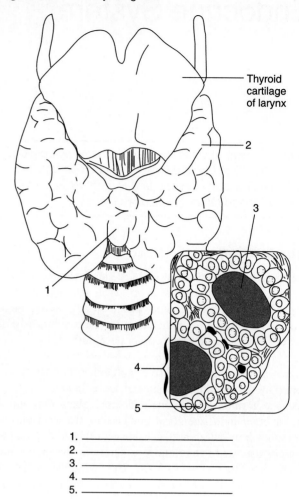

Thyroid cartilage of larynx

1. _____
2. _____
3. _____
4. _____
5. _____

C. Pancreatic Islets

1. Identify, color-code, and label the following regions and structures in figure 26.4:

    Pancreas
       Body
       Tail
    Pancreatic islet (islet of Langerhans)

2. Obtain a histological slide of the pancreas that is stained to show the pancreatic islets, and identify the following:

    Acini (clusters of cells that compose the majority of pancreatic tissue)
    Pancreatic islet
       Alpha cells (red-staining granules)
       Beta cells (blue-staining granules)

D. Adrenal Glands

1. Identify and label the following regions and structures in figure 26.5. Use different colors to distinguish between the adrenal cortex and adrenal medulla.

    Adrenal cortex
       Zona fasciculata
       Zona glomerulosa
       Zona reticularis
    Adrenal gland
    Adrenal medulla

2. Obtain a histological slide of the adrenal gland and identify as many of these regions as possible.

E. Ovaries

1. Label the ovaries in figure 26.6.
2. Identify and label the following structures in figure 26.7:

    Antrum
    Corona radiata
    Granulosa cells
    Ovarian follicle
    Ovum

3. Obtain a histological slide of an ovary and identify these structures.

Figure 26.3 A posterior view of the parathyroid glands.

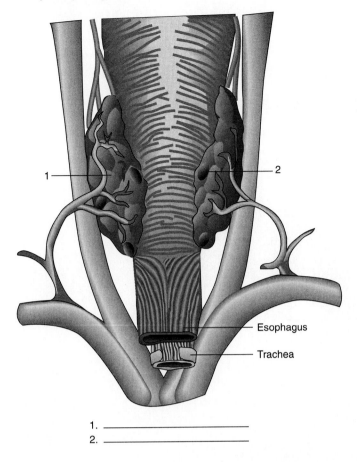

1 ———— 2

2 ————

Esophagus

Trachea

1. _____

2. _____

Figure 26.4 The pancreas and the associated pancreatic islets (of Langerhans).

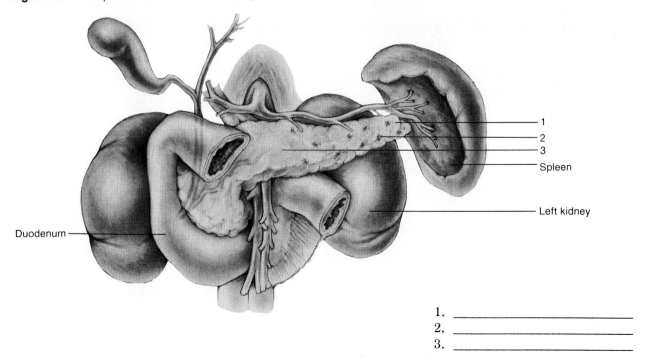

1

2

3

Spleen

Left kidney

Duodenum

1. _____

2. _____

3. _____

Figure 26.5 The structure of the adrenal gland showing the three zones of the adrenal cortex and the adrenal medulla.

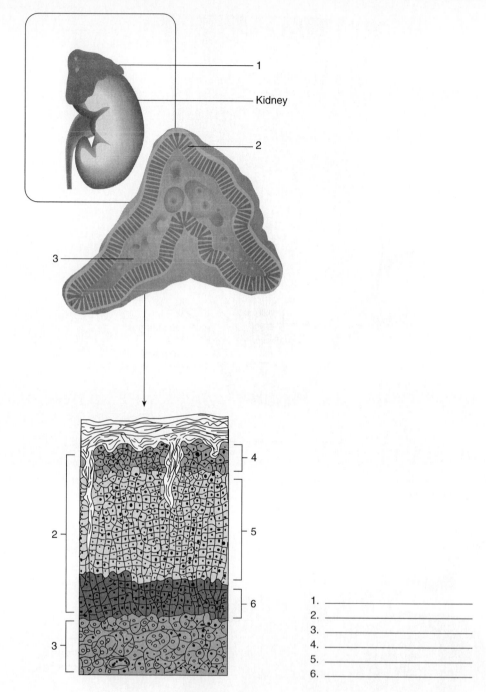

Kidney

1. _____
2. _____
3. _____
4. _____
5. _____
6. _____

Figure 26.6 Various endocrine glands.

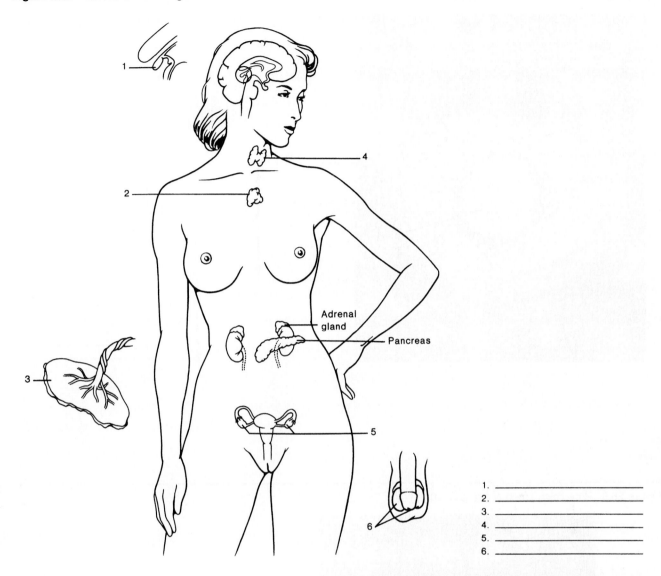

Adrenal gland

Pancreas

1. _____
2. _____
3. _____
4. _____
5. _____
6. _____

F. Testes

1. Label the testes in figure 26.6.
2. Identify and label the following structures in figure 26.8:

   Interstitial (Leydig's) cells
   Seminiferous tubules
   Spermatozoa

3. Obtain a histological slide of a testis and identify these structures.

G. Pineal Gland, Placenta, and Thymus

1. Label the pineal gland, placenta, and thymus in figure 26.6.
2. Locate the pineal gland in a human brain model; find the placenta and thymus on a human torso model.

Figure 26.7 An ovarian follicle with ovum (450×).

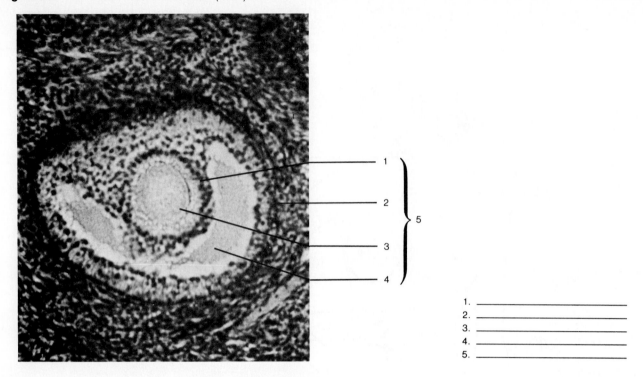

1. _____
2. _____
3. _____
4. _____
5. _____

Figure 26.8 Photomicrograph of a testis (100×).

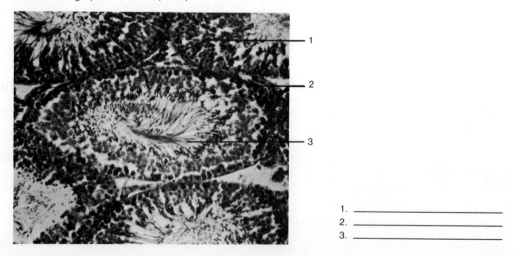

1. _____
2. _____
3. _____

Laboratory Report 26

Name _____

Date _____

Section _____

Endocrine System

Read the assigned section in the textbook before completing the laboratory report.

A. Pituitary Gland

1. List the alternative name for each of the following terms:

   a. Pituitary gland _____

   b. Anterior pituitary gland _____

   c. Posterior pituitary gland _____

2. Why does the anterior pituitary gland have a typical glandular appearance (with many epithelial cells that stain intensely, and numerous capillaries) under the microscope, whereas the posterior pituitary gland does not? What is the functional significance of this difference?

3. Describe how the hypothalamus regulates the hormonal secretion of the anterior pituitary and posterior pituitary glands.

B. Thyroid and Parathyroid Glands

1. List the two major hormones secreted by the thyroid gland:

   _____

   _____

2. Briefly describe the major functions of the thyroid and parathyroid glands.

C and D. Pancreatic Islets and Adrenal Glands

1. What are the differences between an endocrine gland and an exocrine gland? Relate your answer to the structure and functions of the pancreas.

2. Briefly describe the physiological role of insulin and glucagon.

3. List the three major categories of hormones secreted by the adrenal cortex and the two hormones secreted by the adrenal medulla.

   Adrenal cortex _____

   _____

   _____

   Adrenal medulla _____

   _____

4. Identify the embryonic tissues that develop into the adrenal cortex and adrenal medulla. What is the significance of their embryonic origin to the regulation of the two parts of the adrenal gland?

E and F. Ovaries and Testes

1. Describe the structure of an ovarian follicle. Which cells in the follicle secrete estrogen?

2. The testis is said to be composed of two "compartments." What are they? If a man has a vasectomy, will both compartments be affected? Explain.

G. Pineal Gland, Placenta, and Thymus

1. List four hormones secreted by the placenta.

   _____

   _____

   _____

   _____

2. In addition to its role as an endocrine gland, the placenta performs many other functions. List three other organs that perform important functions in addition to their endocrine gland activities.

   _____

   _____

   _____

3. Briefly describe the locations and possible functions of the pineal gland and thymus gland.

Regulation and Maintenance of the Human Body

The following exercises are included in this unit:

These exercises are based on information presented in the following chapters of *Concepts of Human Anatomy and Physiology,* 4th ed.:

The blood carries oxygen (derived from the function of the respiratory system) and nutrients (derived from the function of the digestive system) to the body cells for metabolism. The blood also transports waste products from the body cells: carbon dioxide to the lungs for elimination, and other wastes to the kidneys for excretion as urine. The body systems studied in this section are thus intimately interrelated, and they function together to maintain homeostasis. Since blood flow is integral to these relationships, the function of the heart and blood vessels is central to the proper functioning of all of the other body systems.

Red Blood Cells and Oxygen Transport

Before coming to class, review "Composition of the Blood" in chapter 20 and "Hemoglobin and Oxygen Transport" in chapter 24 of the textbook.

Introduction

Almost all of the oxygen transported by the blood is carried within the **red blood cells (erythrocytes),** where it is attached to hemoglobin. A deoxyhemoglobin molecule binds to four oxygen molecules to form oxyhemoglobin in the lungs, and the oxyhemoglobin dissociates to release its oxygen in the tissue capillaries. Measurements of the oxygen-carrying capacity of the blood include the red blood cell count, hemoglobin concentration, and hematocrit. The condition known as *anemia* is indicated when one or more of these measurements is abnormally low.

Objectives

Students completing this exercise will be able to:

1. Describe the composition of blood.
2. Describe the composition of hemoglobin, and explain how it participates in oxygen transport.
3. Demonstrate the procedures for measuring the red blood cell count, hemoglobin concentration, and hematocrit, and list the normal values for these measurements.
4. Explain how measurements of the oxygen-carrying capacity of blood can be used to diagnose anemia and polycythemia.

Materials

1. Hemocytometer
2. Unopettes (Becton-Dickinson) for manual red blood cell count and hemoglobin measurements
3. Heparinized capillary tubes, clay capillary tube sealant (Seal-ease), microcapillary centrifuge, hematocrit reader
4. Microscope
5. Sterile lancets and 70% alcohol
6. Colorimeter and cuvettes
7. Container for the disposal of blood-containing objects

A. Red Blood Cell Count

The object of this exercise is to determine the number of red blood cells in a cubic millimeter of blood. Since this number is very large, it is practical to dilute a sample of blood with an isotonic solution, count the number of red blood cells in a small volume of this diluted blood, and then multiply by a correction factor. This procedure is accurate only when (1) the blood diluted is a representative fraction of all the blood in the body, (2) the dilution volumes are accurate, and (3) the sample counted is representative of the total volume of diluted blood.

Procedure
Obtaining and Diluting Blood Samples

1. The **Unopette** reservoir (fig. 27.1) contains a premeasured amount of diluting solution. Use the shielded capillary tube (fig. 27.1) to puncture the plastic top of this reservoir. You may leave the shielded capillary tube stuck into the top of the reservoir until it is needed.

Figure 27.1 The Unopette system (Becton-Dickinson), consisting of a reservoir containing the premeasured amount of diluent (left) and a plastic capillary tube within a shield (right) for puncturing the reservoir top and delivering a measured amount of whole blood to the reservoir.

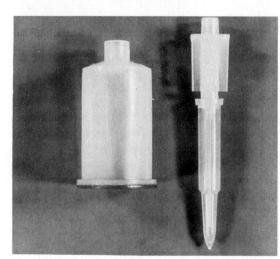

Figure 27.2 The Unopette method for measuring a red blood cell count: (a) the method of filling the plastic capillary pipette with fingertip blood; (b) squeezing of the reservoir to draw blood out of the pipette into diluent within the reservoir.

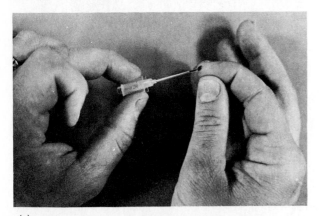

(a)

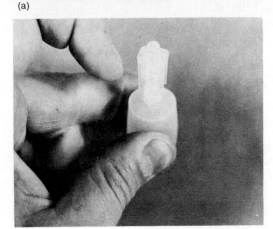

(b)

2. Swing your hand around until your fingers become engorged with blood (hyperemia). Cleanse the tip of your index or third finger with 70% alcohol, and prick it with a sterile lancet.

☞ **Note:** *Because of the danger of exposure to the AIDS virus when handling blood, each student should perform this and other blood exercises with his or her own blood only. Be sure to dispose of all objects that have been in contact with blood in the container designated by the instructor.*

3. Discard the first drop of blood, and hold your finger downward to collect a large droplet of blood. Remove and discard the shield over the capillary pipette of the Unopette, and simply touch the tip of the pipette to the drop of blood. Allow the pipette to fill by capillary action (fig. 27.2a).
4. Squeeze the previously punctured reservoir with the fingers of one hand and, while squeezing, insert the blood-filled pipette into the punctured top of the reservoir. When you release pressure

Figure 27.3 A procedure for filling a hemocytometer with diluted blood for a red blood cell count from a Unopette reservoir. Squeezing of the reservoir places a drop of diluted blood at the edge of the cover slip; the drop of blood then moves under the cover slip by capillary action.

on the reservoir, the blood will be pulled into the premeasured Hayem's solution within the reservoir (fig. 27.2b).

5. Thoroughly shake the blood with the Hayem's solution for approximately 1 minute.

Procedure

Filling the Hemocytometer and Determining Red Blood Cell Count

1. Place a cover slip on the hemocytometer so that it covers one of the silvered areas.
2. Remove the capillary pipette from the reservoir, turn it around, and reinsert it backwards into the reservoir so that the capillary is pointing out of the reservoir (like the needle of a syringe—see fig. 27.3). Discard the first 3 drops of blood from the Unopette onto a piece of cotton or paper, and dispose of this properly into a designated container. Place the next drop of diluted blood at the edge of the cover slip. The diluted blood will be drawn under the cover slip by capillary action (fig. 27.3).
3. Locate the grid on the hemocytometer using the 10× objective. Change to 45× ,and count the total number of red blood cells in the five indicated squares (fig. 27.4).

☞ **Note:** *If a red blood cell lies on the upper or left-hand line, include it in your count. Do not include red blood cells that lie on the lower or right-hand lines.*

4. The central grid of 25 squares is 1 square millimeter (mm^2) in area and 0.10 mm deep. The dilution factor is 1:200. To convert the number of red blood cells that you counted in five squares to the number of red blood cells per mm^3, you must multiply your count by 10,000 ($5 \times 10 \times 200$).

Figure 27.4 The hemocytometer grid: squares 1–5 are used for red blood cell counts, whereas squares A–D are used for white blood cell counts.

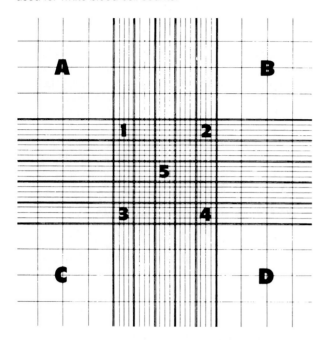

Figure 27.5 A method for filling a capillary tube with blood.

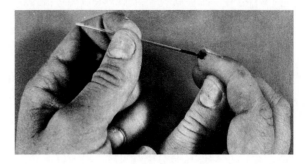

3. Using clay capillary sealant, seal the red-banded end of the capillary tube (this is the fire-polished end) by gently pushing it into Seal-ease and then removing it.
4. Place the sealed capillary tube in a numbered slot on a microcapillary centrifuge, with the sealed end facing outward against the rubber gasket. Screw the top plate onto the centrifuge head, and centrifuge for 3 minutes. At the end of the centrifugation, determine the hematocrit with the hematocrit reader provided, and enter this in your laboratory report.

> The normal hematocrit for an adult male is 47 ± 7, and for an adult female is 42 ± 5. Hematocrit values are percentages.

5. Record your count of the red blood cells in the five squares. Calculate the number of red blood cells in 1 mm^3 of your blood, and enter this number in the laboratory report.
   Red blood cells/5 squares: _____

> The normal red blood cell count per mm^3 for a male is 4.5 million to 6.0 million; for a female, 4.0 million to 5.5 million.

B. Hematocrit

When you centrifuge whole blood, the red blood cells become packed at the bottom of the tube, and the plasma remains at the top. The ratio of packed red blood cell volume to total blood volume is called the *hematocrit*.

Procedure

1. Prick your finger with a sterile lancet to obtain a drop of blood. Discard the first drop onto an alcohol swab, and dispose of this properly in a designated container.
2. Obtain a heparinized capillary tube (heparin is an anticoagulant). Notice that one end of the tube is marked with a red band. Touch the opposite end of the capillary tube to the drop of blood, allowing blood to enter the tube by capillary action and gravity (fig. 27.5). The tube does not have to be completely full (half-full or more is adequate), and the presence of air bubbles is not important (these will disappear during centrifugation).

C. Hemoglobin Concentration

In this exercise, the concentration of hemoglobin in a solution of hemolyzed blood is determined by measuring the solution's color intensity with a spectrophotometer. To do this accurately, the different forms of hemoglobin (oxyhemoglobin, deoxyhemoglobin, and others) must be converted into one form. This is done by oxidizing the Fe^{++} in hemoglobin to Fe^{+++} (forming methemoglobin), and then combining it with cyanide to make it a more stable chemical. The hemoglobin concentration of the unknown sample is determined by comparing its absorbance with that of a standard hemoglobin solution of known concentration.

Procedure

1. A different Unopette will be used for this procedure. This Unopette contains a solution of cyanomethemoglobin reagent (yellow colored), and it has a capillary pipette that delivers twice as much blood as the one used for the red blood cell count. The correct capillary pipette can be identified by the yellow number 20 (for 20 microliters) on its side. Puncture the top of the reservoir with the shielded capillary pipette as in the previous procedure.

2. Clean your finger with 70% alcohol, and puncture it with a sterile lancet. Discard the first drop of blood onto an alcohol swab, and properly dispose of this in a designated container. Fill the plastic capillary pipette in the Unopette with 0.02 milliliters of blood by touching the tip to the drop of blood and allowing the pipette to completely fill by capillary action (fig. 27.2a).

3. Squeeze the reservoir, insert the pipette, and release the reservoir (fig. 27.2b). Squeeze and release the reservoir a few more times to completely wash the pipette. Mix and let stand at room temperature for 10 minutes.

4. Some standard hemoglobin solutions come full strength and must be diluted with cyanomethemoglobin reagent to be at the same dilution as the unknown. Use the procedure in steps 2 and 3 to make this dilution (using a new Unopette).

   Some pre-diluted hemoglobin standards are available and can be used as they are; however, it may be necessary to calculate the blood hemoglobin concentration equivalent to that of the diluted standard solution.

5. Set the colorimeter at a wavelength of 540 nanometers, and standardize the instrument using plain cyanomethemoglobin solution as the blank. Record the absorbance values of the unknown and standard.

   Absorbance of unknown: _____
   Absorbance of standard: _____

6. Calculate the hemoglobin concentration of the unknown, using the formula

$$Concentration_{unknown} = \frac{Concentration_{standard} \times A_{unknown}}{A_{standard}}$$

Enter the hemoglobin concentration of your blood in the laboratory report.

> The normal hemoglobin concentration of an adult male is 14–18 grams per deciliter (g/dl), and of an adult female is 12–16 g/dl.

D. Calculation of Mean Corpuscular Volume (MCV) and Mean Corpuscular Hemoglobin Concentration (MCHC)

An abnormally low hemoglobin, hematocrit, or red blood cell count indicates a condition known as **anemia.** Anemia may have many different causes, such as iron deficiency, vitamin B_{12} deficiency, folic acid deficiencies, bone marrow disease, hemolytic disease (e.g., sickle-cell anemia), blood loss from hemorrhage, and infection. The type of anemia can be more specifically determined by using the hemoglobin, hematocrit, and red blood cell count to derive the **mean corpuscular volume (MCV)** and the **mean corpuscular hemoglobin concentration (MCHC).**

Clinical Significance

Anemia is categorized on the basis of the MCV and MCHC. *Macrocytic anemia* (MCV greater than 94, MCHC within normal range) may be caused by folic acid deficiency and by vitamin B_{12} deficiency associated with the disease *pernicious anemia,* in which the intrinsic factor necessary for vitamin B_{12} absorption is not secreted by the stomach. *Normocytic normochromic anemia* (normal MCV and MCHC) may be due to acute blood loss, hemolysis, aplastic anemia (damage to the bone marrow), and a variety of chronic diseases. *Microcytic hypochromic anemia* (low MCV and MCHC), the most common type, is caused by inadequate dietary intake of iron.

Procedure

1. Calculate the MCV using the formula:

$$MCV = \frac{hematocrit \times 10}{RBC\ count\ (in\ millions\ per\ mm^3\ blood)}$$

Example

$$Hematocrit = 46$$
$$RBC\ count = 5.5\ million$$

$$MCV = \frac{46 \times 10}{5.5} = 84$$

Calculate the MCV of your blood cells, and enter it in the laboratory report.

> Average normal adult (male and female) MCV is 82–92.

2. Calculate your MCHC using the formula:

$$MCHC = \frac{hemoglobin\ (in\ g/dl) \times 100}{hematocrit}$$

Example

$$Hematocrit = 46$$
$$Hemoglobin = 16\ g/dl$$

$$MCHC = \frac{16 \times 100}{46} = 35$$

Calculate the MCHC of your blood, and enter it in the laboratory report.

> Average normal adult (male and female) MCHC is 32–36.

Laboratory Report 27

Name _____

Date _____

Section _____

Red Blood Cells and Oxygen Transport

Read the assigned sections in the textbook before completing the laboratory report.

1. Write your red blood cell count in the space below:
   _____ /mm^3 blood

2. Write your hematocrit in the space below:

   _____

3. Enter your hemoglobin concentration in the space below:
   _____ g/dl blood

4. Enter your MCV and MCHC values in the spaces below:

   MCV _____

   MCHC _____

5. Compare your results to the normal values, and write your conclusions in the space below:

6. One hemoglobin molecule contains _____ heme groups and can thus combine with _____ molecules of oxygen.

7. The hormone _____ stimulates the bone marrow to produce red blood cells.

8. The hormone in question 7 is produced by the _____ .

9. When red blood cells are destroyed, the heme (minus the iron) is coverted into a new pigment called _____ .

10. Accumulations of the pigment in question 9 (due to bile duct obstruction, for example) produce a condition called

    _____ .

11. The ratio of the volume of packed red blood cells to the total volume of a blood sample is called the _____ .

12. The molecule formed by the combination of reduced hemoglobin and oxygen is called _____ .

13. A hemoglobin molecule containing oxidized iron (Fe^{+++}) is called _____ .

14. The molecule formed by hemoglobin and carbon monoxide is called _____ .

15. Define the term *anemia;* describe its causes, and explain its dangers.

16. Newborns, especially premature babies, often have a rapid rate of red blood cell destruction and are jaundiced. Explain the relationship between these two conditions.

Total and Differential White Blood Cell Counts

Before coming to class, review the information about the appearance of white blood cells in chapter 20 of the text book, and the information about the function of leukocytes in chapter 23.

Introduction

Leukocytes (**white blood cells**)—*lymphocytes, monocytes, neutrophils, eosinophils,* and *basophils*—are agents of the immune system. Lymphocytes attack specific antigens: B lymphocytes secrete antibodies that bind to antigens, and T lymphocytes directly destroy antigenic cells. Other leukocytes participate in immunity by phagocytosis. The total white blood cell count and the relative proportion of each type of leukocyte change characteristically in different disease states.

The leukocytes (white blood cells) are divided into two general categories based on histological appearance—*granular* (or *polymorphonuclear*) and *nongranular*. Leukocytes in the former category have lobed or segmented nuclei and cytoplasmic granules, whereas leukocytes in the latter category have unlobed nuclei and lack cytoplasmic granules (plate 25).

Objectives

Students completing this exercise will be able to:

1. Distinguish the different types of leukocytes by the appearance of their nuclei and their cytoplasm.
2. Describe the origin and function of B and T lymphocytes.
3. List the phagocytic white blood cells, and explain their functions in local inflammation.
4. Perform a total and a differential white blood cell count, and explain the importance of this information in the diagnosis of diseases.

Materials

1. Microscopes
2. Unopette system for white blood cell counts
3. Lancets, alcohol swabs
4. Wright's stain (or Diff-Quik, Harleco) for differential count
5. Heparinized capillary tubes, glass slides
6. Container for the disposal of blood-containing objects

A. Total White Blood Cell Count

In this procedure, a small amount of blood is diluted with a solution that lyses the red blood cells and lightly stains the white blood cells (WBC). The white blood cells are counted in the four large corner squares of a hemocytometer (see exercise 27).

The dilution factor is 20, and each of the four squares counted has a volume of 0.1 cubic millimeters (mm^3); therefore, the number of white blood cells per cubic millimeter of blood can be calculated as follows:

$$WBC/mm^3 = \frac{\# \text{ cells} \times 20}{4 \times 0.1 \text{ mm}^3} \text{ or,}$$

$$WBC/mm^3 = \text{cells} \times 50$$

Procedure

1. Obtain the Unopette system for a white blood cell count, and puncture the reservoir as in the previous exercise.
2. Discard the first drop of blood on an alcohol swab and properly dispose it; then fill the capillary pipette with blood.
3. Push the capillary pipette into the previously punctured reservoir while squeezing the reservoir. Release the reservoir to draw the blood into the diluting solution.
4. Shake the pipette for 3 minutes.
5. Discard the first four drops onto a paper towel and properly dispose of it; then fill the hemocytometer as described in exercise 27.

6. Allow the cells to settle for 1 minute; then use the low-power objective and count the number of white blood cells in the four large corner squares (labeled A, B, C, and D in fig. 27.4). Count the cells touching the upper and left-hand lines, but not the cells touching the lower and right-hand lines.
7. Calculate the number of white blood cells per mm^3 of blood, and enter this value in the laboratory report.

> The normal white blood cell count is 5000–10,000/ mm^3 of blood.

B. Differential White Blood Cell Count

It is usually necessary to know not only the total number of white blood cells but also the relative abundance of each type. This knowledge is obtained by determining the number of each type of leukocyte in a total of 100 white blood cells.

Clinical Significance

An increase in the white blood cell count (*leukocytosis*) may be produced by an increase in any one of the leukocyte types: *neutrophil leukocytosis* can result from appendicitis, rheumatic fever, smallpox, diabetic acidosis, and hemorrhage; *lymphocyte leukocytosis,* from infectious mononucleosis and chronic infections, such as syphilis; *eosinophil leukocytosis,* from parasitic diseases (such as trichinosis), psoriasis, bronchial asthma, and hay fever; *basophil leukocytosis,* from hemolytic anemia and chicken pox; and *monocyte leukocytosis,* from malaria, Rocky Mountain spotted fever, bacterial endocarditis, and typhoid fever. In certain cases, an increase in the relative abundance of one type of leukocyte may occur in the absence of an increase in the total white blood cell count as, for example, in lymphocytosis due to pernicious anemia, influenza, infectious hepatitis, German measles, and mumps.

A decrease in the white blood cell count (*leukopenia*) is usually due to a decrease in the number of either neutrophils or eosinophils. A decrease in the number of neutrophils occurs in typhoid fever, measles, infectious hepatitis, rubella, and aplastic anemia. *Eosinopenia* is produced by an elevated secretion of corticosteroids, which occurs in various conditions of stress (such as severe infections and shock) and in adrenal hyperfunction (Cushing's syndrome).

Procedure
Making a Blood Smear

1. Fill a heparinized capillary tube at least one-third full with blood. This can serve as a reservoir of blood for making a number of slides.

☞ **Note:** *Observe the proper procedures for handling blood described in the previous exercise. Handle only your own blood, and dispose of all objects that have been in contact with blood in the containers indicated by your instructor.*

2. Using the capillary tube, apply a small drop of blood on one end of a glass slide that is *absolutely clean* and free of fingerprints (fig. 28.1*a*). Place this slide flat on a laboratory bench.
3. Lower a second clean glass slide at a 30° angle until it lightly touches the first slide in front of the drop of blood (fig. 28.1*b*).
4. Back the second slide into the drop of blood, maintaining the pressure and angle that allows the blood to spread out along the edge of the second slide (fig. 28.1*b*).
5. Keeping the same angle and pressure, push the second slide across the first in a single, rapid, smooth motion. The blood should now be spread in a thin film across the first slide. If this is done correctly, the concentration of blood in the smear should diminish toward the distal end, producing a feathered appearance (fig. 28.1*c* and *d*).

Procedure
Staining a Slide Using Wright's Stain

1. Place the slide on a slide rack, and flood the surface of the slide with Wright's stain. Rock the slide back and forth gently for 1 to 3 minutes.

☞ **Caution:** *The stain is dissolved in methyl alcohol, which quickly evaporates. If any part of the slide should dry during this procedure, the stain will precipitate, ruining the slide.*

2. Drip buffer or distilled water on top of the Wright's stain, being careful not to wash the stain off the slide. Mixing Wright's stain with water is crucial for proper staining; this mixing can be aided by gently blowing on the surface of the stain. Proper

Figure 28.1 A procedure for making a blood smear for a differential white blood cell count.

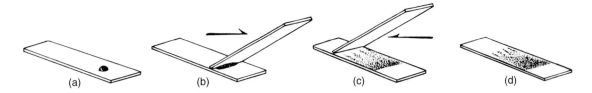

(a)   (b)   (c)   (d)

staining is indicated by a metallic sheen on the surface. The diluted stain should be left on the slide for a full 5 minutes.

3. Wash the stain off the slide with a jet of distilled water from a water bottle, and allow the slide to drain at an angle for a few minutes.

4. Using the oil-immersion objective, count the different types of white blood cells by starting at one point at the feathered distal end of the slide area and systematically scanning the slide until you have counted a total of 100 leukocytes.

5. Keep a running count of the different leukocytes in the table provided on plate 25, and indicate the total of each. Calculate the percentage of each leukocyte, and enter these percentages in your laboratory report.

Procedure
Staining a Slide Using Diff-Quik (Harleco)

1. Dip the slide in fixative solution (light blue) five times, allowing 1 second per dip.

2. Dip the slide in solution 1 (orange) five times, allowing 1 second per dip.

3. Dip the slide in solution 2 (dark blue) five times, allowing 1 second per dip.

4. Rinse the slide with distilled water, and count the white blood cells under a microscope, using the oil-immersion objective. Keep a running count of each different type of leukocyte in the table provided on plate 25. Calculate the percentage of each type of white blood cell, and enter these values in your laboratory report.

Laboratory Report 28

Name _____

Date _____

Section _____

Total and Differential White Blood Cell Counts

Read the assigned sections in the textbook before completing the laboratory report.

A. Total White Blood Cell Count

1. Enter your white blood cell count in the space below:

   _____ WBC/mm^3

2. Compare your results to the normal values, and write your conclusions in the space below:

B. Differential White Blood Cell Count

1. Add the totals of each type of white blood cell to determine your grand total. Use this number to calculate the percentage of each type of white blood cell, and enter these values in the table below:

Leukocyte	Percent
Neutrophils	
Eosinophils	
Basophils	
Lymphocytes	
Monocytes	

2. Compare your results to the normal values, and write your conclusions in the space below:

3. Identify the leukocyte described by each of the following statements:
   a. Polymorphonuclear, with poorly staining granules _____
   b. Agranular with round nucleus, relatively little cytoplasm _____
   c. Granules have affinity for red stain _____
   d. Rarest white blood cell _____
4. Antibodies are produced by _____ lymphocytes; cell-mediated immunity is provided by _____ lymphocytes.
5. White blood cells leave capillaries by a process called _____ .
6. The major phagocytic white blood cell is the _____ .

7. How do phagocytic and antibody-secreting white blood cells cooperate in the fight against infection?

8. Compare the origin and function of B and T lymphocytes.

9. What is a clone of lymphocytes? How is this clone produced?

10. Active immunity occurs when a person is exposed to a pathogen whose virulence (ability to cause disease) has been reduced but whose antigenicity is unaltered. How does this form of immunization protect the person during subsequent exposures to the pathogen?

11. Passive immunity occurs when an individual who has been exposed to a virulent pathogen receives an injection of serum that contains antibodies (*antiserum*) against that pathogen. Antiserum is usually obtained from an animal previously infected with the same pathogen. What are the advantages and disadvantages of passive immunity and of active immunity?

Blood Types

Before coming to class, review "Erythrocyte Antigens" and "Blood Typing" in chapter 20 of the textbook. Also, read about the patterns of inheritance in chapter 30.

Introduction

Red blood cells have characteristic molecules on the surface of their cell membranes. These molecules can function as antigens, which means that they can stimulate the production of specific antibodies and can bind to these specific antibodies. The blood group antigens bind to specific antibodies present in the plasma of a person with a different blood type. The major blood group antigens are the Rh antigen and the antigens of the ABO system.

Objectives

Students completing this exercise will be able to:
1. Explain the meaning of the term *blood type,* and identify the different major blood types.
2. Explain how agglutination occurs, and how this test can be used to determine a person's blood type.
3. Identify the different genotypes that can produce the different blood groups, and describe how different blood types can be inherited.
4. Explain how erythroblastosis fetalis is produced.
5. Explain the dangers of mismatching blood types in blood transfusions.

Materials

1. Glass slides, marking pencils, lancets, and applicator sticks or toothpicks
2. Anti-A, -B, and -Rh sera
3. Slide warmer
4. Container for the disposal of blood-containing objects

When blood from one person is mixed with plasma from another person, the red blood cells will sometimes **agglutinate,** or clump together. This agglutination reaction, which is due to a mismatch of genetically determined blood types, is very important in determining the safety of transfusions.

On the surface of each red blood cell there are a number of molecules that have antigenic properties, and each antibody in the plasma has two combining sites for antigens. In a positive agglutination test, the red blood cells clump together because they are combined through antibody bridges.

A. The Rh Factor

One of the antigens on the surface of red blood cells, the Rh factor (named because it was first discovered in rhesus monkeys), is found on the red blood cells of approximately 85% of the people in the United States. The presence of this antigen on the red blood cells (an Rh positive phenotype) is inherited as a dominant trait and is produced by both the homozygous (*RR*) and the heterozygous (*Rr*) genotypes. Individuals who have the homozygous recessive genotype (*rr*) do not have this antigen on their red blood cells and are said to have the *Rh negative* phenotype.

Suppose an Rh positive man who is heterozygous (*Rr*) mates with an Rh positive woman who is also heterozygous (*Rr*).

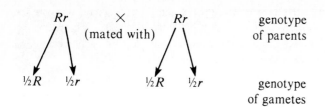

$$Rr \quad \overset{\times}{\underset{\text{(mated with)}}{}} \quad Rr \qquad \text{genotype of parents}$$

½R ½r   ½R ½r   genotype of gametes

Since the mother is Rh positive in this example, her immune system cannot be stimulated to produce antibodies by the presence of an Rh positive fetus; fortunately for the Rh positive mother, an Rh negative fetus does not yet have an immune response. The development of **immunological competence** occurs shortly after birth.

When an Rh negative mother is carrying an Rh positive fetus, some of the Rh antigens may enter her circulation when the placenta breaks at birth (red blood cells do not normally cross the placenta during pregnancy). Since these red blood cells contain an antigen (the Rh factor) that is foreign to the mother, they will stimulate her immune system to produce antibodies that are capable of destroying the red blood cells of subsequent Rh positive fetuses. **Hemolytic disease of the newborn** (or **erythroblastosis fetalis**) can be prevented by the administration of exogenous Rh antibodies (a drug known as *RhoGAM*) to the mother within 72 hours after delivery. These antibodies destroy the fetal red blood cells that have entered the maternal circulation before they can stimulate an immune response.

Procedure

1. Place 1 drop of anti-Rh serum on a clean glass slide.
2. Use a lancet to prick your finger. Add an equal amount of fingertip blood, and mix it with the antiserum (use an applicator stick or a toothpick).
3. Place the slide on a slide warmer (45°–50° C), and rock it back and forth.
4. Examine the slide for agglutination. If no agglutination is observed after a 2-minute period, examine the slide under the low-power objective of the microscope. The presence of grains of agglutinated red blood cells indicates Rh positive blood.

☞ **Caution:** *Handle only your own blood and be sure to discard the slide, toothpicks, and lancet in the container provided by the instructor.*

5. Enter your Rh factor type (positive or negative) in the laboratory report.

B. The ABO Antigen System

Each individual inherits two genes, one from each parent, that control the synthesis of red blood cell antigens of the ABO classification. Each gene contains the information for one of three possible phenotypes: antigen A, antigen B, or no antigen (written O). Thus, an individual may have one of six possible genotypes: *AA, AO, BB, BO, AB, OO.*

An individual who has the genotype *AO* will produce type A antigens just like an individual who has the genotype *AA*; both are therefore said to have **type A** blood. Likewise, an individual with the genotype *BO* and one with the genotype *BB* will both have **type B** blood. Since lack of antigen is a recessive trait, an individual with blood **type O** must have the genotype *OO*.

Unlike the other traits considered, the heterozygous genotype *AB* has a phenotype that is different from either of the homozygous genotypes (*AA* or *BB*). Since there is no dominance (or codominance) between A and B, individuals with the genotype *AB* produce red blood cells with both the A and B antigens (**type AB** blood).

Also, unlike the other immune responses considered, antibodies against the A and B antigens are not induced by prior exposure to these blood types. A person with type A blood, for example, has antibodies in the plasma against type B blood even though that person may never have been exposed to this antigen. A transfusion with type B blood would

Antigen on RBCs	Antibody in Plasma
A	Anti-B
B	Anti-A
O	Anti-A and anti-B
AB	No antibody

be extremely dangerous because the antibodies in the recipient's plasma would agglutinate the red blood cells in the donor's blood. Exactly the same result would occur if the donor were type A and the recipient type B (see plate 26). The most common blood types are type O and type A, whereas type AB is the rarest (table 29.1).

Table 29.1 Incidence of blood types

	Approximate Incidence in U.S. (%)		
Blood Types	**Caucasian**	**Black**	**Asian**
O	45	48	36
A	41	27	28
B	10	21	23
AB	4	4	13

Procedure

1. Draw a line down the center of a clean glass slide with a marking pencil, and label one side *A* and the other *B*.
2. Place a drop of anti-A serum on the side marked *A* and a drop of anti-B serum on the side marked *B*.
3. Add a drop of blood to each antiserum, and mix with a clean applicator stick.
4. Tilt the slide back and forth, and look for agglutination over a 2-minute period. Do not heat the slide on the slide warmer.
5. Enter your ABO blood type in the laboratory report.

Laboratory Report 29

Name _____

Date _____

Section _____

Blood Types

1. Did your blood agglutinate with the anti-Rh serum? _____

2. Are you Rh positive or negative? _____

3. Indicate below (with a yes or no) if your blood agglutinated with the anti-A and anti-B sera:

   Anti-A: _____

   Anti-B: _____

4. What is your blood type? _____

5. What antigens are or are not present on a person's red blood cells if that person is: (a) type A negative; (b) type O positive; and (c) type AB negative.

6. Explain the dangers of giving a person with type A negative blood a transfusion of type B positive blood.

7. Explain how hemolytic disease of the newborn is produced, and how this disease can be prevented.

8. In a paternity suit, a woman (type O) accuses a man (type A) of being the father of her baby (type O). Will the blood types prove or disprove her claim? Explain.

Heart and Arteries

Introduction

The circulatory system is divided into the *cardiovascular system,* which consists of the heart and blood vessels, and the *lymphatic system,* which consists of lymph vessels and lymph nodes. The cardiovascular system comprises *pulmonary circulation,* which involves the heart, lungs, and their connecting vessels, and *systemic circulation,* which involves the heart and the vessels of the rest of the body. The names of individual blood vessels frequently refer to the organs they serve or the body region in which they are found.

Objectives

Students completing this exercise will be able to:

1. Describe the location of the heart; identify the heart's chambers, vessels, and valves; and trace the flow of blood through the heart.
2. Describe the path of the electrical impulse through the conduction system of the heart.
3. Compare the structure and function of arteries, arterioles, capillaries, venules, and veins.
4. Identify the principal arteries of systemic circulation.

Materials

1. Fresh sheep, pig, or beef heart
2. Dissecting instruments and trays
3. Embalmed cat
4. Heart models and charts of the circulatory system
5. Microscopes
6. Prepared slides of stained heart and blood vessels
7. Reference text

A. Heart

Identify and label the following regions and structures in figures 30.1 and 30.2:

Aortic arch
Apex of the heart
Anterior interventricular artery
Atrioventricular bundle (bundle of His)
    Right and left bundle branches
Atrioventricular (AV) node
Circumflex artery
Conduction myofibers (Purkinje fibers)
Great cardiac vein
Interatrial septum
Interventricular septum
Left atrium
    Pulmonary veins: (a) left, (b) right
Left ventricle
    Aortic semilunar valve
    Bicuspid (mitral) valve
Papillary muscle
Pulmonary trunk
Right and left coronary arteries
Right and left pulmonary arteries
Right atrium
    Coronary sinus
    Inferior vena cava
    Superior vena cava
Right ventricle
    Chordae tendineae
    Pulmonary semilunar valve
    Tricuspid valve
Sinoatrial node (SA node)
Thoracic portion of aorta
    Descending thoracic aorta
Trabeculae carneae

B. Dissection of a Mammalian Heart

1. Use cold tap water to rinse excess preservative or blood from the heart. Examine the *pericardium,* which surrounds the heart. This fibroserous membrane may already have been removed, but portions of it may still be attached around the bases of the large vessels. Note the fat deposits on the surface of the heart around the *coronary vessels.*

Figure 30.1 The structure of the heart: (*a*) an anterior view; (*b*) a posterior view; (*c*) an internal view.

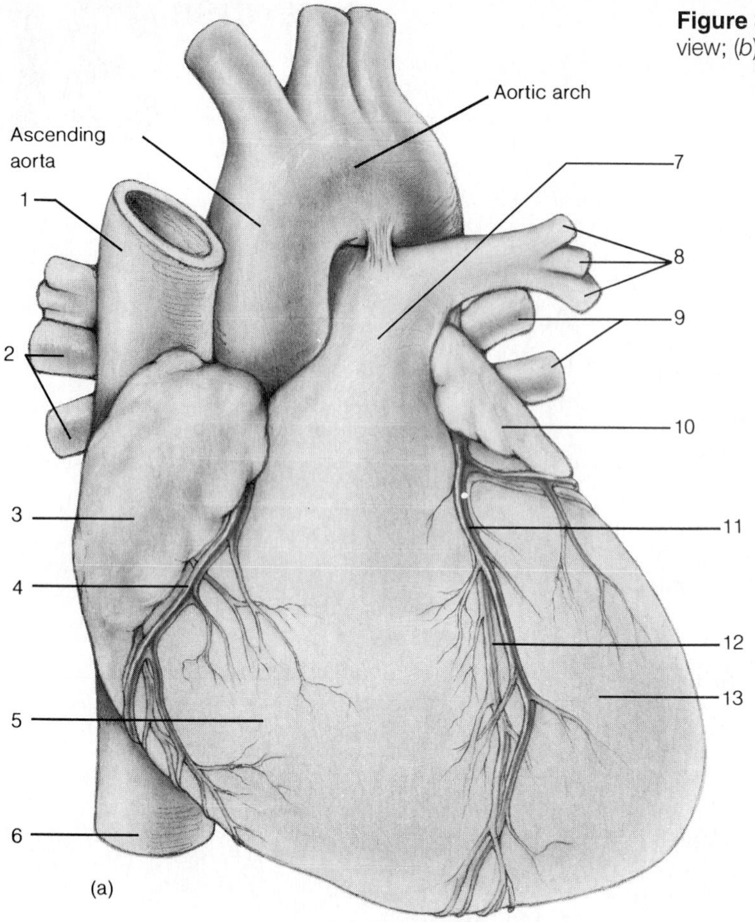

Ascending aorta

Aortic arch

1

2

3

4

5

6

(a)

7

8

9

10

11

12

13

1. _____
2. _____
3. _____
4. _____
5. _____
6. _____
7. _____
8. _____
9. _____
10. _____
11. _____
12. _____
13. _____

Left common carotid artery

Left subclavian artery

Descending aorta

Brachiocephalic trunk

1

2

3

4

5

6

7

8

9

10

11

(b)

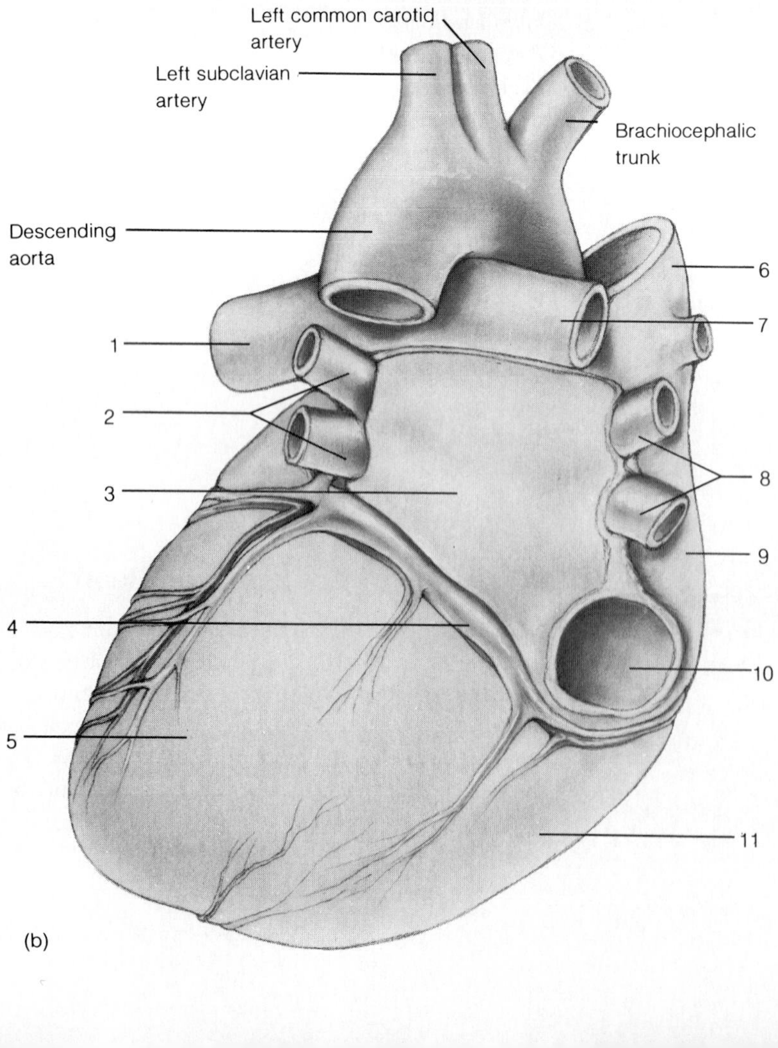

1. _____
2. _____
3. _____
4. _____
5. _____
6. _____
7. _____
8. _____
9. _____
10. _____
11. _____

Figure 30.1   Continued

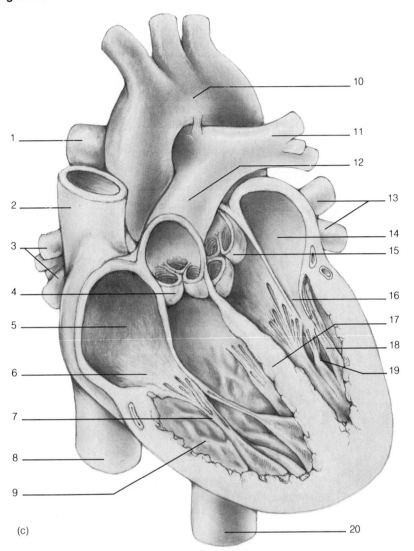

(c)

1. _____
2. _____
3. _____
4. _____
5. _____
6. _____
7. _____
8. _____
9. _____
10. _____
11. _____
12. _____
13. _____
14. _____
15. _____
16. _____
17. _____
18. _____
19. _____
20. _____

Figure 30.2 The conduction system of the heart.

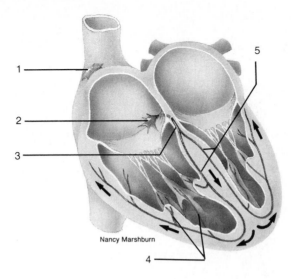

Nancy Marshburn

1. _____

2. _____

3. _____

4. _____

5. _____

2. The next step is to examine the external topography of the heart. All the vessels of the heart enter and exit from the superior portion. The inferiorly pointed portion is called the *apex*. Squeeze the lower part of the heart. The portion that feels noticeably thicker is the *left ventricle* (fig. 30.3). Extending diagonally between the ventricles on the anterior surface of the heart is the *anterior interventricular groove*. Carefully remove the fat from this groove and expose the *coronary vessels*. After the position of the *ventricles* is determined, the two *atria* can be found. Identify the *superior* and *inferior venae cavae*, which enter the right atrium.

3. Insert one blade of the dissecting scissors into the superior vena cava and make a cut through the right atrium to expose the *tricuspid valve* (fig. 30.4). Slowly fill the right ventricle with tap water. Gently squeeze the lower portion of the heart to observe the closing action of this valve as the water pushes against it. At this point, identify the *fossa ovalis,* an oval depression between the *atrial septa* and the *coronary sinus.* The *coronary sinus* is the slitlike opening below the fossa ovalis.

Figure 30.3 Hold the sheep heart so that the ventral surface is toward you. Squeeze the lower portion of the heart, and feel the thicker left ventricle.

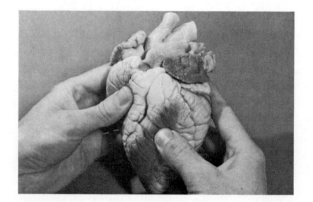

Figure 30.4 Make the initial cut through the wall of the superior vena cava into the right atrium to expose the tricuspid valve.

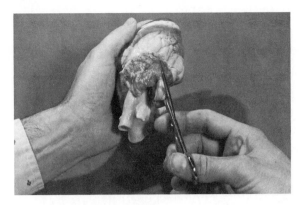

Figure 30.6 The next cut is started in the left atrium and continues down into the left ventricle.

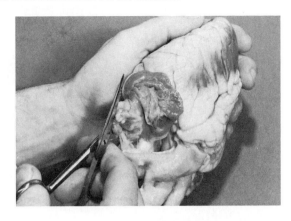

Figure 30.5 Continue the incision throught the wall of the right ventricle, and examine the structures of the right side of the heart.

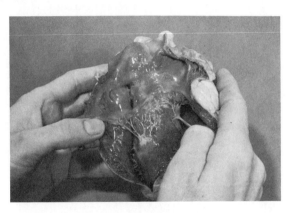

Figure 30.7 Examine the left side of the heart, and note the structural differences between the right side and the left side.

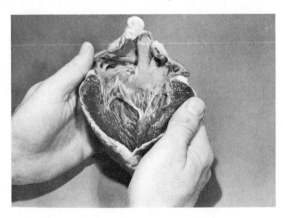

4. Continue cutting the right atrium, and cut through the tricuspid valve to the apex of the heart. Rinse the right ventricle with cold tap water, and examine the interior (fig. 30.5). Identify the *papillary muscles* and the *chordae tendineae,* which are attached to the flaps of the tricuspid valve. The *trabeculae carneae* are fleshy ridges on the interior wall. Insert a probe into the *pulmonary orifice,* which is the opening of the right ventricle into the *pulmonary artery.* Make a cut with the scissors to this point so that the *pulmonary semilunar valve* is exposed. Notice how this valve is structured to prevent the backward flow of blood from the artery into the ventricle.

5. To open the left atrium, insert one blade of the scissors into the lowest of the four *pulmonary veins,* and cut forward to the extremity of the *auricular appendage* (fig. 30.6). Rinse the cavity with tap water, and examine the interior.

6. Extend the cut longitudinally through the wall of the left ventricle. Note the thickness of the *myocardium* in this region of the heart (fig. 30.7). Examine the *bicuspid valve* between the left atrium and the left ventricle. Locate the *aortic semilunar valve* through which blood leaves the heart. Make a longitudinal incision into the aorta, and identify the openings to the *coronary arteries.* These two openings are in the walls of the aorta just above the aortic semilunar valve.

C. Blood Vessels

1. Identify and label the following layers in figure 30.8:

   Tunica externa
   Tunica intima
   Tunica media
   Valve

2. Observe a histological slide of a blood vessel and identify the layers listed above.

Figure 30.8   The structure of (a) an artery and (b) a vein, showing the relative thickness and composition of the tunicas.

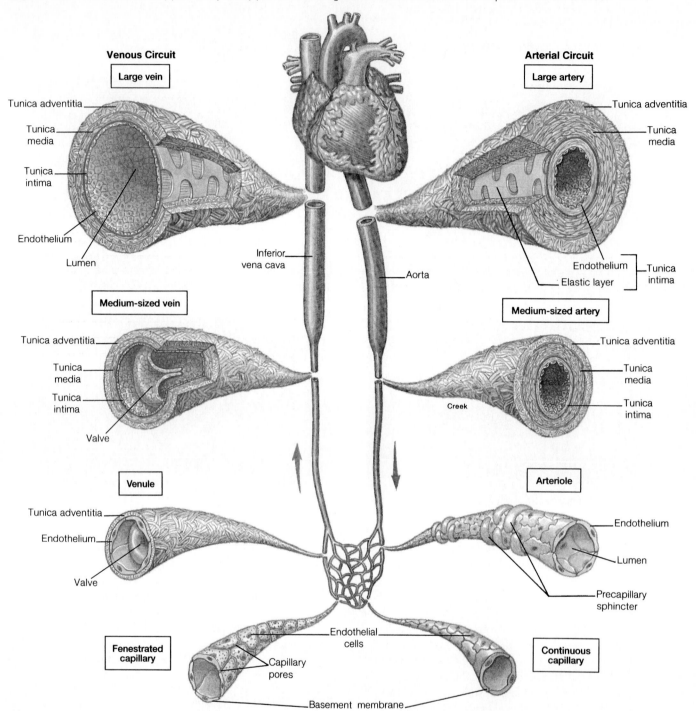

D. Principal Arteries of the Body

Identify and label the following arteries in figure 30.9:

Ascending aorta
Arteries from the aortic arch
    Brachiocephalic trunk
    Left common carotid artery
    Left subclavian artery
Arteries of the neck and head

External carotid artery
Internal carotid artery
Right common carotid arteries
Arteries of the upper extremity
    Axillary artery
    Brachial artery
    Radial artery
    Right subclavian arteries
    Right vertebral artery

Figure 30.9 The principal arteries of the body.

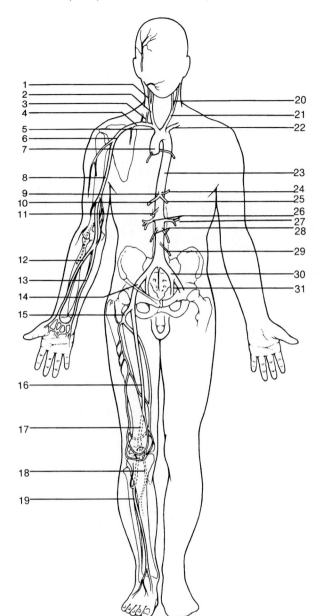

1. _____
2. _____
3. _____
4. _____
5. _____
6. _____
7. _____
8. _____
9. _____
10. _____
11. _____
12. _____
13. _____
14. _____
15. _____
16. _____
17. _____
18. _____
19. _____
20. _____
21. _____
22. _____
23. _____
24. _____
25. _____
26. _____
27. _____
28. _____
29. _____
30. _____
31. _____

Ulnar artery
Branches of the thoracic aorta
   Intercostal arteries
   Coronary arteries
   Phrenic arteries (not shown)
Branches of the abdominal aorta
   Celiac trunk
      Hepatic artery
      Left gastric artery
      Splenic artery
   Common iliac artery
   Inferior mesenteric artery
   Lumbar arteries (not shown)

Ovarian arteries (female) (not shown)
   Renal arteries
      Suprarenal artery
   Superior mesenteric artery
   Gonadal (testicular or ovarian) artery
Arteries of the lower extremities
   Anterior tibial artery
   Deep femoral artery
   External iliac artery
   Femoral artery
   Internal iliac artery
   Popliteal artery
   Posterior tibial artery

Figure 30.10 The cat heart: (*a*) a ventral view; (*b*) a dorsal view.

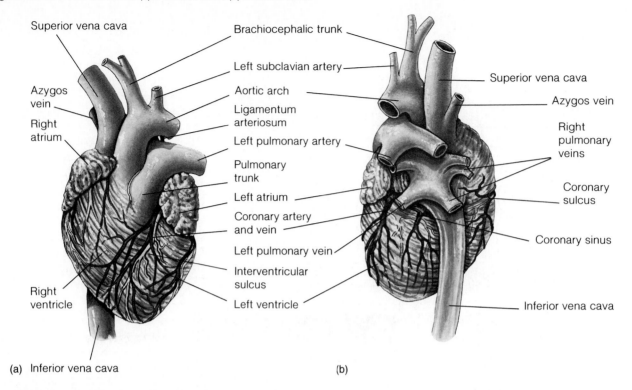

(a)

Superior vena cava

Azygos vein

Right atrium

Right ventricle

Inferior vena cava

Brachiocephalic trunk

Left subclavian artery

Aortic arch

Ligamentum arteriosum

Left pulmonary artery

Pulmonary trunk

Left atrium

Coronary artery and vein

Left pulmonary vein

Interventricular sulcus

Left ventricle

(b)

Superior vena cava

Azygos vein

Right pulmonary veins

Coronary sulcus

Coronary sinus

Inferior vena cava

E. Heart and Arterial System of the Cat: Cat Dissection

Biological suppliers generally inject the arteries of embalmed cats with red latex, and the veins with blue. This not only facilitates vessel differentiation, but it also strengthens the vessels, which otherwise would be easily broken during dissection. When tracing blood vessels, you must free each artery or vein from supporting connective tissue so that it is clearly visible. A sharp probe or dissecting needle should be used for this technical dissection.

Vessels should be learned in the sequence the blood flows through them. In other words, trace arterial blood through arteries away from the heart to the organs being served. Trace the venous return from the organs being drained to the heart. It is important to associate vessels with the particular organs they serve.

Exposing the Visceral Organs

Open the entire thoracic and abdominal cavities of the cat by making a longitudinal incision with a sharp scalpel or scissors along the ventral surface of the animal. The incision should pass about 5 centimeters (cm) to the right or left of the midline so that costal cartilages are cut instead of the sternum or ribs. The cut should extend from the apex of the thorax posteriorly to the pubic region. In addition, make lateral cuts immediately posterior to the diaphragm and in the pelvic region.

The contents of the thorax can be observed by spreading apart the walls. Greater exposure can be achieved by cutting the ribs about halfway down with heavy-duty scissors. It would be helpful at this point to identify the visceral organs. Note the meshlike covering over much of the abdominal viscera. This is the *greater omentum,* which stores lipids and protects the viscera. It must be lifted out of the way to expose the other visceral organs. Referring to the color plates of cat dissections and other figures, identify the lungs, thyroid gland, diaphragm, spleen, stomach, liver, small intestine, large intestine, kidneys, urinary bladder, and reproductive organs. The heart is not yet visible; it is enveloped by a tough pericardium.

Heart. All mammals have a remarkably similar four-chambered heart (fig. 30.10). Identify the *superior (cranial) vena cava* and the *inferior (caudal) vena cava,* which return venous blood to the right atrium. The *pulmonary veins,* which carry oxygenated blood from the lungs to the heart, can be seen where they enter the left atrium.

Identify the two large arteries that transport blood away from the heart. The more ventral vessel is the *pulmonary trunk,* which arises from the right ventricle and transports blood to the lungs. The pulmonary trunk bifurcates (splits) into *right* and *left pulmonary arteries,* which pass to the right and left lungs. The thicker, more dorsal vessel is the *ascending aorta,* which arises from the left ventricle. As the aorta arches to the left to course through

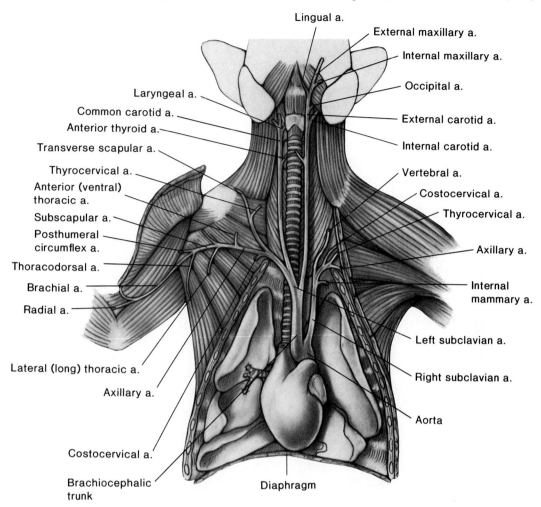

Lingual a.
External maxillary a.
Internal maxillary a.
Occipital a.
External carotid a.
Internal carotid a.
Vertebral a.
Costocervical a.
Thyrocervical a.
Axillary a.
Internal mammary a.
Left subclavian a.
Right subclavian a.
Aorta

Laryngeal a.
Common carotid a.
Anterior thyroid a.
Transverse scapular a.
Thyrocervical a.
Anterior (ventral) thoracic a.
Subscapular a.
Posthumeral circumflex a.
Thoracodorsal a.
Brachial a.
Radial a.
Lateral (long) thoracic a.
Axillary a.
Costocervical a.
Brachiocephalic trunk

Diaphragm

the thorax, it is known as the *aortic arch.* The three large vessels that arise from the aortic arch will be discussed in the following section.

Aortic arch. The *right* and *left coronaries* serve the myocardium of the heart; they are the only branches on the ascending aorta. The aorta then arches to the left and slightly dorsally over the pulmonary trunk as the aortic arch. Three vessels arise from the aortic arch: the *brachiocephalic* (innominate) *trunk,* the *left common carotid artery,* and the *left subclavian artery.* The common carotid arteries transport blood to the neck and head, while the subclavian arteries supply the upper extremities (fig. 30.11).

Arteries of the neck and head. The common carotid arteries course cranially in the neck of the cat along either lateral side of the trachea. Several small vessels arise from the common carotid artery to supply blood to the larynx, thyroid, anterior neck muscles, and lymph nodes of the neck. The common carotid artery bifurcates into the *internal* and *external carotid arteries* slightly behind the angle of the mandible.

Arteries of the upper extremity. The *right subclavian artery* branches off of the brachiocephalic *trunk,* while the *left subclavian artery* arises directly from the aorta arch. The subclavian artery passes laterally dorsal to the clavicle, carrying blood toward the upper extremity.

From each subclavian artery, four vessels arise: (1) a *vertebral artery,* which carries blood to the brain through the transverse foramina of the cervical vertebrae; (2) a *costocervical artery,* which supplies blood to the deep muscles of the back and neck; (3) an *internal mammary artery,* which originates on the ventral surface of the subclavian artery and supplies blood to the ventral body wall and mammary glands; and (4) a *thyrocervical artery,* which becomes the *transverse scapular artery* in the shoulder region and supplies blood to the neck and shoulder.

The subclavian artery becomes the *axillary artery* as it passes through the axillary region. Three vessels arise from the axillary artery: (1) a *ventral thoracic artery,* which branches from the ventral side of the axillary artery and supplies the pectoral muscles; (2) a long *thoracic artery,* which arises a short distance from the ventral thoracic artery and passes posteriorly to serve the pectoralis and latissimus dorsi muscles; (3) a *subscapular artery,* which gives rise to the *thoracodorsal artery,* which supplies blood to the latissimus dorsi and other back muscles.

The axillary artery becomes the *brachial artery* as it enters the brachial region. At the elbow, the brachial artery bifurcates into the *radial* and *ulnar arteries,* which supply the lower forelimb and foot (paw).

Branches of the thoracic aorta. The thoracic aorta is a continuation of the aortic arch posteriorly through the thoracic cavity to the diaphragm. This large vessel gives off branches (not illustrated) to the viscera and muscles of the thoracic region: *pericardial arteries* go to the pericardium; *bronchial arteries* provide systemic circulation to the lungs; *esophageal arteries* go to the esophagus where blood passes through the mediastinum; *intercostal arteries* serve the intercostal muscles and structures of the wall of the thorax; and *phrenic arteries* supply blood to the diaphragm.

Branches of the abdominal aorta. The *abdominal aorta* is the segment of the dorsal aorta between the diaphragm and the lower lumbar region. In the lower lumbar region, it divides into the *right* and *left external iliac arteries* (no common iliac arteries exist in the cat). The first branch of the abdominal aorta is called the *celiac trunk* (fig. 30.12). The celiac trunk divides immediately into three arteries: the *splenic artery,* going to the spleen, the *left gastric artery,* going to the stomach, and the *hepatic artery,* going to the liver. The *superior mesenteric artery* is also an unpaired vessel; it arises ventrally from the abdominal aorta just below the celiac trunk. The superior mesenteric artery supplies blood to the small intestine and to portions of the large intestine.

Paired *suprarenal arteries* arise from the abdominal aorta about 2 cm caudal to the superior mesenteric artery. These vessels serve the adrenal (suprarenal) glands, di-

aphragm, and muscles of the body wall. The next major vessels that arise from the abdominal aorta are the paired renal arteries, which conduct blood to the kidneys. The *spermatic (testicular) arteries* in the male, and the *ovarian arteries* in the female, are small, paired vessels that come off of the abdominal aorta just below the renal arteries and, as their name implies, serve the gonads. The *inferior mesenteric artery* is the last single offshoot of the abdominal aorta. It branches into the *left colic* and *superior hemorrhoidal arteries,* which supply blood to the terminal portion of the large intestine and to the rectum. Several *lumbar arteries* branch dorsally from the abdominal aorta and serve the muscles and the spinal cord in the lumbar region. The *external iliac arteries* are large arteries that pass through the body wall and become the femoral arteries in the hind legs. The *internal iliac arteries* are the most caudal paired vessels to arise from the abdominal aorta. These arteries serve the gluteal muscles and rectum, and the uterus of the female. The *caudal artery* is an extension of the abdominal aorta; it traverses along the median ventral surface of the sacrum and into the tail.

Arteries of the hind limb. The external iliac artery passes out of the cat's pelvic cavity ventrally to the inguinal ligament, after which it becomes the *femoral artery* (fig. 30.12). Several vessels arise from the femoral artery to serve the thigh region, one of which—the *deep femoral*—is larger and passes posteriorly to serve as the hamstring muscles. The femoral artery becomes the *popliteal artery* when it passes across the back of the knee. The popliteal artery bifurcates into the *anterior tibial artery* and the *posterior tibial artery,* which serve the leg and foot.

Figure 30.12 A ventral view of the abdomen and hind limbs of a cat showing the arteries posterior to the diaphragm.

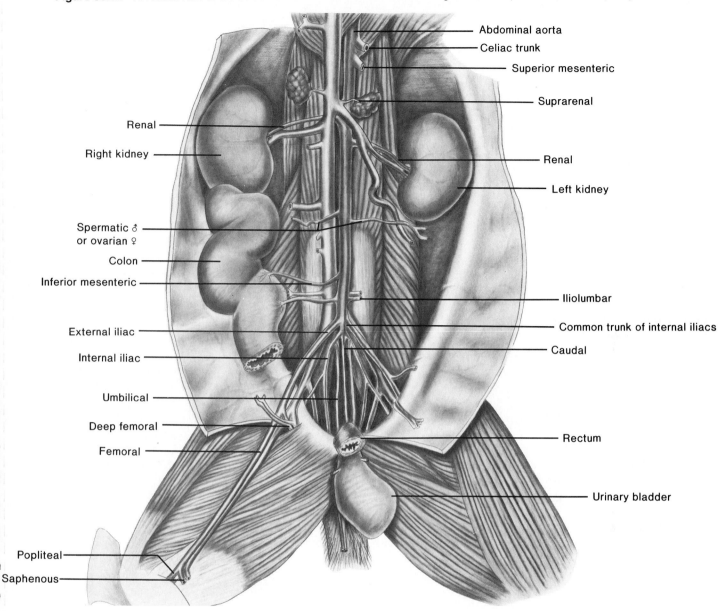

Abdominal aorta

Celiac trunk

Superior mesenteric

Suprarenal

Renal

Right kidney

Renal

Left kidney

Spermatic ♂
or ovarian ♀

Colon

Inferior mesenteric

Iliolumbar

External iliac

Common trunk of internal iliacs

Internal iliac

Caudal

Umbilical

Deep femoral

Rectum

Femoral

Urinary bladder

Popliteal

Saphenous

Laboratory Report 30

Name _____

Date _____

Section _____

Heart and Arteries

Read the assigned sections in the textbook before completing the laboratory report.

A and B. Heart, and Dissection of a Mammalian Heart

1. Contrast the right and left sides of the heart in the table below:

	Right Side	Left Side
Structure of the atrioventricular valves	_____	_____
Thickness of the myocardium	_____	_____
Oxygen content of the blood	_____	_____
Number of vessels going into the atria	_____	_____

2. Trace, in proper sequence, the passage of blood through the heart. Name all the chambers, vessels, and valves that the blood contacts as it travels from the right atrium to the aorta.

3. Trace, in proper sequence, the passage of a cardiac impulse through the conduction system of the heart.

C. Blood Vessels

1. In the adult body, all arteries except one contain oxygenated blood. Which one? In the adult body, all veins except one contain deoxygenated blood. Which one? Why is this so?

2. Using the following criteria, compare the structure of arteries and veins.

	Artery	Vein
Size of lumen		
Individual variation		
Direction of blood flow		
Thickness of tunicas		
Relative position—superficial or deep		

D. Principal Arteries of the Body

1. Name the arteries that supply blood to the following organs or regions:
   a. Myocardium _____
   b. Kidney _____
   c. Side of face _____
   d. Gluteal muscles _____
   e. Diaphragm _____
   f. Brain _____
   g. Ovary _____
   h. Stomach _____
   i. Rectum _____

2. Identify the arteries referred to in the following descriptions:
   a. Used to determine the pulse at the wrist _____
   b. Nonpaired, short truck immediately below diaphragm, which has three branches _____
   c. Arises from the aortic arch and continues up the left side of the neck _____
   d. Found in the popliteal region of the leg _____
   e. First major vessel to arise from the aortic arch, immediately bifurcates _____
   f. Used to determine arterial blood pressure _____
   g. Important pressure point in groin area _____

3. List the vessels that are not symmetrical; that is, those without a counterpart on the opposite side of the body.

Veins, Lymphatics, and Fetal Circulation

Before coming to class, review the following sections in chapter 21 of the textbook: "Principal Veins of the Body" and "Fetal Circulation." Also review "Lymphatic System" in chapter 23.

Introduction

In the venous portion of the systemic circulation, blood flows from smaller vessels into larger ones; thus a vein receives smaller tributaries, instead of branching off smaller vessels as does an artery. There are more veins than arteries, and veins are both superficial and deep. Superficial veins are clinically important for blood drawing and injection administration. Deep veins are near the principal arteries and are usually named similarly.

The fetal circulation is adapted to transport blood to and from the placenta, where oxygen and nutrients are obtained, and where carbon dioxide and other products are eliminated.

The lymphatic vessels transport interstitial fluid (initially formed as a blood filtrate) back to the bloodstream. These vessels also transport absorbed fat from the small intestine to the blood. In addition, the lymphatic system comprises lymph nodes and lymphoid organs, which produce lymphocytes and function as part of the immune system to protect the body from disease.

Objectives

Students completing this exercise will be able to:

1. Identify the principal veins of systemic circulation in a human.
2. Identify the principal veins of systemic circulation in a cat.
3. Describe the components and functions of the lymphatic system, and explain the relationship between the lymphatic system and the vascular system.

4. Describe the unique structures of fetal circulation, explain their function, and contrast the circulation of a fetus with the circulation of a newborn.

Materials

1. Colored pencils
2. Embalmed cat, dissecting tray, and dissecting instruments
3. Reference text

A. Principal Veins of the Body

Identify and label the following vessels in figures 31.1, 31.2, and 31.3:

Veins draining the head and neck
External jugular vein
Inferior thyroid veins
Internal jugular vein
Right brachiocephalic vein
Subclavian vein
Superior vena cava
Veins of the upper extremity
Axillary vein
Basilic vein (not shown)
Brachial vein
Cephalic vein
Median cubital vein
Radial vein
Ulnar vein
Veins of the thorax
Accessory hemiazygos vein
Azygos vein
Communicating veins
Hemiazygos vein
Intercostal vein

Figure 31.1 Principal veins of the body: the superficial veins are on the left extremities and the deep veins are on the right extremities.

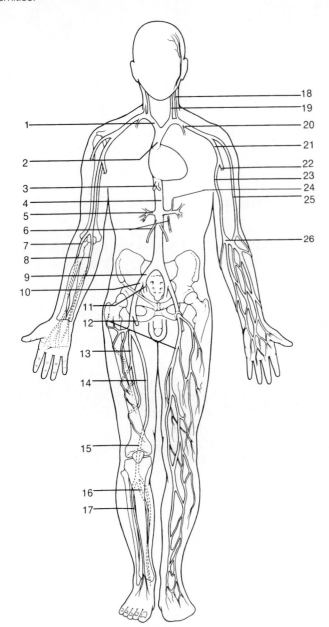

1. _____
2. _____
3. _____
4. _____
5. _____
6. _____
7. _____
8. _____
9. _____
10. _____
11. _____
12. _____
13. _____
14. _____
15. _____
16. _____
17. _____
18. _____
19. _____
20. _____
21. _____
22. _____
23. _____
24. _____
25. _____
26. _____

Figure 31.2 Veins of the thoracic region (the lungs and heart have been removed).

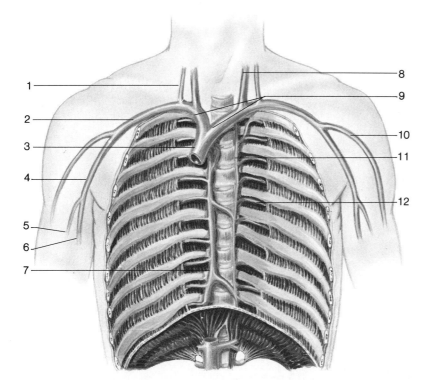

1. _____
2. _____
3. _____
4. _____
5. _____
6. _____
7. _____
8. _____
9. _____
10. _____
11. _____
12. _____

Veins of the lower extremity
   Anterior tibial vein
   Common iliac vein
   Deep femoral vein (not shown)
   External iliac vein
   Femoral vein
   Great saphenous vein
   Internal iliac vein
   Popliteal vein
   Posterior tibial vein
Veins of the abdominal region
   Hepatic vein
   Inferior vena cava
   Lumbar vein (not shown)

Ovarian vein (female) (not shown)
Renal vein
Suprarenal vein
Testicular (internal spermatic) vein (male)
Hepatic portal system
   Cystic vein (not shown)
   Gastric vein: (a) left, (b) right
   Gastroepiploic vein: (a) left, (b) right
   Hepatic vein (not shown)
   Hepatic portal vein
   Inferior mesenteric vein
   Pancreatic vein (not shown)
   Splenic vein
   Superior mesenteric vein

Figure 31.3 The hepatic portal system.

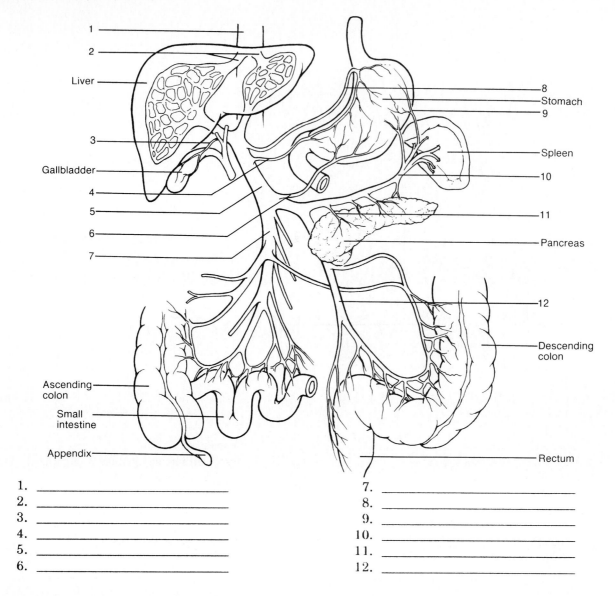

1. _____   7. _____
2. _____   8. _____
3. _____   9. _____
4. _____   10. _____
5. _____   11. _____
6. _____   12. _____

B. Fetal Circulation

1. Identify and label the following structures in figure 31.4:

    Ductus arteriosus
    Ductus venosus
    Foramen ovale
    Inferior vena cava
    Placenta (not shown)
    Umbilical arteries
    Umbilical cord
    Umbilical vein

2. In figure 31.4, use arrows to indicate the pathway of blood through the fetal circulatory system. Use red to color the vessel that transports highly oxygenated blood; use blue for vessels that carry oxygen-depleted blood. Use both red and blue to indicate "mixed blood."

C. Lymphatic System

Identify and label the following structures in figure 31.5:

    Capillary bed
    Cisterna chyli
    Lymph nodes: (a) axillary, (b) inguinal, (c) lumbar
    Lymphoid organs
        Spleen
        Thymus
        Tonsils
    Right lymphatic duct
    Thoracic duct

Figure 31.4 Fetal circulation.

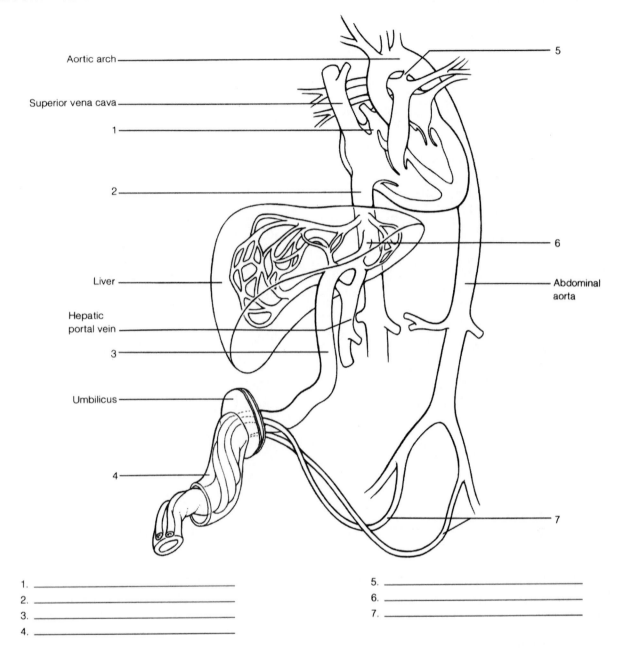

Aortic arch

Superior vena cava

1

2

Liver

Hepatic portal vein

3

Umbilicus

4

5

6

Abdominal aorta

7

1. _____
2. _____
3. _____
4. _____

5. _____
6. _____
7. _____

D. Venous System of the Cat

Veins Draining the Head, Neck, and Upper Extremities

Venous blood from the head, neck, and upper extremities drains into the right atrium of the cat heart via the *superior (anterior) vena cava.* The superior vena cava is formed by the convergence of *right* and *left brachiocephalic veins* (fig. 31.6). The brachiocephalics, in turn, are formed by the union of the *external jugular vein* and the *subclavian vein* on each side.

Blood from the head and portions of the neck is drained by the external jugular vein, which is formed by three tributaries. The most lateral of the three is the *posterior facial vein,* which drains blood from the side of the head. Next, the *anterior facial vein* drains blood from the buccal (mouth) area, which includes the lips and teeth. Finally, the *transverse vein* forms a connection between the two external jugular veins at the base of the chin.

As the external jugular vein descends in the neck, it is joined by two additional veins. The large *transverse scapular vein* drains blood from the shoulder to the external jugular

Figure 31.5 Lymphatic vessels: (*a*) a magnified view of the upper right quadrant showing the lymph drainage of the right breast; (*b*) major lymph drainage of the body.

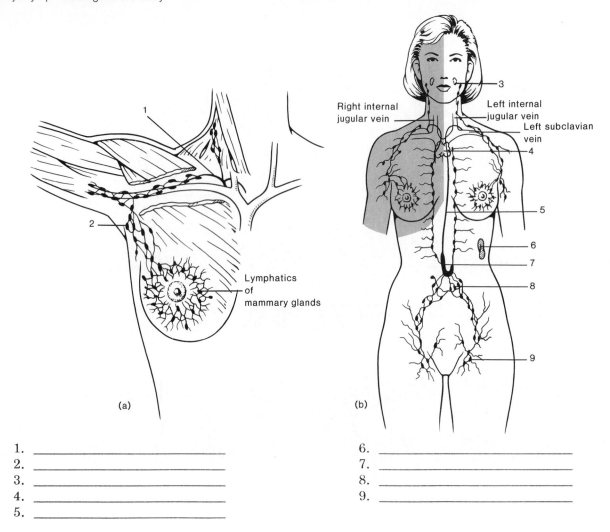

Right internal jugular vein

Left internal jugular vein

Left subclavian vein

Lymphatics of mammary glands

(a)

(b)

1. _____
2. _____
3. _____
4. _____
5. _____

6. _____
7. _____
8. _____
9. _____

vein ventromedially to the scapula. Most of the blood flowing through the transverse scapular vein comes from the lateral surface of the forelimb through the *cephalic vein.* In addition to the transverse scapular vein, the external jugular receives blood from the small *internal jugular vein.* One of the paired internal jugular veins may occasionally be absent. The internal jugular vein parallels the common carotid artery and enters the external jugular vein near its convergence with the subclavian vein.

The venous return from the forelimb of the cat is through the superficial cephalic vein (already described) and the deep *brachial vein.* The brachial vein begins at the convergence of the *ulnar vein* and *radial vein* (not illustrated) and then runs parallel to the brachial artery along the medial side of the humerus. At the cubital fossa of the elbow, a short *median cubital vein* connects the cephalic and brachial veins.

Near the head of the humerus, the brachial vein converges with the *subscapular vein* to become the *axillary*

vein, which runs through the axillary region. The subscapular vein has two tributaries: the *thoracodorsal vein,* from the teres major and subscapularis muscles, and the *humeral circumflex vein,* draining the deep muscles of the upper brachium. A ventral (anterior) *thoracic vein* drains blood from the pectoral muscles to the axillary vein.

Veins of the thorax. In addition to receiving the two brachiocephalic veins that return blood from the head, neck, and upper extremities, the superior vena cava receives three venous tributaries from the thoracic region (fig. 31.6). The *azygos vein* empties blood into the superior vena cava near its entrance to the right atrium. Force the heart and lungs of the cat to the left to locate the azygos vein. *Intercostal, esophageal,* and *bronchial veins* empty into the azygos vein. The second vessel is the *internal mammary (sternal) vein,* which drains the medial pectoral region and enters the superior vena cava at the level of the third rib. The third vessel to empty into the superior vena cava is the *right*

Figure 31.6 A ventral view of the thorax, neck, and brachium of the cat showing the veins anterior to the diaphragm.

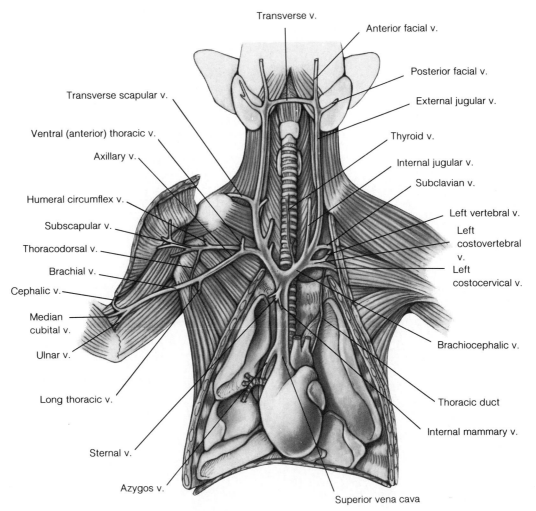

Transverse v.
Anterior facial v.
Posterior facial v.
External jugular v.
Thyroid v.
Internal jugular v.
Subclavian v.
Left vertebral v.
Left costovertebral v.
Left costocervical v.
Brachiocephalic v.
Thoracic duct
Internal mammary v.
Superior vena cava

Transverse scapular v.
Ventral (anterior) thoracic v.
Axillary v.
Humeral circumflex v.
Subscapular v.
Thoracodorsal v.
Brachial v.
Cephalic v.
Median cubital v.
Ulnar v.
Long thoracic v.
Sternal v.
Azygos v.

vertebral vein (not illustrated), draining blood from the brain. The left vertebral vein enters the left brachiocephalic vein rather than the superior vena cava.

Veins of the lower extremity. The blood of the lower hind limb of the cat drains through a superficial *saphenous vein* along the medial aspect and a deep *popliteal vein,* which passes through the popliteal region at the back of the knee (fig. 31.7). The saphenous and popliteal veins converge to become the *femoral vein,* which runs parallel to the femoral artery along the medial aspect of the thigh. As the femoral vein passes through the abdominal wall, it receives the *deep femoral vein* from a medial position. The deep femoral vein carries blood from the *posterior epigastric vein,* which drains the posterior abdominal wall and small tributaries from the urinary bladder and external genitalia.

The *external iliac vein* is formed at the convergence of the femoral and deep femoral veins. The external iliac vein courses cranially to the level of the sacroiliac joint. There it merges with the *internal iliac vein* of the pelvic and genital regions to become the *common iliac vein.* The right and left common iliacs then converge to become the large *inferior vena cava.* A small, unpaired *caudal vein* drains blood from the tail into the right common iliac vein.

Veins of the abdominal region. The inferior vena cava parallels the dorsal aorta; it extends through the abdominal cavity of the cat, penetrates the diaphragm and enters the right atrium. As the inferior vena cava passes through the abdominal cavity, it receives tributaries from veins that correspond in name and position to arteries previously identified.

The *iliolumbar veins* drain the posterior abdominal wall and enter the inferior vena cava at the level of the iliolumbar arteries. Several small, unpaired *lumbar veins* drain into the inferior vena cava from the lumbar spinal region. In the male cat, the *spermatic (genital) veins* receive blood from the testes. The right spermatic vein drains directly into the inferior vena cava, and the left spermatic vein generally empties into the left renal vein. The drainage pathway is similar in the female, except the vessels are called *ovarian veins* and they receive blood from the ovaries. The *renal veins* drain blood from the kidneys and ureters into the

Figure 31.7 A ventral view of the abdomen and lower extremities of the cat showing the veins below the diaphragm.

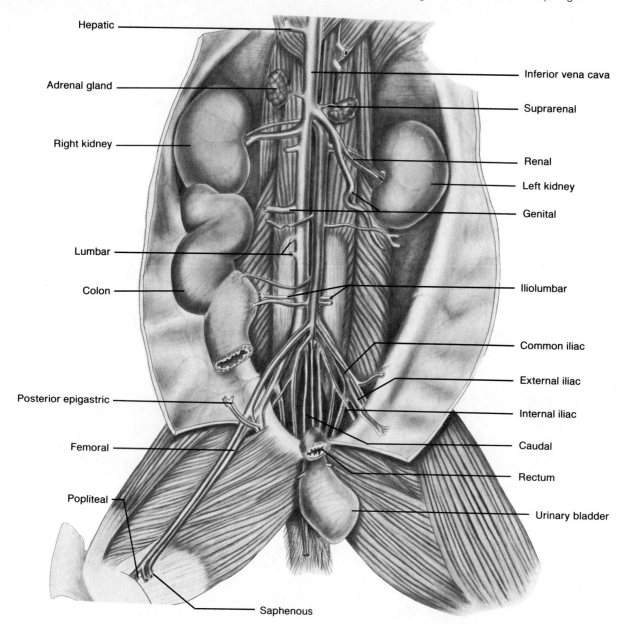

Hepatic

Adrenal gland

Right kidney

Lumbar

Colon

Posterior epigastric

Femoral

Popliteal

Saphenous

Inferior vena cava

Suprarenal

Renal

Left kidney

Genital

Iliolumbar

Common iliac

External iliac

Internal iliac

Caudal

Rectum

Urinary bladder

inferior vena cava. Small *suprarenal veins* drain the adrenal (suprarenal) glands. The *inferior phrenic veins* (not illustrated) receive blood from the inferior side of the diaphragm and drain into the inferior vena cava. *Right* and *left hepatic veins* originate in the liver and empty into the inferior vena cava immediately below the diaphragm.

Hepatic portal system. The hepatic portal system refers to those vessels that drain blood from the abdominal digestive organs and spleen to the liver. This system allows transport of nutrient-rich blood from the capillary network of the

digestive organs to the capillary network of the liver. Thus the liver—the principal organ that processes digested food—has immediate access to absorbed nutrients.

The hepatic portal vein is the large vessel entering the inferior side of the cat's liver (fig. 31.8). It is formed by the union of the *superior (anterior) mesenteric vein,* which drains nutrient-rich blood from the small intestine, and the *gastrosplenic vein,* which drains blood from the stomach and spleen. The *inferior (posterior) mesenteric vein* is a large vessel draining the colon via the *left colic vein* and the rectum via the *superior (anterior) hemorrhoidal vein.* The

Figure 31.8 The hepatic portal system of the cat.

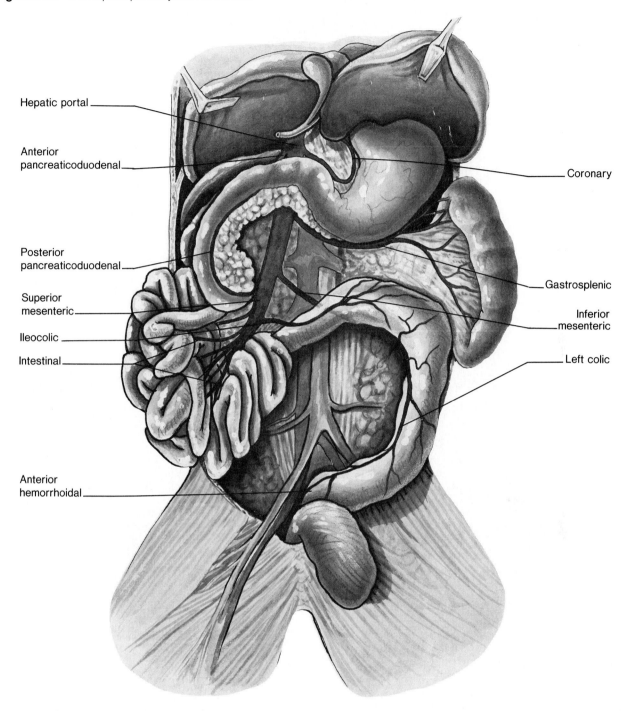

Hepatic portal

Anterior
pancreaticoduodenal

Posterior
pancreaticoduodenal

Superior
mesenteric

Ileocolic

Intestinal

Anterior
hemorrhoidal

Coronary

Gastrosplenic

Inferior
mesenteric

Left colic

inferior mesenteric vein parallels the inferior mesenteric artery and drains into the superior mesenteric vein. A *coronary vein,* which drains blood from the lesser curvature of the stomach, also empties into the superior mesenteric vein. The *anterior pancreaticoduodenal vein* drains the anterior portion of the pancreas and duodenum, and it enters the hepatic portal vein at the level of entrance of the coronary veins. The *posterior pancreaticoduodenal vein* empties into the superior mesenteric vein.

After the venous hepatic portal blood flows through the sinusoids of the liver, it enters the hepatic veins, which in turn drain into the inferior vena cava.

Laboratory Report 31

Name _____

Date _____

Section _____

Veins, Lymphatics, and Fetal Circulation

Read the assigned sections in the textbook before completing the laboratory report.

A. Principal Veins of the Body

1. Trace the path of a glucose injection from the median cubital vein of the arm to the brain, and list, in proper sequence, all the blood vessels and chambers of the heart through which it passes. You may have to review the arterial pathways learned in the previous exercise.

2. Define a *portal system*. What is the significance of the hepatic portal system?

3. Name the vein or veins that drain blood from each of the following organs or regions:
   a. Ovary _____
   b. Brain _____
   c. Superficial thigh _____
   d. Liver _____
   e. Descending colon and rectum _____
   f. Forearm _____
   g. Posterior abdominal wall and spinal cord _____
   h. Diaphragm _____
   i. Gallbladder _____
   j. Pancreas _____
   k. Kidney _____
   l. Stomach _____

B. Fetal Circulation

1. What is significant about the pattern of blood flow through the fetal heart?

2. What are five fetal structures or vessels that cease their vascular function in the postnatal infant?

3. What are the differences between congenital heart disorders and acquired heart problems? List examples of each.

C. Lymphatic System

1. What might a physician conclude if, during a physical examination, he or she discovered enlarged cervical lymph nodes?

2. Why are the thymus and spleen considered lymphoid organs? How are these organs similar, and how are they different?

3. If cancer cells from the right breast were to metastasize, which lymph nodes would probably be involved?

Effects of Drugs on the Frog Heart

Before coming to class, review the following sections in chapter 21 of the textbook: "Pressure Changes during the Cardiac Cycle" and "Electrical Activity of the Heart."

Introduction

A *drug* is a substance that affects some aspect(s) of physiology. Drugs may be identical to substances found in the body, such as minerals, vitamins, and hormones, or they may be molecules uniquely produced by particular plants or fungi. Many drugs marketed by pharmaceutical companies are natural products whose chemical structure has been slightly modified to alter the biological activity of the native compounds.

The biological effects of endogenous compounds (those compounds normally found in the body) vary with their concentration. A normal blood potassium concentration, for example, is necessary for health, but too high a concentration can be fatal. Similarly, the actions exhibited by many hormones at abnormally high concentrations may not occur when the hormones are at normal concentrations. It is important, therefore, to distinguish between the *physiological effects* (normal effects) of these substances and their *pharmacological effects* (those that occur when the substances are administered as drugs). A study of the pharmacology of various substances can, however, reveal much about the normal physiology of the body.

In this exercise, the effects of various pharmacological agents will be tested on the heart of a pithed frog. Although heart muscle, like skeletal muscle, is striated, it differs from skeletal muscles in several respects. The heartbeat is *automatic;* unlike skeletal muscle, heart muscle needs no nerve or electrode stimulation to contract. This is because action potentials begin spontaneously in the *pacemaker region* (called the sinoatrial, or *SA, node*) in the right atrium, and spread through the ventricles in an automatic, rhythmic cycle. In the exposed frog heart, the atria can be seen to contract before the ventricle contracts. (Unlike mammals, frogs have only one ventricle.)

When a thread connects the frog heart to the recording equipment, contractions of the atria and ventricle produce two successive peaks in the recordings. The strength of contraction is related to the amplitude of these peaks, and the rate of the heartbeat can be determined by the distance between the ventricular peaks (if the chart speed is known). The rate of impulse conduction between the atria and ventricle is related to the distance between the atrial and ventricular peaks. Thus the effects of various drugs on the strength of contraction, rate of contraction, and rate of impulse conduction from the atria to the ventricle can be determined.

Objectives

Students completing this exercise will be able to:

1. Describe the pattern of contraction in the frog heart.
2. Describe the effect of various drugs on the heart, and explain their mechanisms of action.

Materials

1. Frogs, dissecting instruments, and dissecting trays
2. Copper wire and thread
3. Recording apparatus: Physiograph, transducer coupler, and myograph transducer (Narco); or kymograph, kymograph paper, and kerosene burner
4. Ringer's solution (see exercise 14—all drugs should be added to the Ringer's solution, which is used as the solvent); calcium chloride (2.0 g/dl); digitalis (0.2 g/dl); pilocarpine (2.5 g/dl); atropine (2.5 g/dl); potassium chloride (2.0 g/dl); epinephrine (0.01 g/dl); caffeine (saturated; 0.2 g/dl)); and nicotine (1.0 g/dl)

Procedure

1. Double pith a frog and expose its heart (fig. 32.1). Skewer the apex of the heart muscle with a short length of thin copper wire, being careful not to let the wire enter the ventricle.
2. Bend the copper wire into a loop, and tie one end of cotton thread to this loop (see enlarged insert in fig. 32.2).
3. Procedure for **kymograph** recording:
   a. Tie the other end of the thread to a heart lever. The thread and heart should be pulled fairly tight so that contractions of the heart produce movements of the lever.
   b. Attach kymograph paper (shiny side out) to the kymograph drum, and rotate the drum

Figure 32.1 A procedure for exposing the frog heart: (*a*) first the skin is cut; (*b*) then the body cavity is exposed by cutting through the muscles to the sternum. The sternum will next be split to expose the heart.

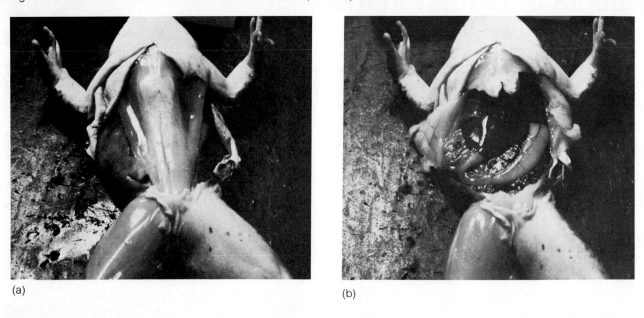

(a)

(b)

Figure 32.2 A frog heart setup: the contractions of the heart pull a lever that writes on a moving chart (kymograph).

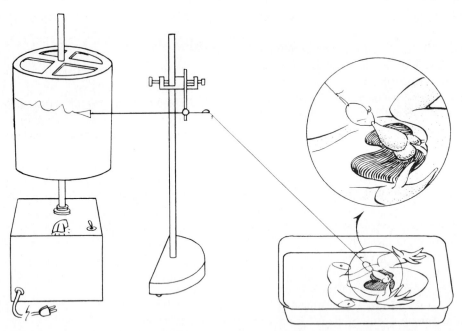

slowly over a kerosene burner until the paper is uniformly blackened. Arrange the heart lever so that it lightly drags across the smoked paper. (Too much pressure of the writing stylus against the kymograph will prevent movement of the heart lever.) See figure 32.2 for the proper setup.

4. Procedure for **myograph** recording:
   a. Tie the other end of the thread to the hook below the myograph transducer. (Make sure the myograph is plugged into the transducer

coupler on the Physiograph.) The heart should be positioned directly below the myograph. Adjust the height of the myograph on its stand so that the heart is pulled out of the chest cavity (fig. 32.3).
   b. Make sure the Physiograph is properly balanced, and set the paper speed at 0.5 centimeters per second. Push the *record* button *in*, and set the *paper advance* button *out* when you are ready to record.

Figure 32.3 A procedure for setting up to record the frog's heart contractions: (a) a small length of thin copper wire is passed through the tip of the ventricle; (b) this wire is then twisted together to form a loop; (c) the loop is tied by a cotton thread to the hook in the myograph transducer.

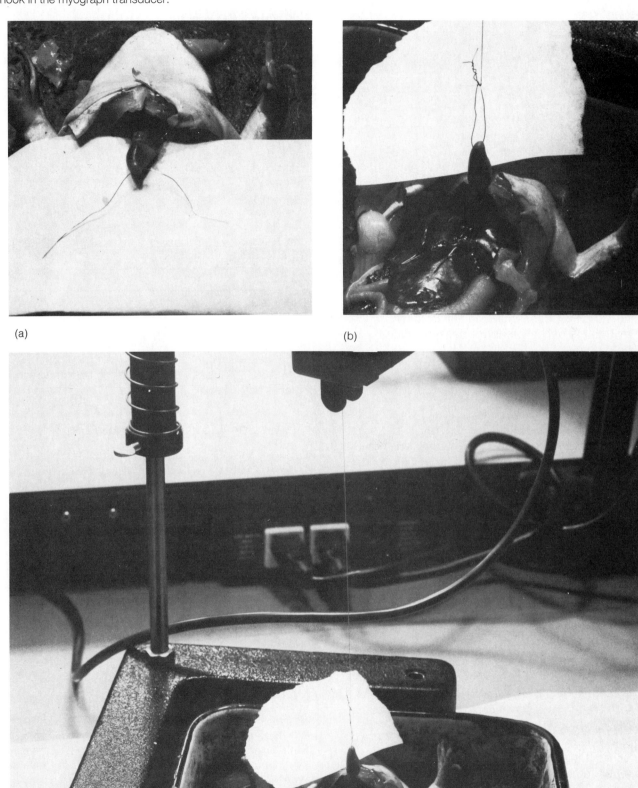

(a)

(b)

(c)

Figure 32.4 A recording of frog heart contractions on a Physiograph recorder. The arrow points to a recording of a smaller atrial contraction, which is followed by a recording of a larger ventricular contraction.

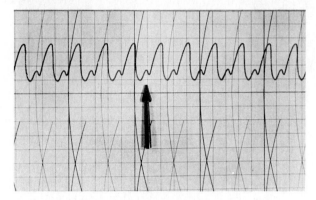

5. Observe the pattern of the heartbeat prior to the addition of drugs (see example in fig. 32.4). Be sure that you can distinguish atrial and ventricular beats, and that you can measure the rate (in beats per minute) and strength (in millimeters deflection above baseline) of the heartbeat.

A. Effect of Calcium Ions on the Heart

In addition to the intracellular role calcium plays in coupling excitation to contraction, the extracellular Ca^{++}/Mg^{++} ratio also affects impulse transmission at cholinergic neuromuscular junctions. An increase in the extracellular concentration of calcium (above the normal concentration of 4.5–5.5 milliequivalents per liter) affects both the electrical properties and the contractility of muscle.

The heart is thus affected in a number of ways by an increase in extracellular calcium: (1) increased force of contraction, (2) decreased cardiac rate, and (3) the appearance of ectopic pacemakers in the ventricles, producing abnormal rhythms (extrasystoles and idioventricular rhythm).

Procedure

1. Obtain a record of the normal heartbeat. Then use a dropper to bathe the heart in 2.0% calcium chloride ($CaCl_2$). On the recording paper, indicate the time at which calcium was added. Observe the effects of the calcium solution for a few minutes.
2. On the table in your laboratory report, tape the recording (or draw a facsimile) of the heartbeat before (normal) and after the calcium solution bath.

3. Rinse the heart thoroughly with Ringer's solution until the normal heartbeat returns.

B. Effect of Digitalis on the Heart

It is thought that digitalis inhibits the Na^+/K^+–ATPase pump, thus causing an influx of Na^+, an efflux of K^+, and an enhanced uptake of Ca^{++}. The effects of digitalis and of increased extracellular calcium on the heart are thus very similar.

Clinical Significance

Digitalis glycosides, such as digoxin, are frequently used to treat congestive heart failure, atrial flutter, and atrial fibrillation. Digitalis benefits these conditions by: (1) increasing the force of contraction, (2) decreasing the cardiac rate (it directly inhibits the SA node and it stimulates the vagus nerve, which also then inhibits the SA node), and (3) decreasing the conduction rate in the atrioventricular bundle.

Procedure

1. Obtain a record of the normal heartbeat, and then bathe the heart in a 2.0% digitalis solution.
2. On the table in your laboratory report, tape the recording (or draw a facsimile) of the heartbeat after adding digitalis.

C. Effect of Pilocarpine on the Heart

Pilocarpine is a *parasympathomimetic* drug; that is, it mimics the effects of parasympathetic nerve stimulation. Pilocarpine facilitates the release of the neurotransmitter *acetylcholine* from the vagus nerves, resulting in a decrease in the cardiac rate.

Procedure

1. Thoroughly rinse the heart with Ringer's solution until the normal heartbeat returns.
2. Bathe the heart in a 2.5% pilocarpine solution, and tape the recording (or draw a facsimile) on the table in your laboratory report.

D. Effect of Atropine on the Heart

Atropine is an alkaloid drug derived from the nightshade plant *Atropa belladonna* (the species name, *belladonna*, is often used as the drug name). Atropine blocks the effects of acetylcholine and inhibits the effects of parasympathetic activity on the heart, smooth muscles, and glands. If the cardiac rate is decreased as a result of vagus nerve stimulation, the administration of atropine will increase this rate.

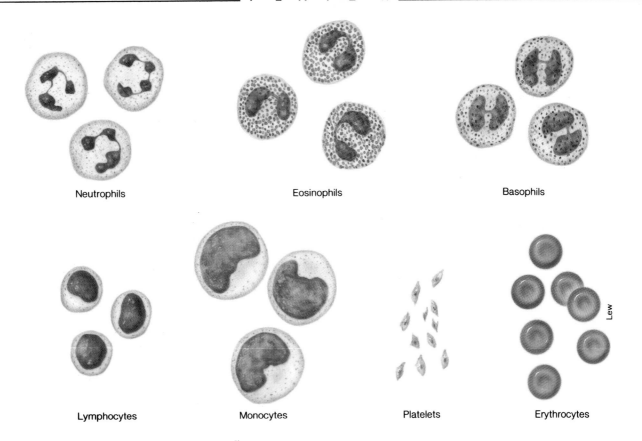

Neutrophils   Eosinophils   Basophils

Lymphocytes   Monocytes   Platelets   Erythrocytes

The average differential count in the normal adult is as follows:

Leukocyte	Percentage
Neutrophils	55%–75%
Eosinophils	2%–4%
Basophils	0.5%–1%
Lymphocytes	20%–40%
Monocytes	3%–8%

Leukocyte	Cells Counted	Total
Neutrophils		
Eosinophils		
Basophils		
Lymphocytes		
Monocytes		

▲ **Exhibit 25**

Formed elements of blood.

Anti-B   Anti-A

Type A

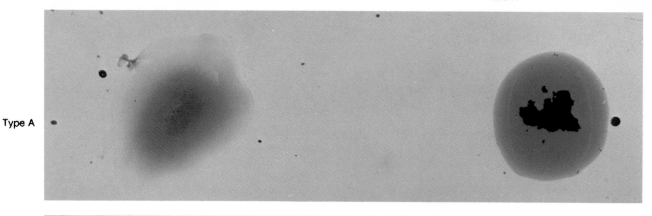

Type B

(a)

Type AB

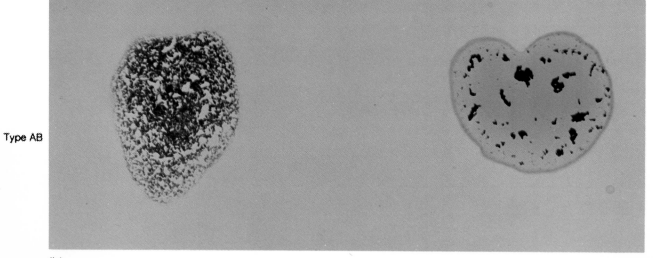

(b)

▲ **Exhibit 26**

Results of blood typing; (a) type A blood agglutinating
(clumping) with antiserum A (top right), and type B blood
agglutinating with antiserum B (bottom left); type O blood would
not agglutinate with either antiserum; (b) type AB blood
agglutinating with both antiserum B (left) and antiserum A
(right).

Clinical Significance

The ability of atropine to block the effects of parasympathetic nervous stimulation is useful in a variety of clinical situations. For example, it is used in ophthalmology to dilate the eyes (parasympathetic nerves cause constriction of the pupils), and in surgery to dry the mouth, pharynx, and trachea (glandular secretions that wet these mucous membranes are stimulated by parasympathetic nerves).

Procedure

1. Bathe the heart in a 5.0% atropine solution while it is still under the influence of pilocarpine.
2. On the table in your laboratory report, tape the recording (or draw a facsimile) of the effects of atropine.

E. Effect of Potassium Ions on the Heart

Since the resting membrane potential partly depends on the maintenance of a higher K^+ concentration inside the cell than outside, an increase in the concentration of extracellular K^+ results in a *decrease in the resting membrane potential*. This, in turn, produces a decrease in the force of contraction and the conduction rate. In extreme *hyperkalemia* (high blood potassium), the conduction rate may be so depressed that ectopic pacemakers appear in the ventricles and fibrillation may develop.

Procedure

1. Rinse the heart in Ringer's solution until the beat returns to normal, and then bathe the heart in 2.0% potassium chloride (KCl).
2. On the table in your laboratory report, tape the recording (or draw a facsimile) of the effects of K^+ on the heartbeat.

F. Effect of Epinephrine on the Heart

Epinephrine is a hormone secreted by the adrenal medulla that increases the strength and rate of cardiac contraction. The sympathetic neurotransmitter norepinephrine has similar actions. Exogenous epinephrine is a *sympathomimetic drug* because it mimics the effect of sympathetic nerve stimulation.

Procedure

1. Rinse the heart with Ringer's solution until the normal heartbeat returns, and then bathe the heart in epinephrine (adrenaline).
2. On the table in your laboratory report, tape the recording (or draw a facsimile) of the effect of epinephrine on the heartbeat.

G. Effect of Caffeine on the Heart

Caffeine is a mild central nervous system stimulant that also acts directly on the myocardium to increase both the strength of contraction and the cardiac rate. It has this effect because it inhibits activity of the enzyme phosphodiesterase, which breaks down cyclic AMP. As a result, there is an increased amount of cyclic AMP in the heart cells, which duplicates the action of epinephrine.

Caffeine's usefulness as a central nervous system stimulant is limited because, in high doses, it can promote the formation of *ectopic pacemakers* (*foci*), resulting in major arrhythmias.

Procedure

1. Bathe the heart with Ringer's solution until the normal heartbeat returns, and then bathe the heart with a saturated solution of caffeine.
2. On the table in your laboratory report, tape the recording (or draw a facsimile) of the effect of caffeine on the heartbeat.

H. Effect of Nicotine on the Heart

Nicotine promotes transmission at the autonomic ganglia by stimulating particular "nicotinic" receptors for acetylcholine in the postganglionic neurons. When applied directly to the heart, the major effect of nicotine will be stimulation of parasympathetic ganglia located within the epicardium. Activation of postganglionic parasympathetic neurons, in turn, will cause slowing of the heart rate. When nicotine is administered systemically, it can act at the sympathetic ganglia and adrenal medulla to stimulate the sympathoadrenal system; this causes an increase in the heart rate.

Procedure

1. Rinse the heart in Ringer's solution until the normal heartbeat returns, and then bathe the heart in a 0.2% solution of nicotine.
2. Tape the recording (or draw a facsimile) on the table provided in your laboratory report.
3. Analyze your data, and record your results for parts A through H in the laboratory report.

Laboratory Report 32

Name _____

Date _____

Section _____

Effects of Drugs on the Frog Heart

Read the assigned sections in the textbook before completing the laboratory report.

1. Enter your data in the following table:

Condition	Effect (Tape the Recording or Draw a Facsimile)
Normal	
Ca^{++}	
Digitalis	
Pilocarpine	
Atropine	
K^+	
Epinephrine	
Caffeine	
Nicotine	

2. Analyze your recordings and enter your results in the following table:

Condition	Rate (beats/min)	Strength (mm above Baseline)	Distance (mm) between Atrial and Ventricular Peaks	Conclusions about Drug Effects
Normal				
Ca^{++}				
Digitalis				
Pilocarpine				
Atropine				
K^{+}				
Epinephrine				
Caffeine				
Nicotine				

3. Match the following:

_____(1) Endogenous substance that makes
      the beat stronger and faster

_____(2) Substance that makes the beat slower
      and stronger

_____(3) Substance that facilitates the release
      of acetylcholine from parasympathetic
      nerve endings

_____(4) Substance that mimics the action of
      epinephrine by inhibiting the action of
      phosphodiesterase

_____(5) Substance that stimulates the acetylcholine
      receptors of autonomic ganglia

_____(6) Substance that blocks the acetylcholine
      receptors of the target cells of postganglionic
      parasympathetic neurons

(a) digitalis
(b) nicotine
(c) caffeine
(d) epinephrine
(e) atropine
(f) pilocarpine

4. What are the effects of hyperkalemia on the heart? How are these effects produced?

5. What effect did Ca^{++} have on the amplitude of the recording of the heartbeat? Explain why.

6. Explain the effect of digitalis on the heart and the clinical uses of this drug.

Electrocardiogram

Before coming to class, review the following sections in chapter 21 of the textbook: "Electrical Activity of the Heart," "The Electrocardiogram," "ECG Leads," "Arrhythmias Detected by the Electrocardiograph," and "Ischemic Heart Disease."

Introduction

The regular pattern of impulse production and conduction in the heart results in contraction of the myocardium, and in the cardiac cycle of systole and diastole. These events can be monitored by an electrocardiogram (ECG, or EKG), which can also reveal abnormal patterns associated with cardiac arrhythmias. The ECG can also be used to observe the effects of exercise on the cardiac cycle.

Objectives

Students completing this exercise will be able to:

1. Describe the normal pattern of impulse production and conduction in the heart, and identify the conducting tissues of the heart.
2. Describe the normal ECG, and explain how it is produced.
3. Obtain an ECG using the limb leads, identify the waves, determine the P-R interval, and measure the cardiac rate.
4. Describe common abnormalities that can easily be seen in an ECG.
5. Explain how the cardiac cycle changes during exercise.

Materials

1. Electrocardiograph
2. Electrode plates, rubber straps, electrolyte gel or paste
3. Alternative equipment: *Cardiocomp* (Intelitool, Inc.), with Apple or IBM (8086 processor) computer

A. Electrocardiogram at Rest

Since the concentration of electrolytes in body fluids is high, the electrical activity generated during heart contraction travels throughout the body and can easily be monitored by placing electrodes on different areas of the skin (fig. 33.1). A graphic representation of this electrical activity is called an *electrocardiogram* (*ECG,* or *EKG*), and the instrument producing this record is called an *electrocardiograph.*

In the following exercises the *standard limb leads* I, II, and III will be used. These leads record the difference in potential (that is, the voltage) between two electrodes placed on the arms and legs (fig. 33.1). In clinical electrocardiography, however, *unipolar leads* are also used. These are the *AVR* (right arm), *AVL* (left arm), *AVF* (left leg), and the chest leads labeled V_1 to V_6.

Procedure

1. The subject should lie comfortably on a cot. The examiner rubs a small amount of electrolyte gel on the medial surface of the arms and legs (about 2 inches above the wrists and ankles) so that the gel covers an area about the size of a silver dollar. Attach electrode plates to these four locations, using the rubber straps provided.
2. Attach the four ECG leads to the appropriate plates.
3. ☞ **Note:** *the specific instructions for obtaining an ECG vary with the type of instrument used. Your instructor will demonstrate the recording equipment used in your lab. The following instructions apply only to the use of a single-channel electrocardiograph.*

   a. Turn on the power switch.
   b. Set the speed selector switch to 25 millimeter (mm) per second (sec).
   c. Set the sensitivity to *1*.
   d. Set the lead selector switch to the first dot to the left of the *STD* position.
   e. Turn the control knob to the *run* position.

4. Turn the position knob until the stylus is centered on the ECG paper.

Figure 33.1 The placement of the bipolar limb leads of an electrocardiogram (ECG). (RA = right arm, LA = left arm, LL = left leg.)

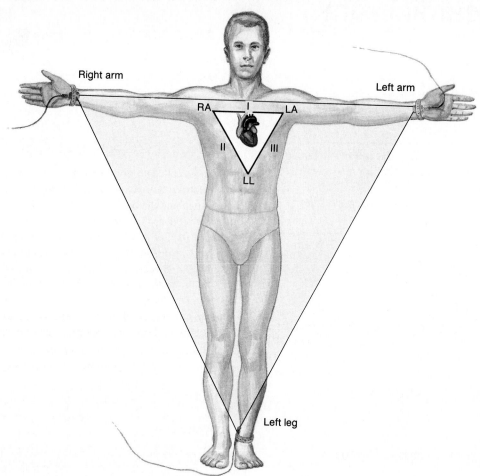

5. Turn the lead selector switch to the *1* position; this ECG measures the voltage between the right and left arms. As the paper is running, depress the *mark* button; this makes a single dash at the top of the chart to indicate that this is the recording from lead I. Allow enough time so that each member of the subject's group can cut off a sample of the recording, then turn the lead selector switch to the dot above the *1* position. The dots are the *rest* positions, and they stop the movement of the chart between recordings of different leads.

6. Turn the lead selector switch to the *2* position; this ECG measures the voltage between the right arm and left leg. As the chart is running, depress the *mark* button twice; two dashes thus indicate that this is the recording from lead II. Turn the lead selector switch to the dot above the *2* position.

7. Repeat this procedure with lead III; this measures the voltage between the left arm and the left leg.

8. After recordings from leads I, II, and III have been obtained, turn the lead selector switch to the *STD* position. This will run the recording out of the machine.

9. Remove the electrode plates from the subject's skin, and thoroughly wash the electrolyte gel from both the plate and the skin.

10. Tape samples of the recordings in your laboratory report, and label all the waves.

11. Determine the P-R interval of lead II. This is done by counting the number of small boxes between the beginning of the P and the Q and multiplying this number by 0.04 second.

☞ **Note:** *If you use a multichannel recorder (e.g., a Physiograph) instead of an electrocardiograph, the calculation will be different. If you use a Cardiocomp, the P-R interval will be provided automatically.*

_____second

| The normal P-R interval is 0.12–0.20 second. |

12. Determine the cardiac rate by the following methods:

   a. Count the number of QRS complexes in a 3-second interval (the distance between two vertical lines at the top of the ECG paper), and multiply by 20.

☞ **Note:** *If you use a multichannel recorder (e.g., a Physiograph), the amount of chart paper corresponding to a given time interval must be calculated and will vary with the paper speed. If you use a Cardiocomp, the cardiac rate will be provided automatically.*

Beats per minute = _____

b. Count the number of QRS complexes in a 6-second interval, and multiply by 10.

Beats per minute = _____

c. At a chart speed of 25 mm/sec, the time interval between one light vertical line and the next is 0.04 second. The time interval between heavy vertical lines is 0.20 second. The cardiac rate in beats per minute can be calculated if one knows the time interval between two R waves in two successive QRS complexes.

For example, suppose that the time interval from one R wave to the next is exactly 0.60 second. Therefore,

$$\frac{1 \text{ beat}}{0.60 \text{ sec}} = \frac{x \text{ beats}}{60 \text{ sec}}$$

$$x = \frac{1 \text{ beat} \times 60 \text{ sec}}{0.60 \text{ sec}}$$

$$x = 100 \text{ beats per minute}$$

Beats per minute = _____

d. The values obtained by method c can be approximated by counting the number of heavy vertical lines between one R wave and the next according to the memorized sequence: 300, 150, 100, 75, 60, 50.

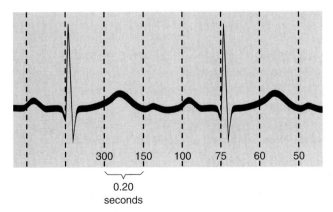

300   150   100   75   60   50

0.20
seconds

The cardiac rate in the above example is 100 beats per minute.

Beats per minute = _____

B. Electrocardiogram following Exercise

The heart is innervated by both sympathetic and parasympathetic fibers. At the beginning of exercise, the activity of the parasympathetic fibers that innervate the SA node decreases. Since these fibers have an inhibitory effect on the pacemaker, a decrease in their activity results in an increase in cardiac rate. As exercise becomes more intense, the activity of sympathetic fibers that innervate the SA node increases. This has an excitatory effect on the SA node and causes even greater increases in cardiac rate.

Sympathetic fibers also innervate the conducting tissues of the heart and the ventricular muscle. Through these innervations, sympathetic stimulation may increase the velocity of both impulse conduction and ventricular contraction. These effects are most evident at high cardiac rates and contribute only slightly to the increased cardiac rate during exercise. Thus, the increased cardiac rates are mainly due to a shortening of the ventricular diastole and only secondarily due to a shortening of ventricular systole.

Clinical Significance

A portion of cardiac muscle may receive insufficient blood flow because of a clot in a coronary artery (*coronary thrombosis*) or because of narrowing of the vessel due to *atherosclerosis*. The condition of insufficient blood flow—**ischemia**—may, however, be relative. The rate of blood flow may be adequate to meet the aerobic requirements of the heart at rest, but it may be inadequate for the increased metabolic energy demand of the heart during exercise. Exercise tests help diagnose this kind of ischemia, which may be revealed by a change in the S-T segment of the ECG (fig. 33.2).

Procedure

1. After the resting ECG has been recorded, unplug the electrode leads from the electrocardiograph.
2. The subject should hold the lead wires and exercise by walking up and down stairs or by hopping.

☞ **Caution:** *The intensity of exercise should be monitored so that 80% of the maximum cardiac rate is not exceeded. This activity should be performed only by students who are in good health.*

3. Immediately after exercising, the subject should lie down. Plug the electrode leads into the electrocardiograph and record lead II. Wait 2 minutes and record lead II again.
4. Calculate the cardiac rate and P-R interval as previously described. Also, determine the duration of ventricular diastole by measuring the time period from the middle of a T wave to the Q of the next cycle. Enter these data in the table in your laboratory report.

Figure 33.2 During myocardial ischemia, the S-T segment of the electrocardiogram may be depressed, as illustrated in this figure.

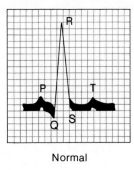

Normal

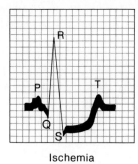

Ischemia

Laboratory Report 33

Name _____

Date _____

Section _____

Electrocardiogram

Read the assigned sections in the textbook before completing the laboratory report.

A. Electrocardiogram at Rest

1. Tape your recording in the spaces below:

   Lead I

   Lead II

   Lead III

2. The cells with the fastest spontaneous cycle of depolarization-contraction are located in the _____.

3. Indicate the electrical events that produce each of the following waves:

    a. P wave _____

    b. QRS wave _____

    c. T wave _____

4. An occasional extra beat, which can be seen as an ectopic QRS complex, is called a(n) _____.

5. An abnormally long P-R interval indicates a condition called _____.

6. A condition in which the ventricles are unable to contract and work as an effective pump, and in which a circus rhythm of electrical activity may be present, is known as _____.

7. Explain why the SA node functions as the normal pacemaker.

8. Compare ventricular tachycardia with paroxysmal supraventricular tachycardia in terms of etiology, ECG pattern, and danger.

9. On initial examination, a patient is found to have a P-R interval of 0.24 seconds. This patient is examined again a year later and found to have a resting pulse of 40 beats per minute with very little increase after exercise. Explain what happened.

B. Electrocardiogram following Exercise

1. Enter your data in the table below:

	Cardiac Rate (beats/min)	P-R interval (beginning of P to Q)	Ventricular Diastole (middle of T to next Q)
Resting ECG			
Immediate postexercise ECG			
2-minute postexercise ECG			

2. The ECG wave that occurs at the beginning of ventricular systole is the _____ wave.

3. The ECG wave that occurs at the end of systole and beginning of diastole is the _____ wave.

4. The ECG wave that occurs at the end of ventricular diastole is the _____ wave.

5. Describe the regulatory mechanisms that produce an increase in cardiac rate during exercise. Explain how these changes affect the ECG.

6. Explain why a person may have a normal ECG at rest but may show evidence of myocardial ischemia after moderate exercise.

Mean Electrical Axis of the Ventricles

Before coming to class, review "The Electrocardiogram" in chapter 21 of the textbook.

Introduction

Depolarization waves spread through the heart in a characteristic pattern. Depolarization begins at the sinoatrial (SA) node and spreads from this pacemaker through the entire mass of both atria. This produces the *P wave* in an electrocardiogram (ECG). After the atrioventricular (AV) node is excited, the impulses spread through the atrioventricular bundle (bundle of His), and the interventricular septum becomes depolarized. Since, at this moment, the lateral walls of the ventricles maintain their original polarity, there is a difference in electrical potential (voltage) between the septum and the ventricle wall, which produces the *R wave*. When the entire mass of the ventricles is depolarized, the difference in potential between them and the septum—the voltage—returns to zero (completing the *QRS complex*).

The direction of the depolarization waves depends on the orientation of the heart and on the particular instant of the cardiac cycle being considered. It is clinically useful, however, to determine the *mean axis* (average direction) of depolarizaton during the cardiac cycle. This can be done by observing the voltages of the QRS complex from two different perspectives using two different leads. Lead I provides a horizontal axis of observation (from left arm to right arm); lead III has an axis of about 120° (from left arm to left leg). Using the recordings from leads I and III, one finds that the normal mean electrical axis of the ventricles is about 59°. This is shown in figure 34.1.

Clinical Significance

Hypertrophy of one ventricle shifts the mean axis of depolarization toward the hypertrophied ventricle, because it takes longer for the larger ventricle to depolarize. A left axis deviation thus occurs when the left ventricle is hypertrophied (this may occur as a result of hypertension or narrowing of the aortic semilunar valve). A right axis deviation occurs when the right ventricle hypertrophies. This may occur as a result of narrowing of the pulmonary semilunar valve, or a congenital abnormality, such as a septal defect or tetralogy of Fallot.

The depolarization wave normally spreads through both ventricles at the same time. However, in the presence of a conduction block in one of the branches of the atrioventricular bundle—**bundle branch block**—depolarization is much slower in the blocked ventricle. In left bundle branch block, for example, depolarization occurs more slowly in the left ventricle than in the right ventricle, and the mean electrical axis deviates to the left. On the other hand, in right bundle branch block, there is a right axis deviation. Deviations of the electrical axis also occur in varying degrees as a result of myocardial infarction.

Objectives

Students completing this exercise will be able to:

1. Describe the electrical changes in the heart that produce the ECG waves.
2. Determine the mean electrical axis of the ventricles in a test subject, and explain the clinical significance of this measurement.

Materials

1. Electrocardiograph, or multichannel recorder (e.g., Physiograph) with ECG module; alternatively, a Cardiocomp (Intelitool, Inc.) may be used with an Apple or IBM (8086 processor) computer
2. ECG plates, straps, and gel

Procedure

☞ **Note:** *If a Cardiocomp-7 or -12 is used, the mean electrical axis of the ventricles can be determined by examining the QRS loop of the vectorgram displayed on the computer screen. Alternatively, the following procedure can be used by examining leads I and III that were previously obtained.*

1. Begin by using lead I (with a sensitivity setting of *1*).
    a. Find a QRS complex, and count the number of millimeters (small boxes) that it projects above the top of the baseline. Enter this value in the following space.

    + _____

Figure 34.1 (a) The convention by which the axis of depolarization is measured. The bottom half of the circle (with the heart at the center) is considered the positive pole (LA = left arm, RA = right arm, LL = left leg); (b) when the interventricular septum is depolarized, the surface of the septum is electrically negative compared with the walls of the ventricles, which have not yet become depolarized; the average normal direction of depolarization, or mean electrical axis of the ventricles, is about 59° (RV = right ventricle, LV = left ventricle).

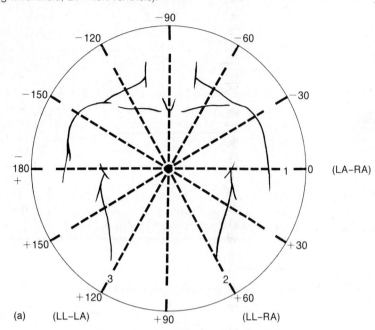

(a) (LL–LA) (LL–RA)

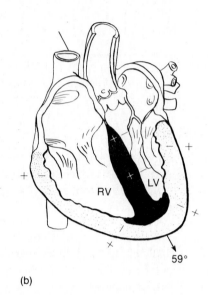

(b)

b. Count the millimeters that the Q and S waves (or the R wave, if it is inverted) project below the top of the baseline. Add these measurements of downward deflections and enter this sum in the following space.

_____

c. Add the two values from parts a and b (keep the negative sign if the sum is negative), and enter this value in the laboratory report.

2. Next, using lead III (with a sensitivity setting of *1*).

a. Find a QRS complex and count the millimeters it projects above the top of the baseline. Enter this value in the following space.

+ _____

b. Count the millimeters that the Q and S waves (or the R wave, if it is inverted) project below the top of the baseline. Add these measurements of downward deflections and enter this sum in the following space.

– _____

c. Add the two values from parts a and b (keep the negative sign if the sum is negative), and enter this value in the laboratory report.

3. On the grid chart in your laboratory report, use a straightedge to make a line on the axis of lead I that corresponds to the sum you obtained in step 1c.

4. Make a line on the axis of lead III that corresponds to the sum you obtained in step 2c.

5. Draw an arrow from the center of the grid chart to the intersection of the two lines drawn in steps 3 and 4. Extend this arrow to the edge of the circle around the grid chart, and record the mean electrical axis of the ventricles.

Example (fig. 34.2)

Lead I of sample ECG

Upward deflection:	+7 mm
Downward deflections:	–1 mm
	6

Lead III of sample ECG	+14 mm
Upward deflection:	–2 mm
Downward deflections:	12

A line perpendicular to the axis of lead I that corresponds to position 6 on the scale is drawn. Then a line perpendicular to the axis of lead III that corresponds to position 12 on the scale is drawn.

An arrow is then drawn from the center of the circle through the intersection of two lines previously drawn (fig. 34.3). In this example, the mean electrical axis of the ventricles is +71°.

Figure 34.2 Sample electrocardiograms of leads I and III used in the example for determination of the mean electrical axis of the heart.

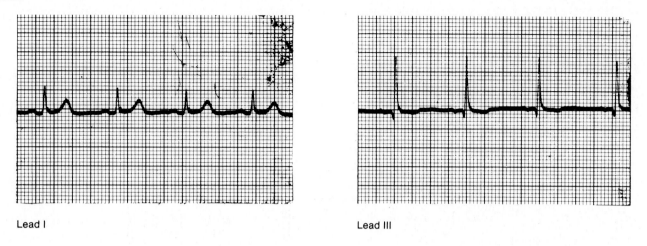

Lead I   Lead III

Figure 34.3 An example of the method used to determine the mean electrical axis of the heart, using data from leads I and III in the sample ECG provided in figure 34.2. In this example, the mean electrical axis is +71°.

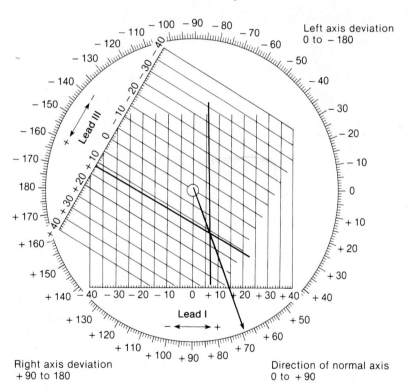

Laboratory Report 34

Name _____

Date _____

Section _____

Mean Electrical Axis of the Ventricles

Read the assigned section in the textbook before completing the laboratory report.

1. Enter the value from step 1c in the space below. Remember to indicate whether it is a positive or negative number.

   _____

2. Enter the value from step 2c in the space below. Be sure to indicate whether it is a positive or negative number.

   _____

3. Use your data and the grid chart to determine the mean electrical axis of the ventricles as described in steps 3, 4, and 5. Mean electrical axis of the ventricles:

Grid chart

Direction of the electrical axis of the heart as determined by the human electrocardiogram

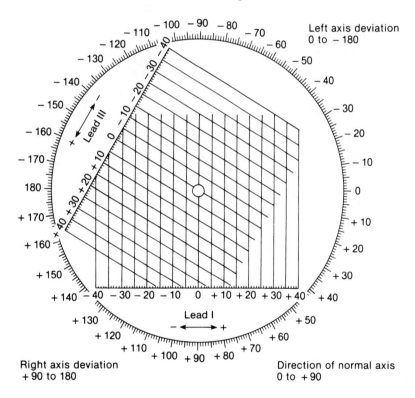

4. Explain the electrical events in the heart that produce a QRS wave. Why does the tracing go up from Q to R and then back to baseline from R to S?

5. Explain how the normal pattern of depolarization is affected by ventricular hypertrophy and by bundle branch block.

Regulation and Maintenance of the Human Body

Heart Sounds and Measurement of Blood Pressure

Before coming to class, review "Pressure Changes during the Cardiac Cycle" and "Heart Sounds" in chapter 21 of the textbook, and "Measurement of Blood Pressure" and "Pulse Pressure and Mean Arterial Pressure" in chapter 22.

Introduction

Pressure changes, which occur during contraction and relaxation of the ventricles, cause the one-way heart valves to open and close. This generates sounds that allow monitoring of the cardiac cycle and that aid in the diagnosis of structural abnormalities of the heart.

Contraction of the ventricles at systole produces a rise in intraventricular pressure. This causes the atrioventricular (AV) valves to close and the semilunar valves to open, so that blood is directed into the arteries. The closing of the AV valves results in the first sound of the heart ("lub"), and the ejection of blood into the arterial system produces a rise in arterial pressure to the systolic level. Relaxation of the ventricles in diastole produces a fall in intraventricular pressure. This causes the semilunar valves to close, resulting in the second sound of the heart ("dub"), and also results in a fall in arterial blood pressure to the diastolic level.

Objectives

Students completing this exercise will be able to:

1. Describe the causes of the heart sounds.
2. List some of the causes of abnormal heart sounds.
3. Correlate the heart sounds with the waves of the electrocardiogram and the events of the cardiac cycle.
4. Describe how the sounds of Korotkoff are produced, and explain why measurements of systolic and diastolic pressure correspond to the first and last sounds of Korotkoff.
5. Describe how pulse pressure and mean arterial pressure are calculated.

Materials

1. Stethoscope
2. Sphygmomanometer

A. Heart Sounds

Careful **auscultation** (listening) may reveal two components to each of the two heart sounds. The *splitting* of the heart sounds is more evident during inhalation than it is during exhalation. During inhalation, the first heart sound may be split into two sounds because the tricuspid valve and mitral valve close at different times. The second heart sound may also split during inspiration because the pulmonary and aortic semilunar valves close at different times.

Clinical Significance

Auscultation of the chest is a valuable aid in the diagnosis of a variety of cardiac conditions, including **heart murmurs.** These murmurs may be caused by an irregularity in a valve, a septal defect, or the persistence after birth of the fetal opening (*foramen ovale*) between the right and left atria, resulting in a regurgitation of blood in the reverse direction of normal flow. Abnormal splitting of the first and second heart sounds occurs as a result of a variety of conditions, including heart block, septal defects, aortic stenosis, and hypertension.

Procedure

1. To best hear the first heart sound, auscultate the *apex* beat of the heart by placing the diaphragm of the stethoscope on the fifth left intercostal space (fig. 35.1).
2. To best hear the second heart sound, place the stethoscope at the second intercostal space to the right or left of the sternum.
3. Compare the heart sounds heard at these three stethoscope positions during quiet breathing, during slow, deep inhalation, and during slow exhalation.

Figure 35.1 Stethoscope positions for auscultation of heart sounds.

Figure 35.2 The use of a sphygmomanometer to measure blood pressure.

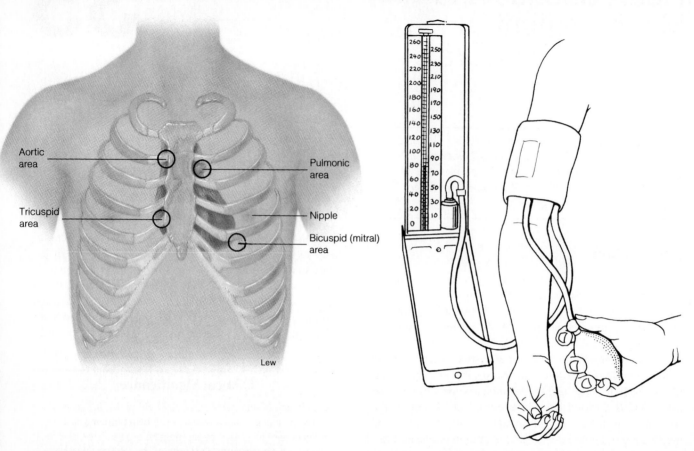

B. Measurement of Blood Pressure

The arterial blood pressure is routinely measured indirectly with a *sphygmomanometer*. This device consists of an inflatable rubber bag connected by rubber hoses to a bulbous hand pump and to a pressure gauge (manometer) graduated in millimeters of mercury. The rubber bag is wrapped within a cuff around the upper arm and inflated to a pressure greater than the suspected systolic pressure, thus occluding the brachial artery. The examiner auscultates the brachial artery by placing the bell of a stethoscope in the cubital fossa; the pressure in the rubber bag is allowed to fall gradually by opening a screw valve next to the hand pump (fig. 35.2).

When the subject is at rest, the blood flows through the arteries in a *laminar flow*—that is, the fluid in the central axial stream moves faster than the material in the peripheral layers, but there is very little transverse flow or mixing between these axial layers. Under these conditions, the artery is silent when auscultated.

When the sphygmomanometer cuff is inflated to a pressure above the systolic blood pressure, the flow is stopped and the artery is again silent. As the pressure in the cuff gradually drops below the systolic (but above the diastolic) pressure, the blood pushes through the compressed artery walls in a *turbulent flow*. Under these conditions, ed-

dies flow at right angles to the axial stream, layers of blood are mixed, and the turbulence causes vibrations in the artery that can be heard with the stethoscope. These sounds are known as the **sounds of Korotkoff** (named after the man who first described them).

The cuff pressure at which the first sound is heard is the **systolic pressure.** The cuff pressure at which the sound becomes muffled (or the pressure at which the sound disappears) is the **diastolic pressure.** Although the pressure at which the sounds disappear is closer to the true diastolic pressure than the pressure at which they muffle, the beginning of muffling is easier to detect, and the results are more reproducible. It is often recommended that both measurements of diastolic pressure be recorded; for example, 120/81/76. Frequently, however, this blood pressure is simply recorded as 120/76.

Clinical Significance

A normal blood pressure measurement for an individual depends on age, gender, heredity, and environment. Considering these factors, chronically elevated blood pressure measurements may indicate an unhealthy state called **hypertension,** a major contributing factor in heart disease and stroke.

Hypertension may be divided into two general categories. *Primary hypertension* (95% of all cases) refers to hypertension

Table 35.1 Normal Arterial Blood Pressure at Different Ages

Age	Systolic Males	Systolic Females	Diastolic Males	Diastolic Females	Age	Systolic Males	Systolic Females	Diastolic Males	Diastolic Females
1 day	70				16 years	118	116	73	72
3 days	72				17 years	121	116	74	72
9 days	73				18 years	120	116	74	72
3 weeks	77				19 years	122	115	75	71
3 months	86				20–24 years	123	116	76	72
6–12 months	89	93	60	62	25–29 years	125	117	78	74
1 year	96	95	66	65	30–34 years	126	120	79	75
2 years	99	92	64	60	35–39 years	127	124	80	78
3 years	100	100	67	64	40–44 years	129	127	81	80
4 years	99	99	65	66	45–49 years	130	131	82	82
5 years	92	92	62	62	50–54 years	135	137	83	84
6 years	94	94	64	64	55–59 years	138	139	84	84
7 years	97	97	65	66	60–64 years	142	144	85	85
8 years	100	100	67	68	65–69 years	143	154	83	85
9 years	101	101	68	69	70–74 years	145	159	82	85
10 years	103	103	69	70	75–79 years	146	158	81	84
11 years	104	104	70	71	80–84 years	145	157	82	83
12 years	106	106	71	72	85–89 years	145	154	79	82
13 years	108	108	72	73	90–94 years	145	150	78	79
14 years	110	110	73	74	95–106 years	145	149	78	81
15 years	112	112	75	76					

Source: *Documenta Geigy Scientific Tables*, 7th ed., edited by K. Diem and C. Lentner. Copyright © 1970 CIBA-GEIGY AG, Basle, Switzerland.

of unknown etiology. There are also two kinds of primary hypertension: benign hypertension (also known as *essential hypertension*) and malignant hypertension. *Secondary hypertension* refers to hypertension for which the causative pathology is known.

Procedure

1. The subject should sit and rest the right or left arm on a table at the level of the heart. Wrap the cuff of the sphygmomanometer around the arm about 2.5 centimeters above the elbow.
2. Palpate the brachial artery in the cubital fossa, and place the stethoscope where the pulse is felt. Close the screw valve, and pump up the cuff pressure about 20 millimeters of mercury (mm Hg) above the point where sounds disappear, or about 20 mm Hg above the point where the radial pulse can no longer be felt.
3. Open the screw valve to allow the pressure in the cuff to fall at a rate of about 2 or 3 mmHg per second.
4. Record the systolic pressure and the two measurements of diastolic pressure (beginning of muffling and disappearance of sounds). Enter these values in your laboratory report, and compare your pressure with the range of normal values listed in table 35.1.
5. Calculate the subject's pulse pressure (systolic minus diastolic pressure). Enter this value in your laboratory report.
6. Calculate the subject's mean arterial pressure. This is equal to the diastolic pressure plus one-third of the pulse pressure. Enter this in the laboratory report.
7. Repeat all these measurements when the same subject is reclining (arms at sides); and a few minutes later, when the subject is standing (arms down).

Laboratory Report 35

Name _____

Date _____

Section _____

Heart Sounds and Measurement of Blood Pressure

Read the assigned sections in the textbook before completing the laboratory report.

A. Heart Sounds

1. Defects of the heart valves can be detected by auscultation but not by electrocardiography. Explain why.

2. What is meant by the "splitting" of the heart sounds? What causes the heart sounds to be split?

3. Why do the first and second heart sounds correlate with the QRS and T waves, respectively?

B. Measurement of Blood Pressure

1. Enter your data in the table below:

	Sitting	Reclining	Standing
Systolic pressure			
Diastolic pressure			
Pulse pressure			
Mean arterial pressure			

2. When blood pressure measurements are taken, the first sound of Korotkoff occurs when the cuff pressure equals the _____ , and the last sound occurs when the cuff pressure equals the _____.

3. Suppose a person's blood pressure is 165/110.
   a. What is his or her systolic pressure? _____
   b. What is his or her diastolic pressure? _____
   c. What is his or her pulse pressure? _____
   d. What is his or her mean arterial pressure? _____

4. What condition does the person in question 3 have? Explain the dangers of this condition.

5. What causes the sounds of Korotkoff? Why can't they normally be heard in the brachial artery before the cuff is inflated?

6. What effect does arm position have on the measurement of blood pressure? Explain.

Cardiovascular System and Physical Fitness

Before coming to class, review the following sections in chapter 22 of the textbook: "Aerobic Requirements of the Heart," "Regulation of Coronary Blood Flow," "Regulation of Blood Flow through Skeletal Muscles," and "Changes in Cardiac Output during Exercise."

Introduction

Physical fitness is dependent on adaptations of the cardiovascular system that include increased stroke volume, decreased resting cardiac rate, and changed cardiovascular responses to exercise. In addition, endurance training increases the ability of the trained muscles to extract oxygen from the blood, producing a higher aerobic capacity. The physically fit thus require a slower rate of increase of the cardiac rate with exercise, and they have a faster return to the resting cardiac rate after exercise. Physical fitness, therefore, involves not only muscular development but also the ability of the cardiovascular system to adapt to changes in demand.

Clinical Significance

Exercise tests have proved extremely useful in the diagnosis of heart disease, particularly *myocardial ischemia* (inadequate blood flow to the heart). People who, at rest, seem normal and have normal electrocardiograms (ECGs) may, with exercise, develop *angina pectoris* and abnormal ECGs. In these tests, a standard exercise procedure is performed (e.g., the Harvard one-step, the Master two-step, the treadmill, or the bicycle ergometer) for the length of time determined to yield a fraction (such as 90%) of the maximum cardiac rate for the patient's age. At the end of the test, irregularities in the ECG are noted, such as depressed or elevated S-T segments, which indicate myocardial ischemia.

Objectives

Students completing this exercise will be able to:

1. Describe the relationship between age and the maximum cardiac rate.
2. Describe the cardiovascular changes that occur during exercise.
3. Compare the effects of exercise in people who are physically fit and in those not physically fit.
4. Explain how controlled exercise can be used to detect heart disease.

Materials

1. Sphygmomanometer
2. Stethoscope
3. Chair, stair, or platform 18 inches high

Procedure

1. Measure your reclining pulse by placing your fingertips (not the thumb) on the radial artery in the ventrolateral region of the wrist.[3] Count the number of pulses in 30 seconds, and multiply by 2. Score points as indicated.

Reclining Pulse

Rate	Points
50–60	3
61–70	3
71–80	2
81–90	1
91–100	0
101–110	−1

Score: _____

[3] The procedure for this exercise is from E. C. Schneider, "A Cardiovascular Rating as a Measure of Physical Fatigue and Efficiency." *JAMA* 74 (1920): 1507. Copyright 1920, American Medical Association.

2. After measuring the reclining pulse, stand and measure the pulse rate immediately upon standing.

Standing Pulse Rate

Rate	Points
60–70	3
71–80	3
81–90	2
91–100	1
101–110	1
111–120	0
121–130	0
131–140	−1

Score: _____

3. Subtract the pulse rate of step 1 from the pulse rate of step 2 to get the pulse rate increase on standing.

Pulse Rate Increase on Standing

Reclining Pulse	0–10 Beats	11–18 Beats	19–26 Beats	27–34 Beats	35–43 Beats
50–60	3	3	2	1	0
61–70	3	2	1	0	−1
71–80	3	2	0	−1	−2
81–90	2	1	−1	−2	−3
91–100	1	0	−2	−3	−3
101–110	0	−1	−3	−3	−3

Score: _____

4. Place your right foot on a chair or stair 18 inches high. Raise your body so that your left foot comes to rest by your right foot. Return your left foot to the original position. Repeat this exercise five times, allowing 3 seconds for each step up. Immediately upon completion of this exercise, measure the pulse for 15 seconds and multiply by 4. Record this pulse rate.

Pulse: _____

Measure the pulse for 15 seconds at 30, 60, 90, and 120 seconds after completion of the exercise, and multiply each value by 4. Record the time that it takes for the pulse to return to normal standing level (step 2). Score points as indicated.

Return of Pulse to Standing Normal after Exercise

Seconds	Points
0–30	4
31–60	3
61–90	2
91–120	1
After 120	0

Score: _____

5. Subtract your normal standing pulse rate (step 2) from your pulse rate immediately after exercise (step 4).

Pulse Rate Increase Immediately after Exercise

Standing Pulse	0–10 Beats	11–20 Beats	21–30 Beats	31–40 Beats	41+ Beats
60–70	3	3	2	1	0
71–80	3	2	1	0	−1
81–90	3	2	1	−1	−2
91–100	2	1	0	−2	−3
101–110	1	0	−1	−3	−3
111–120	1	−1	−2	−3	−3
121–130	0	−2	−3	−3	−3
131–140	0	−3	−3	−3	−3

Score: _____

6. Calculate the change in systolic blood pressure as you go from a reclining to a standing position.

Change in Systolic Pressure from Reclining to Standing

Change (mm Hg)	Points
Rise of 8 or more	3
Rise of 2–7	2
No rise	1
Fall of 2–5	0
Fall of 6 or more	−1

Score: _____

7. Determine your total score for all the tests and evaluate this score on the following basis. Enter your score and rating in the laboratory report.

Excellent	18–17
Good	16–14
Fair	13–8
Poor	7 or less

Laboratory Report 36

Name _____

Date _____

Section _____

Cardiovascular System and Physical Fitness

Read the assigned sections in the textbook before completing the laboratory report.

1. Write your total score in the space below, and indicate if this score is excellent, good, fair, or poor according to the rating scale in the procedure.

   Total Score: _____

   Rating: _____

2. Does a person's aerobic capacity change with age? Explain.

3. In a physically fit person, what factors improve the delivery of oxygen to muscles?

4. How does the increase in blood pressure and pulse rate after exercise compare in people who are and who are not physically fit?

5. What are the consequences of inadequate delivery of oxygen to the heart? How does exercise help in the diagnosis of heart disease?

Respiratory System

Before coming to class, review the following sections in chapter 24 of the textbook: "Conducting Division" and "Alveoli, Lungs, and Pleurae."

Introduction

Breathing, or *ventilation*, moves air into and out of the respiratory system. The respiratory system is frequently divided into the *conducting zone* and the *respiratory zone*. The conducting zone includes all of the cavities and structures that transport gases to and from alveoli. The alveoli constitute the respiratory zone where gas exchange between the air and blood occurs. This gas exchange is termed *external respiration*, and it is distinguished from the gas exchange of *internal respiration* that occurs between the blood and tissues.

Objectives

Students completing this exercise will be able to:

1. Identify the structural components of the respiratory system.
2. Explain the functions of each structure and cavity of the respiratory system.
3. Discuss the dual role the laryngeal region plays in digestion and respiration.
4. Identify the anatomical features of the larynx that are associated with sound production and respiration.
5. Describe the surface anatomy of the lungs in relation to the thoracic cavity.
6. Describe the role of alveoli in the diffusion of respiratory gases.

Materials

1. Models and charts of the respiratory system
2. Fresh pluck (connected thoracic organs) of a sheep
3. Embalmed cat
4. Dissecting tray and dissecting instruments
5. Reference text
6. Colored pencils

A. Structures of the Conducting Zone

Identify and label the following regions and structures in figures 37.1 and 37.2:

> Conducting passages
>> Bronchial tree
>>> Right and left primary bronchus
>>>> Carina
>>> Secondary bronchi
>>> Segmental bronchi
>>> Terminal bronchioles
>> Larynx
>>> Arytenoid cartilages (not shown)
>>> Cricoid cartilage
>>> Epiglottis
>>> Glottis
>>> Thyroid cartilage
>>> Laryngeal prominence (Adam's apple)
>>> Vocal cords (true and false)
>> Nasal cavity
>>> Meatus
>>> Nostril
>>> Nasal conchae (turbinate bones):
>>>> (a) inferior, (b) middle, (c) superior
>>> Paranasal sinuses
>>> Frontal sinus
>>> Sphenoidal sinus

Figure 37.1 A sagittal section of the head showing the structures of the upper respiratory tract.

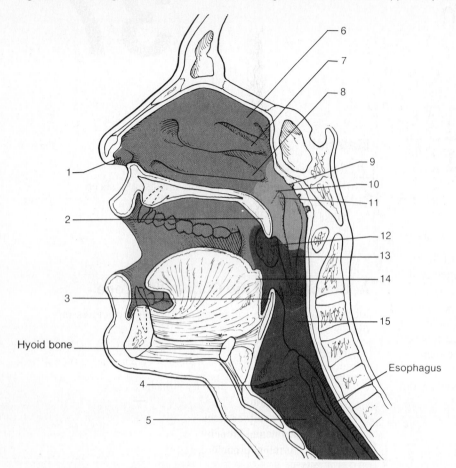

Hyoid bone

Esophagus

1. _____
2. _____
3. _____
4. _____
5. _____
6. _____
7. _____
8. _____
9. _____
10. _____
11. _____
12. _____
13. _____
14. _____
15. _____

Figure 37.2 The larynx, trachea, and bronchial tree.

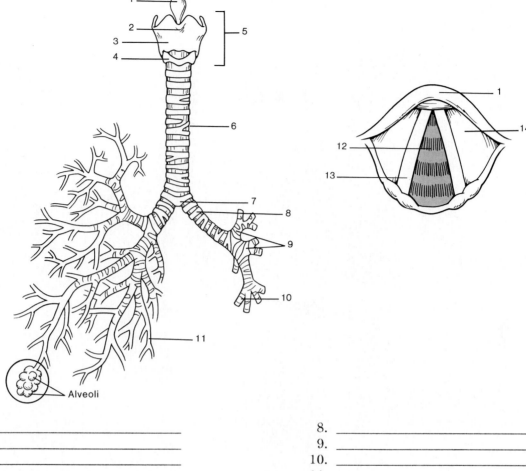

1. _____
2. _____
3. _____
4. _____
5. _____
6. _____
7. _____

8. _____
9. _____
10. _____
11. _____
12. _____
13. _____
14. _____

Pharynx
    Choana
    Laryngopharynx
    Nasopharynx
    Pharyngeal tonsils
    Auditory (eustachian) tube
    Oropharynx
    Hard palate
    Soft palate
    Uvula
    Palatine tonsil
    Lingual tonsil
Trachea

B. Alveoli, Lungs, and Pleura

1. Identify and label the following structures in figures 37.3 and 37.4:

Lungs
    Left lung
        Cardiac notch
        Inferior lobe
        Superior lobe
    Right lung
        Inferior lobe
        Middle lobe
        Superior lobe

Figure 37.3 An anterior view of the lower respiratory system.

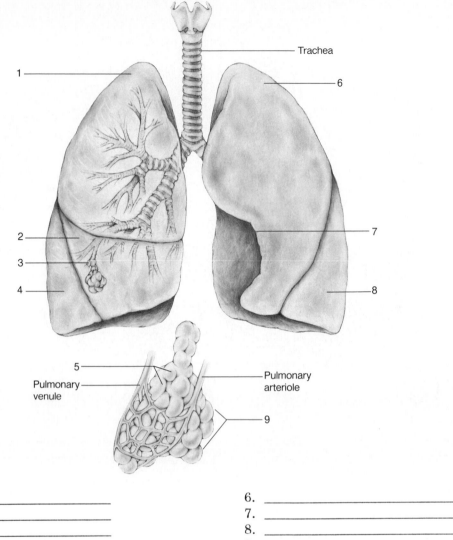

Trachea

Pulmonary venule

Pulmonary arteriole

1. _____
2. _____
3. _____
4. _____
5. _____

6. _____
7. _____
8. _____
9. _____

Pleurae
    Mediastinum: (a) anterior, (b) posterior
    Parietal pleura
    Pleural cavity
    Visceral pleura
  Respiratory sacs
    Alveolar duct
    Alveolar sac
    Alveoli
2. Using colored pencils, color the visceral pleura and the parietal pleura in different colors.

C. Examination of a Sheep Pluck

Examine fresh sections of trachea and lung tissue, and iden-tify as many structures as possible. Pinch a portion of lung tissue and note its texture. Cut off a small portion of lung tissue and place it in a beaker of water. Observe that the alveolar air will cause the lung tissue to float even when cut into small pieces.

Figure 37.4 A cross section of the thoracic cavity showing the mediastinum and pleural membranes.

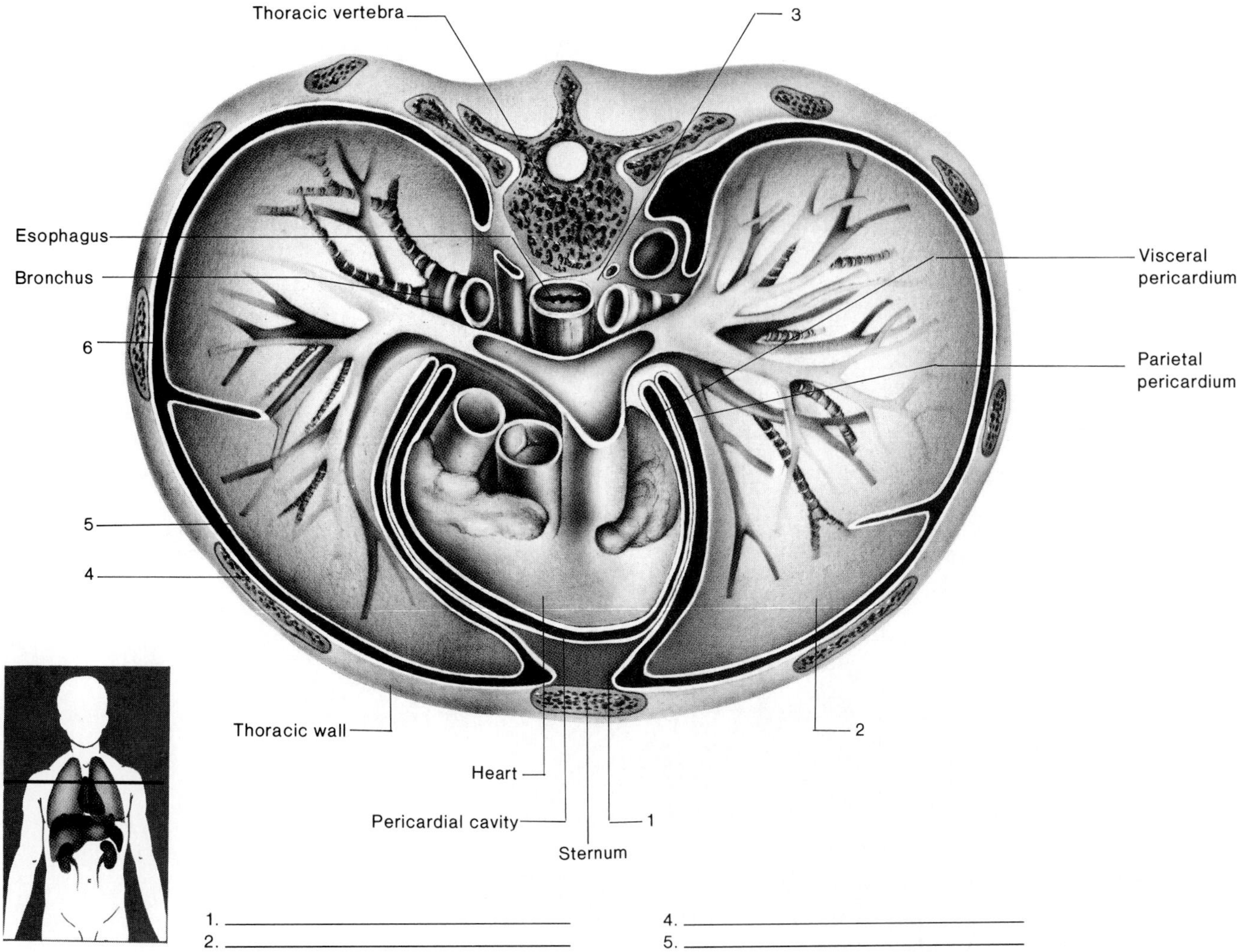

Thoracic vertebra — 3

Esophagus

Bronchus

6

5

4

Thoracic wall

Heart

Pericardial cavity — 1

Sternum

Visceral pericardium

Parietal pericardium

2

1. _____
2. _____
3. _____
4. _____
5. _____
6. _____

D. Respiratory System of a Cat

Exposing the respiratory system of the cat will require some dissection. The amount of dissection will depend on what systems on the specimen have been previously studied and to what extent you desire to have the respiratory system exposed. Many structures of the upper respiratory tract have already been studied. Review the skeletal structure of the *nasal cavity* at this time to prepare for the exercise on the cat respiratory system.

If a sagittal section of the cat head is desired, use a mechanical bone saw or an electric autopsy saw. The head *must* be held securely by one person while another does the cutting, but the saw must *never* be in a position that could injure the hand of the person holding the specimen. Once the sagittal cut is made, the head should be washed of loose debris. The thoracic cavity should already have been opened when the circulatory system was studied. Refer to figure 37.1 for a sagittal view of a mammalian head.

Expose the *oral* (*buccal*) *cavity* (mouth) and *pharynx* (throat) by making a dorsolateral cut from the right corner of the mouth toward the angle of the jaw. Using bone shears, cut through the mandible at the angle of the jaw. Then use a scalpel to cut caudally through the wall of the pharynx to a joint just cranial to the thyroid gland (fig. 37.5). Free the jaw on the left side by cutting through the body of the mandible, and then reflecting the cut position caudally. Examine the dissected cat and identify as many structures of the respiratory system as possible by referring to figure 37.5.

Figure 37.5 The respiratory system of the cat (a ventrolateral view of the head, oral cavity, and pharynx after reflection of the mandible).

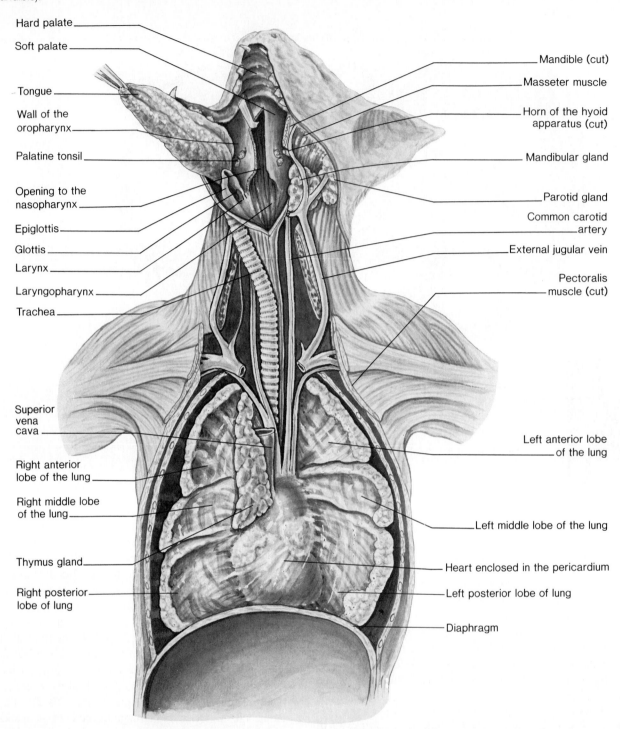

Hard palate

Soft palate

Tongue

Wall of the oropharynx

Palatine tonsil

Opening to the nasopharynx

Epiglottis

Glottis

Larynx

Laryngopharynx

Trachea

Superior vena cava

Right anterior lobe of the lung

Right middle lobe of the lung

Thymus gland

Right posterior lobe of lung

Mandible (cut)

Masseter muscle

Horn of the hyoid apparatus (cut)

Mandibular gland

Parotid gland

Common carotid artery

External jugular vein

Pectoralis muscle (cut)

Left anterior lobe of the lung

Left middle lobe of the lung

Heart enclosed in the pericardium

Left posterior lobe of lung

Diaphragm

Laboratory Report 37

Name _____

Date _____

Section _____

Respiratory System

Read the assigned sections in the textbook before completing the laboratory report.

1. What is the primary function of the respiratory system, and what is the basic functional unit?

2. What are the advantages of breathing through the nose rather than the mouth?

3. Explain the close relationship between the respiratory system and the circulatory system.

4. Differentiate between the various regions of the pharynx, both in structure and function.

5. Why is it advantageous for the larynx, trachea, and bronchi to be cartilaginous?

6. Diagram the structure of the larynx, and describe its location. What is meant by food or fluid "going down the wrong pipe"?

7. What are the advantages of having two separate pleural cavities lined with visceral and parietal pleural membranes?

8. Describe the surface features of the lungs and the position of the lungs within the thoracic cavity.

Measurements of Pulmonary Function

Before coming to class, review the following sections in chapter 24 of the textbook: "Physical Aspects of Ventilation," "Inspiration and Expiration," "Pulmonary Function Tests," and "Common Respiratory Disorders."

Introduction

Spirometry is a technique used to measure lung volumes and capacities and to measure ventilation as a function of time. A Collins respirometer can be used for this purpose (fig. 38.1). As the subject exhales into a mouthpiece, the oxygen bell rises, which causes a pen to move downward on a moving chart (kymograph). Since this is a closed system, soda lime is provided to remove CO_2 from exhaled air. As the subject inhales, the oxygen bell moves downward, causing the pen to move upward on the moving chart. The *y* axis of the chart is graduated in milliliters (ml), and the *x* axis is graduated in millimeters (mm). Since the speed at which the chart moves is known, a graph of the volume (in ml) of air moved into and out of the lungs in a given time interval can be obtained.

Alternatively, a computerized setup using a Phipps and Bird spirometer and a program for analyzing the data (Spirocomp) may be used (fig. 38.2). In this case, the treated data are displayed on a computer screen.

Objectives

Students completing this exercise will be able to:

1. Identify the muscles involved in inspiration and expiration, and explain the mechanics of breathing.
2. Define the different lung volumes and capacities, and determine the values of these measurements in a spirogram.
3. Describe and perform the forced expiratory volume and maximum breathing capacity tests, and determine these measurements in a spirogram.

4. Explain how pulmonary function tests are used in the diagnosis of restrictive and obstructive pulmonary disorders.

Materials

1. Collins, Inc. 9-liter *respirometer*
2. Disposable mouthpieces and nose clamp
3. Alternative equipment: *Spirocomp* (Intelitool, Inc.) with Apple or IBM computer and wet 6L spirometer (Phipps and Bird).
4. Reference textbook

A. Measurement of Simple Lung Volumes and Capacities

Spirometry enables you to easily visualize, define, and measure many important aspects of pulmonary function.

The **total lung capacity (TLC)** is the amount of gas in the lungs after a maximum (forced) inhalation.

The **vital capacity (VC)** is the maximum amount of gas that can be exhaled after a maximum inhalation.

The **tidal volume (TV)** is the volume of gas inspired or expired during each normal (unforced) ventilation cycle.

The **inspiratory capacity (IC)** is the maximum amount of gas that can be inhaled after a normal (unforced) exhalation.

The **inspiratory reserve volume (IRV)** is the maximum amount of gas that can be forcefully inhaled after a normal inhalation.

The **expiratory reserve volume (ERV)** is the maximum amount of gas that can be forcefully exhaled after a normal exhalation.

The **functional residual capacity (FRC)** is the amount of gas left in the lungs after a normal (unforced) exhalation.

The **residual volume (RV)** is the amount of gas left in the lungs after a maximum (forced) exhalation.

Figure 38.1 The Collins 9-liter respirometer.

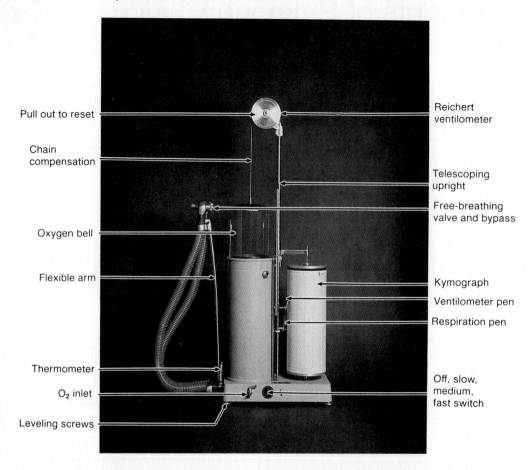

Pull out to reset

Chain compensation

Oxygen bell

Flexible arm

Thermometer

O_2 inlet

Leveling screws

Reichert ventilometer

Telescoping upright

Free-breathing valve and bypass

Kymograph

Ventilometer pen

Respiration pen

Off, slow, medium, fast switch

Procedure (for the Spirocomp)

1. Press the "T" key on the computer and the words "Breathe Normal Cycles" will appear on the computer screen. After three normal tidal volume cycles, the data will appear on the screen.
2. Press the "E" key on the computer and the words "Breathe Normal Cycles" will appear on the screen. At the third breathing cycle the words "Stop After Normal Exhale" will appear.
3. After the subject's breathing pauses, the words "Exhale Forcefully" will appear on the screen. Exhale all you can at this point.
4. Press the "V" key and the words "Inhale Max Then Press V Exhale Fully" will appear on the screen. After inhaling all you can, press "V" and exhale all the air you can as fast as you can.
5. Record the data displayed on the screen and use it to complete your laboratory report.

Procedure (for the Collins Respirometer)

1. Raise and lower the oxygen bell (fig. 38.1) several times to get fresh air into the spirometer. Notice that as the bell moves up and down, one of the pens moves a corresponding distance down and up. Adjust the height of the oxygen bell to position this pen so that it will begin writing in the middle of the chart paper. This pen (the ventilometer pen) usually has black ink; the other (respiration) pen usually has red ink and will not be used for this exercise. (It can be rocked away from the paper.)
2. With the free-breathing valve set to the *open* position, place the mouthpiece in the oral cavity (as in breathing through a snorkel), and go through several ventilation cycles to become accustomed to the apparatus. (When the free-breathing valve is open, you will breathe room air.) If a disposable cardboard mouthpiece is

Figure 38.2 Equipment involved in using the Spirocomp program.

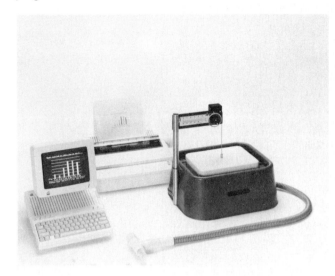

used, be particularly careful to prevent air leakage from the corners of the mouth. Breathing through the nostrils can be prevented by a nose clamp or by pinching the nose tightly with the thumb and forefinger.

3. Turn the respirometer to the *slow* position (32 mm/minute [min]), and close the free-breathing valve so that the oxygen bell and the pen go up and down with each ventilation cycle.

4. Breathe in a normal, relaxed manner for 1–2 minutes. The breaths should appear relatively uniform, and the slope should go upward (see fig. 38.3). A downward slope indicates that there is an air leak; in this event, tighten the grip of the mouth on the mouthpiece and the nose clamp on the nose, and begin again.

☞ **Note:** *At this speed (32 mm/min), the distance between vertical lines on the chart is traversed in 1 minute.*

This procedure measures tidal volume—the amount of air inhaled or exhaled in each resting ventilation cycle.

5. When the tidal volume procedure has been completed, perform a test for vital capacity—the maximum amount of air that can be exhaled after a maximum inhalation. At the end of a normal exhalation, inhale as much as possible, then exhale to the fullest possible extent. At the completion of this exercise, the chart should resemble the one shown in figure 38.3.

6. Remove the chart from the kymograph drum. Notice that the chart is marked horizontally in milliliters.

☞ **Note:** *Since the temperature and pressure of the respirometer are different from those of the body, the volume that the air occupies in the respirometer is different from the actual volume it occupies in the lung—it is subject to changes in ambient (room) conditions. To standardize the volumes measured in spirometry, we multiply these measured volumes by a correction factor known as the BTPS factor (body temperature, atmospheric pressure, saturated with water vapor). Since the BTPS factor is very close to 1.1 at normal room temperatures, we will use this figure in the calculations.*

Calculations

Obtain the measured tidal volume from the chart by subtracting the milliliters corresponding to the trough from the milliliters corresponding to the peak of a typical resting ventilation cycle.

Example (from fig. 38.4)

Step 1
$$\begin{array}{r} 3700 \text{ ml (inhalation)} \\ -3250 \text{ ml (exhalation)} \\ \hline 450 \text{ ml} \end{array}$$

Step 2
$$\begin{array}{r} 450 \text{ ml (measured tidal volume)} \\ \times\, 1.1 \quad \text{(BTPS factor)} \\ \hline 495 \text{ ml} \end{array}$$

Procedure

1. Enter the corrected, measured tidal volume (TV) in the *Measured* column of the table in your laboratory report.

2. Obtain the measured inspiratory capacity from the chart. To do this, subtract the milliliters corresponding to the last normal exhalation before performing the vital capacity maneuver from the milliliters corresponding to the maximum inhalation peak.

Example (from fig. 38.5)

Step 1
$$\begin{array}{r} 6650 \text{ ml} \ \ \text{(maximum inhalation)} \\ -3650 \text{ ml} \ \ \text{(normal exhalation)} \\ \hline 3000 \text{ ml} \end{array}$$

Step 2
$$\begin{array}{r} 3000 \text{ ml (measured inspiratory capacity)} \\ \times\, 1.1 \ \ \text{(BTPS factor)} \\ \hline 3300 \text{ ml} \end{array}$$

Enter the corrected, measured inspiratory capacity (IC) in the *Measured* column of the table in your laboratory report.

Figure 38.3 A spirometry chart of tidal volume, inspiratory capacity, expiratory reserve volume, and vital capacity.

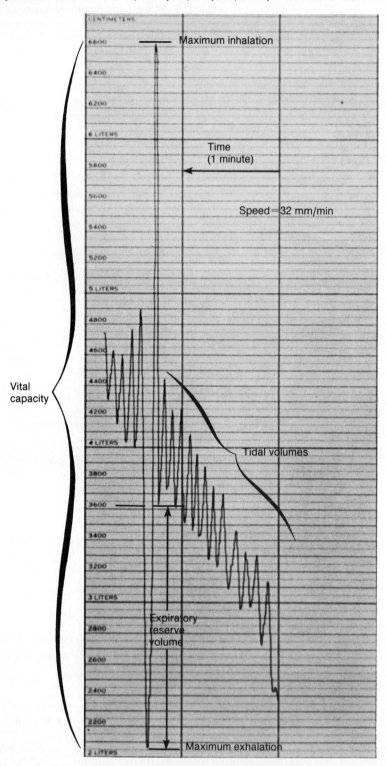

Figure 38.4   A close-up of tidal volume measurements on a spirometry chart.

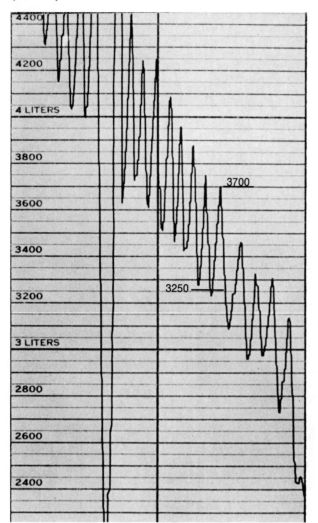

Figure 38.5   A close-up of the inspiratory capacity measurement on a spirometry chart.

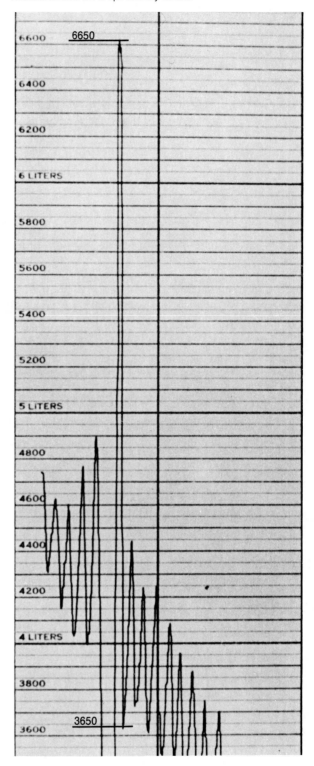

3. Obtain the measured expiratory reserve volume by subtracting the milliliters corresponding to the trough for maximum exhalation from the milliliters corresponding to the last normal exhalation before the vital capacity maneuver (the same value used for step 2).

Example (from fig. 38.6)

Step 1
$$\begin{array}{r} 3650 \text{ ml (normal exhalation)} \\ -2050 \text{ ml (maximum exhalation)} \\ \hline 1600 \text{ ml} \end{array}$$

Step 2
$$\begin{array}{r} 1600 \text{ ml (measured expir. reserve volume)} \\ \times\ 1.1 \text{ (BTPS factor)} \\ \hline 1760 \text{ ml} \end{array}$$

Enter the corrected expiratory reserve volume (ERV) in the *Measured* column of the table in your laboratory report.

4. Obtain the measured vital capacity by either (1) adding the corrected inspiratory capacity (from step 2) and the corrected expiratory reserve volume (step 3) (since these values have already been BTPS-standardized, an additional correction step is unnecessary); or (2) subtracting the milliliters corresponding to maximum exhalation from the milliliters corresponding to maximum inhalation (this value must then be multiplied by the BTPS factor).

Example

Method 1
$$\begin{array}{r} 3300 \text{ ml (corrected inspiratory capacity)} \\ +\ 1760 \text{ ml (corrected expir. reserve volume)} \\ \hline 5060 \text{ ml (corrected vital capacity)} \end{array}$$

Method 2
$$\begin{array}{r} 6650 \text{ ml (maximum inhalation)} \\ -2050 \text{ ml (maximum exhalation)} \\ \hline 4600 \text{ ml} \end{array}$$

$$\begin{array}{r} 4600 \text{ ml (measured vital capacity)} \\ \times\ 1.1 \text{ (BTPS factor)} \\ \hline 5060 \text{ ml (corrected capacity)} \end{array}$$

Enter the corrected vital capacity (VC) in the *Measured* column of the table in your laboratory report.

5. Obtain the predicted vital capacity for the subject's sex, age, and height from tables 38.1 and 38.2. The height in centimeters can be most conveniently obtained by multiplying height in inches by 2.54 cm/in.

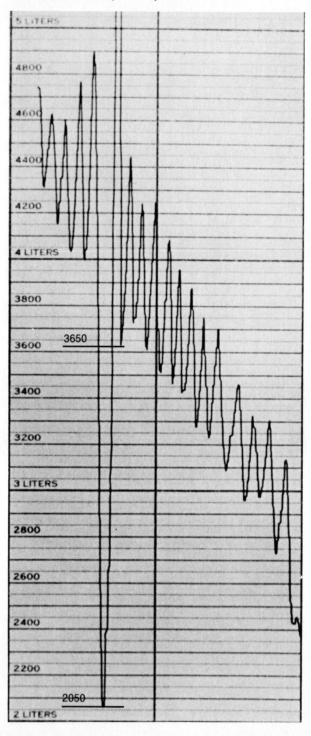

Figure 38.6 A close-up of the expiratory reserve volume measurement on a spirometry chart.

Table 38.1 Predicted Vital Capacities, Females (Milliliters)

Height in Centimeters

Age	146	148	150	152	154	156	158	160	162	164	166	168	170	172	174	176	178	180	182	184	186	188	190	192	194
16	2950	2990	3030	3070	3110	3150	3190	3230	3270	3310	3350	3390	3430	3470	3510	3550	3590	3630	3670	3715	3755	3800	3840	3880	3920
17	2935	2975	3015	3055	3095	3135	3175	3215	3255	3295	3335	3375	3415	3455	3495	3535	3575	3615	3655	3695	3740	3780	3820	3860	3900
18	2920	2960	3000	3040	3080	3120	3160	3200	3240	3280	3320	3360	3400	3440	3480	3520	3560	3600	3640	3680	3720	3760	3800	3840	3880
20	2890	2930	2970	3010	3050	3090	3130	3170	3210	3250	3290	3330	3370	3410	3450	3490	3525	3565	3605	3645	3695	3720	3760	3800	3840
22	2860	2900	2940	2980	3020	3060	3095	3135	3175	3215	3255	3290	3330	3370	3410	3450	3490	3530	3570	3610	3650	3685	3725	3765	3800
24	2830	2870	2910	2950	2985	3025	3065	3100	3140	3180	3220	3260	3300	3335	3375	3415	3455	3490	3530	3570	3610	3650	3685	3725	3765
26	2800	2840	2880	2920	2960	3000	3035	3070	3110	3150	3190	3230	3265	3300	3340	3380	3420	3455	3490	3530	3570	3610	3650	3685	3725
28	2775	2810	2850	2890	2930	2965	3000	3040	3070	3115	3155	3190	3230	3270	3305	3345	3380	3420	3460	3495	3535	3570	3610	3650	3685
30	2745	2780	2820	2860	2895	2935	2970	3010	3045	3085	3120	3160	3195	3235	3270	3310	3345	3385	3420	3460	3495	3535	3570	3610	3645
32	2715	2750	2790	2825	2865	2900	2940	2975	3015	3050	3090	3125	3160	3200	3235	3275	3310	3350	3385	3425	3460	3495	3535	3570	3610
34	2685	2725	2760	2795	2835	2870	2910	2945	2980	3020	3055	3090	3130	3165	3200	3240	3275	3310	3350	3385	3425	3460	3495	3535	3570
36	2655	2695	2730	2765	2805	2840	2875	2910	2950	2985	3020	3060	3095	3130	3165	3205	3240	3275	3310	3350	3385	3420	3460	3495	3530
38	2630	2665	2700	2735	2770	2810	2845	2880	2915	2950	2990	3025	3060	3095	3130	3170	3205	3240	3275	3310	3350	3385	3420	3455	3490
40	2600	2635	2670	2705	2740	2775	2810	2850	2885	2920	2955	2990	3025	3060	3095	3135	3170	3205	3240	3275	3310	3345	3380	3420	3455
42	2570	2605	2640	2675	2710	2745	2780	2815	2850	2885	2920	2955	2990	3025	3060	3100	3135	3170	3205	3240	3275	3310	3345	3380	3415
44	2540	2575	2610	2645	2680	2715	2750	2785	2820	2855	2890	2925	2960	2995	3030	3060	3095	3130	3165	3200	3235	3270	3305	3340	3375
46	2510	2545	2580	2615	2650	2685	2715	2750	2785	2820	2855	2890	2925	2960	2995	3030	3060	3095	3130	3165	3200	3235	3270	3305	3340
48	2480	2515	2550	2585	2620	2650	2685	2715	2750	2785	2820	2855	2890	2925	2960	2995	3030	3060	3095	3130	3160	3195	3230	3265	3300
50	2455	2485	2520	2555	2590	2625	2655	2690	2720	2755	2785	2820	2855	2890	2925	2955	2990	3025	3060	3090	3125	3155	3190	3225	3260
52	2425	2455	2490	2525	2555	2590	2625	2655	2690	2720	2755	2790	2820	2855	2890	2925	2955	2990	3020	3055	3090	3125	3155	3190	3220
54	2395	2425	2460	2495	2530	2560	2590	2625	2655	2690	2720	2755	2790	2820	2855	2885	2920	2950	2985	3020	3050	3085	3115	3150	3180
56	2365	2400	2430	2460	2495	2525	2560	2590	2625	2655	2690	2720	2755	2790	2820	2855	2885	2920	2950	2980	3015	3045	3080	3110	3145
58	2335	2370	2400	2430	2460	2495	2525	2560	2590	2625	2655	2690	2720	2750	2785	2815	2850	2880	2920	2945	2975	3010	3040	3075	3105
60	2305	2340	2370	2400	2430	2460	2495	2525	2560	2590	2625	2655	2685	2720	2750	2780	2810	2845	2875	2915	2940	2970	3000	3035	3065
62	2280	2310	2340	2370	2405	2435	2465	2495	2525	2560	2590	2620	2655	2685	2715	2745	2775	2810	2840	2870	2900	2935	2965	2995	3025
64	2250	2280	2310	2340	2370	2400	2430	2465	2495	2525	2555	2585	2620	2650	2680	2710	2740	2770	2805	2835	2865	2895	2925	2955	2990
66	2220	2250	2280	2310	2340	2370	2400	2430	2460	2495	2525	2555	2585	2615	2645	2675	2705	2735	2765	2800	2825	2860	2890	2920	2950
68	2190	2220	2250	2280	2310	2340	2370	2400	2430	2460	2490	2520	2550	2580	2610	2640	2670	2700	2730	2760	2795	2820	2850	2880	2910
70	2160	2190	2220	2250	2280	2310	2340	2370	2400	2425	2455	2485	2515	2545	2575	2605	2635	2665	2695	2725	2755	2785	2810	2840	2870
72	2130	2160	2190	2220	2250	2280	2310	2335	2365	2395	2425	2455	2480	2510	2540	2570	2600	2630	2660	2685	2715	2745	2775	2805	2830
74	2100	2130	2160	2190	2220	2245	2275	2305	2335	2360	2390	2420	2450	2475	2505	2535	2565	2590	2620	2650	2680	2710	2740	2765	2795

Source: Warren E. Collins, Inc., Braintree, MA.

Table 38.2 Predicted Vital Capacities, Males (Milliliters)

Height in Centimeters

Age	146	148	150	152	154	156	158	160	162	164	166	168	170	172	174	176	178	180	182	184	186	188	190	192	194
16	3765	3820	3870	3920	3975	4025	4075	4130	4180	4230	4285	4335	4385	4440	4490	4540	4590	4645	4695	4745	4800	4850	4900	4955	5005
18	3740	3790	3840	3890	3940	3995	4045	4095	4145	4200	4250	4300	4350	4405	4455	4505	4555	4610	4660	4710	4760	4815	4865	4915	4965
20	3710	3760	3810	3860	3910	3960	4015	4065	4115	4165	4215	4265	4320	4370	4420	4470	4520	4570	4625	4675	4725	4775	4825	4875	4930
22	3680	3730	3780	3830	3880	3930	3980	4030	4080	4135	4185	4235	4285	4335	4385	4435	4485	4535	4585	4635	4685	4735	4790	4840	4890
24	3635	3685	3735	3785	3835	3885	3935	3985	4035	4085	4135	4185	4235	4285	4330	4380	4430	4480	4530	4580	4630	4680	4730	4780	4830
26	3605	3655	3705	3755	3805	3855	3905	3955	4000	4050	4100	4150	4200	4250	4300	4350	4395	4445	4495	4545	4595	4645	4695	4740	4790
28	3575	3625	3675	3725	3775	3820	3870	3920	3970	4020	4070	4115	4165	4215	4265	4310	4360	4410	4460	4510	4555	4605	4655	4705	4755
30	3550	3595	3645	3695	3740	3790	3840	3890	3935	3985	4035	4080	4130	4180	4230	4275	4325	4375	4425	4470	4520	4570	4615	4665	4715
32	3520	3565	3615	3665	3710	3760	3810	3855	3905	3950	4000	4050	4095	4145	4195	4240	4290	4340	4385	4435	4485	4530	4580	4625	4675
34	3475	3525	3570	3620	3665	3715	3760	3810	3855	3905	3950	4000	4045	4095	4140	4190	4225	4285	4330	4380	4425	4475	4520	4570	4615
36	3445	3495	3540	3585	3635	3680	3730	3775	3825	3870	3920	3965	4010	4060	4105	4155	4200	4250	4295	4340	4390	4435	4485	4530	4580
38	3415	3465	3510	3555	3605	3650	3695	3745	3790	3840	3885	3930	3980	4025	4070	4120	4165	4210	4260	4305	4350	4400	4445	4495	4540
40	3385	3435	3480	3525	3575	3620	3665	3710	3760	3805	3850	3900	3945	3990	4035	4085	4130	4175	4220	4270	4315	4360	4410	4455	4500
42	3360	3405	3450	3495	3540	3590	3635	3680	3725	3770	3820	3865	3910	3955	4000	4050	4095	4140	4185	4230	4280	4325	4370	4415	4460
44	3315	3360	3405	3450	3495	3540	3585	3630	3675	3725	3770	3815	3860	3905	3950	3995	4040	4085	4130	4175	4220	4270	4315	4360	4405
46	3285	3330	3375	3420	3465	3510	3555	3600	3645	3690	3735	3780	3825	3870	3915	3960	4005	4050	4095	4140	4185	4230	4275	4320	4365
48	3255	3300	3345	3390	3435	3480	3525	3570	3615	3655	3700	3745	3790	3835	3880	3925	3970	4015	4060	4105	4150	4190	4235	4280	4325
50	3210	3255	3300	3345	3390	3430	3475	3520	3565	3610	3650	3695	3740	3785	3830	3870	3915	3960	4005	4050	4090	4135	4180	4225	4270
52	3185	3225	3270	3315	3355	3400	3445	3490	3530	3575	3620	3660	3705	3750	3795	3835	3880	3925	3970	4010	4055	4100	4140	4185	4230
54	3155	3195	3240	3285	3325	3370	3415	3455	3500	3540	3585	3630	3670	3715	3760	3800	3845	3890	3930	3975	4020	4060	4105	4145	4190
56	3125	3165	3210	3255	3295	3340	3380	3425	3465	3510	3550	3595	3640	3680	3725	3765	3810	3850	3895	3940	3980	4025	4065	4110	4150
58	3080	3125	3165	3210	3250	3290	3335	3375	3420	3460	3500	3545	3585	3630	3670	3715	3755	3800	3840	3880	3925	3965	4010	4050	4095
60	3050	3095	3135	3175	3220	3260	3300	3345	3385	3430	3470	3500	3555	3595	3635	3680	3720	3760	3805	3845	3885	3930	3970	4015	4055
62	3020	3060	3110	3150	3190	3230	3270	3310	3350	3390	3440	3480	3520	3560	3600	3640	3680	3720	3770	3810	3850	3890	3930	3970	4020
64	2990	3030	3080	3120	3160	3200	3240	3280	3320	3360	3400	3440	3490	3530	3570	3600	3650	3690	3730	3770	3810	3850	3900	3930	3980
66	2950	2990	3030	3070	3110	3150	3190	3230	3270	3310	3350	3390	3430	3470	3510	3550	3600	3640	3680	3720	3760	3800	3840	3880	3920
68	2920	2960	3000	3040	3080	3120	3160	3200	3240	3280	3320	3360	3400	3440	3480	3520	3560	3600	3640	3680	3720	3760	3800	3840	3880
70	2880	2930	2970	3010	3050	3090	3130	3170	3210	3250	3290	3330	3370	3410	3450	3480	3520	3560	3600	3640	3680	3720	3760	3800	3840
72	2860	2900	2940	2980	3020	3060	3100	3140	3180	3210	3250	3290	3330	3370	3410	3450	3490	3530	3570	3610	3650	3680	3720	3760	3800
74	2820	2860	2900	2930	2970	3010	3050	3090	3130	3170	3200	3240	3280	3320	3360	3400	3440	3470	3510	3550	3590	3630	3670	3710	3740

Source: Warren E. Collins, Inc., Braintree, MA.

Table 38.3 Factors for obtaining the predicted residual volume and total lung capacity

Age	Residual Volume: Vital Capacity × Factor	Total Lung Capacity: Vital Capacity × Factor
16–34	0.250	1.250
35–49	0.305	1.305
50–69	0.445	1.445

Example

Sex: male
Age: 34
Height: 174 cm
Predicted vital capacity: 4140 ml

Enter the subject's predicted vital capacity from the tables of normal values in the *Predicted* column in your laboratory report.

6. To obtain an estimate of the predicted residual volume and the predicted total lung capacity, refer to table 38.3.

☞ **Note:** *These values cannot be measured by spirometry because residual volume cannot be exhaled; total lung capacity equals the vital capacity plus residual volume.*

7. Obtain the percent predicted value of all the measurements in the following way:
Percent predicted =

$$\frac{\text{corrected measured value}}{\text{predicted value}} \times 100\%$$

Example

Measured vital capacity = 5060 ml
(BTPS corrected)
Predicted vital capacity = 4140 ml
(from table 38.1 or 38.2)

$$\% \text{ predicted} = \frac{5060 \text{ ml}}{4140 \text{ ml}} \times 100\%$$

$$= 122\%$$

Enter the percent predicted values in the appropriate places in the table in your laboratory report.

Measurements of vital capacity that are repeatedly and consistently less than 80% of the predicted value suggest the presence of restrictive lung disease, such as emphysema.

B. Measurement of Forced Expiratory Volume

The ability to ventilate the lungs in a given time interval is often of greater diagnostic value than measurements of simple lung volumes and capacities. One of the measurements that involves time intervals is that of the *forced expiratory volume (FEV)*.

In the forced expiratory volume test, the subject performs a vital capacity maneuver (inhales maximally and then exhales maximally), while the operator of the respirometer sets the kymograph speed at its fastest setting (1920 mm/min). This fast speed stretches out the exhalation tracing, because the distance between vertical lines is now traversed in 1 second. From the recording, the percent of the total vital capacity that is exhaled in the first second ($FEV_{1.0}$), second second ($FEV_{2.0}$), and third second ($FEV_{3.0}$) can be determined. A sample record of the forced expiratory volume is shown in figure 38.7.

Clinical Significance

A measured vital capacity less than the normal range is diagnostic of *restrictive lung disorders,* such as pulmonary fibrosis, pulmonary edema, and emphysema. A person who has asthma or bronchitis may have a normal vital capacity, but not the ability to exhale this volume at a normal rate. Such *obstructive disorders* are detected by an abnormally low forced expiratory volume ($FEV_{1.0}$) test. Emphysema is a lung disease that is usually both restrictive and obstructive.

Procedure

1. The subject breathes normally into the respirometer, with the kymograph set to 32 mm/min.
2. After a normal (unforced) exhalation, the operator instructs the subject to take a deep, forceful inhalation. The subject is instructed to hold this inhalation momentarily.
3. The operator sets the kymograph at the fast speed (1920 mm/min) and instructs the subject to exhale as rapidly and forcefully as possible.

☞ **Note:** *At a speed of 1920 mm/min, the distance between two vertical lines on the chart is equal to 1 second.*

Calculations: Forced Expiratory Volume ($FEV_{1.0}$)

1. Measure the vital capacity from the chart by subtracting the exhalation trough (which is flat, because no more air could be expelled) from the inhalation peak (which is also flat, because the subject's breath is held). You do not have to multiply this value by the BTPS factor.

Figure 38.7 A recording of the forced expiratory volume (FEV), or timed vital capacity, test.

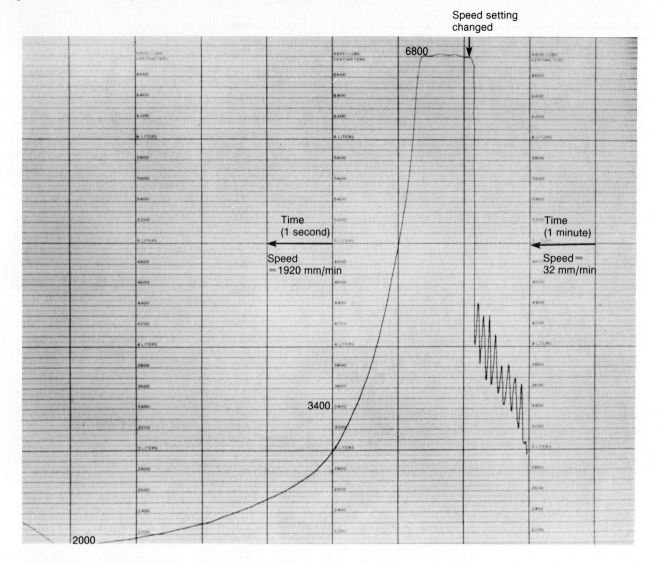

Example (from fig. 38.7)

> 6800 ml (maximum inhalation)
> −2000 ml (maximum exhalation)
>
> 4800 ml (vital capacity)

Enter the uncorrected vital capacity in the following space:

_____ ml

2. Measure the amount of air exhaled in the first second by subtracting the milliliters corresponding to the exhalation line after 1 second from the milliliters of the inhalation peak. Remember that the distance between vertical lines is passed in 1 second. If the subject does not begin to exhale exactly on a vertical line, use a ruler to measure 3.2 cm from the beginning of exhalation. (This is the distance between vertical lines and is equivalent to 1 second at this chart speed.)

Example (from fig. 38.7)

> 6800 ml (maximum inhalation)
> −3400 ml (exhalation line after 1 second)
>
> 3400 ml (amount exhaled in first second)

Enter the amount exhaled in the first second in the following space:

_____ ml

3. Calculate the percent of the vital capacity exhaled in the first second (the $FEV_{1.0}$).

Example

$$FEV_{1.0} = \frac{3400 \text{ ml (from step 2)}}{4800 \text{ ml (from step 1)}} \times 100\%$$

$$= 70.8\%$$

Enter the $FEV_{1.0}$ in the laboratory report. Refer to table 38.4, and enter the predicted percentage for the $FEV_{1.0}$ in the laboratory report.

Table 38.4 Percent of Vital Capacity Exhaled during First Second

Age	Percent of VC
18–29	82–80
30–39	78–77
40–44	75.5
45–49	74.5
50–54	73.5
55–64	72–70

Source: Edward Gaensler and George W. Wright, *Archives of Environmental Health,* Vol. 12:146, 1966.

Laboratory Report 38

Name _____

Date _____

Section _____

Measurements of Pulmonary Function

Read the assigned sections in the textbook before completing the laboratory report.

A. Measurement of Simple Lung Volumes and Capacities

1. Enter your (BTPS-corrected) data under the *Measured* column, and enter your calculated *Percent Predicted,* in the table below:

	Measured	Predicted	Percent Predicted
TV		500 ml (avg. normal)	
IC		2800 ml (avg. normal)	
ERV		1200 ml (avg. normal)	
VC		(from tables)	
RV	not measured	(VC × factor)	cannot calculate
TLC	not measured	(VC × factor)	cannot calculate

B. Measurement of Forced Expiratory Volume

1. Enter your measured and predicted $FEV_{1.0}$ in the spaces below:

   Measured = _____ %   Predicted range = _____ %

2. Compare your values to the normal ranges given in table 38.4, and enter your conclusions in the space below:

3. Identify the following lung volumes and capacities:
   a. Maximum amount of air that can be expired after a maximum expiration _____
   b. Maximum amount of air that can be inspired after a normal inspiration _____
   c. Maximum amount of air that can be inspired after a normal expiration _____
   d. The amount of air left in the lungs after a maximum expiration _____

4. Calculate the following values for the spirogram shown below (correct values to BTPS):

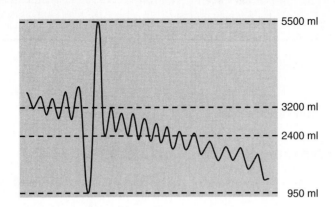

a. Tidal volume _____

b. Inspiratory capacity _____

c. Expiratory reserve volume _____

d. Vital capacity _____

5. Calculate the $FEV_{1.0}$ value for the spirogram shown below:

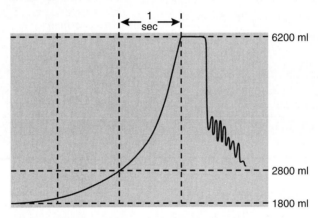

$FEV_{1.0}$ _____ %

6. Pulmonary disorders in which the alveoli are normal but resistance to airflow is abnormally high are categorized as _____ disorders.

7. An example of one of the above disorders is _____ .

8. One pulmonary test for the above disorder is the _____ .

9. Does your chest expand because your lungs inflate, or do your lungs inflate because your chest expands? Explain.

10. Distinguish between obstructive and restrictive pulmonary diseases. How does spirometry aid in their diagnosis?

Effects of Exercise on the Respiratory System

Before coming to class, review the following sections in chapter 24 of the textbook: "Regulation of Breathing" and "Ventilation during Exercise."

Introduction

The **total minute volume,** which is the tidal volume multiplied by the number of breaths per minute, increases during exercise. It is currently believed that the increase in total minute volume that occurs during exercise is due in part to an increase in CO_2 production, although concentrations of arterial CO_2 during exercise are not usually increased. Sensory feedback from the exercising muscles may also contribute to increased breathing during exercise.

Oxygen consumption and the total minute volume remain elevated immediately after exercise. This extra oxygen consumption (over resting levels) following exercise is called the **oxygen debt** and is used to oxidize lactic acid produced by anaerobic respiration of the muscles and to support an increased metabolism of the warmed muscles.

Objectives

Students completing this exercise will be able to:

1. Define the term *total minute volume,* and explain how it is obtained.
2. Describe how the rate of oxygen consumption is measured, and explain how this measurement is related to the metabolic rate.
3. Describe the relationship between the total minute volume and the rate of oxygen consumption, and explain how and why these measurements change during exercise.
4. Explain why oxygen consumption and total minute volume remain high for a time after exercise has ceased.

5. Describe the pH scale, and define the terms *acid* and *base.*
6. Explain how carbonic acid and bicarbonate are formed in the blood, and explain their functions.
7. Define the terms *acidosis* and *alkalosis,* and explain how they relate to *hypoventilation* and *hyperventilation.*
8. Explain how ventilation is adjusted to help maintain acid-base balance.

Materials

1. Spirometer (Collins 9-liter respirometer)
2. Disposable mouthpieces
3. Droppers, beakers, and straws
4. 0.10N NaOH, phenolphthalein solution (saturated)

A. Total Minute Volume and Oxygen Consumption

The increased rate of metabolism during exercise is usually matched by an increased total minute volume. If the match is perfect, the arterial carbon dioxide level (P_{CO_2}) does not change. Under these conditions, the high total minute volume is termed **hyperpnea.** If the total minute volume is excessively high for a given metabolic rate, the arterial P_{CO_2} decreases below normal; this indicates **hyperventilation.** An inadequate total minute volume will increase the arterial P_{CO_2} above normal, indicating **hypoventilation.**

The body's cells consume oxygen during aerobic respiration; thus the amount of air trapped within the oxygen bell of the respirometer decreases as the subject breathes through the mouthpiece. (The exhaled carbon dioxide is eliminated from the air in the respirometer by soda lime.) Oxygen consumption results in an upward slope of the tidal volume measurements, and the amount of oxygen consumed can be determined by the difference in milliliters (ml) before and after one minute of resting ventilation (fig. 39.1).

Figure 39.1 A spirogram showing tidal volume and vital capacity measurements. Note the rising slope of the tidal volume measurements, which indicates oxygen consumption.

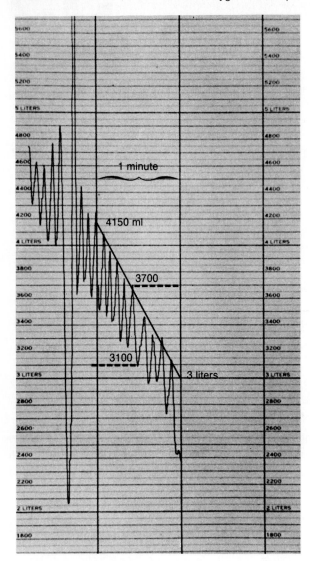

☞ **Note:** *The tidal volume measurements must have an upward slope, indicating that oxygen is being consumed within the air trapped in the oxygen bell. If the slope is downward, air is leaking into the system, usually through the corners of the mouth or the nose. In this event, reposition the mouthpiece, check the nose clamp, and begin the measurements again.*

3. Remove the chart from the kymograph drum, and determine the frequency of ventilation (number of breaths per minute) and the BTPS-corrected tidal volume.

Example (from fig. 39.1)

3700 ml	(inhalation peak of tidal volume)
−3100 ml	(exhalation trough of tidal volume)
600 ml	(uncorrected tidal volume)

600 ml	(BTPS factor)
× 1.1	(corrected tidal volume)
660 ml	

Frequency = 10 breaths/min (from fig. 39.1).

Enter the subject's corrected tidal volume in the following space:

_____ ml

Enter the subject's frequency of ventilation in the following space:

_____ breaths/min

> The average frequency is 14 breaths per minute.

4. Determine the subject's *total minute volume* at rest by multiplying the frequency of ventilation by the tidal volume. Enter this value in the data table in your laboratory report.

Example (from fig. 39.1)

660 ml/breath × 10 breaths/min
= 6600 ml/min total minute volume

> The average total minute volume is 6600 ml/min.

5. Use a straightedge to draw a line that averages either the peaks or the troughs of the tidal volume measurements. Determine the *oxygen consumption* (per minute) by subtracting the milliliters where this line intersects two vertical chart lines at the beginning and at the end of 1 minute.

Procedure

1. Set the Collins respirometer to a speed of 32 millimeters per minute (mm/min). (At this speed the distance between two vertical lines is traversed in 1 minute.) Position the mouthpiece securely in the mouth, purse the lips tightly against it, and clamp the nostrils closed.

2. Under resting conditions, breathe normally into the respirometer for 1 minute (that is, perform the procedure for measuring tidal volume—see exercise 38).

Example (from fig. 39.1)

Using a line that averages the peaks:

4150 ml (at end of 1 min)
−3000 ml (at beginning of 1 min)
1150 ml (oxygen consumption)

Enter the resting oxygen consumption in the data table in your laboratory report.

6. The subject should now perform light exercise, such as five to ten jumping jacks. Then repeat the respirometer measurements and data calculation described in steps 1–5.

☞ **Note:** *If breathing into the respirometer becomes difficult after exercise, the subject should stop the procedure. Results obtained in less than a minute can then be extrapolated to 1 minute. Alternatively, the bell can be filled with 100% oxygen to prevent the possible occurrence of hypoxia.*

Enter the corrected tidal volume after exercise in the following space:

_____ ml

Enter the frequency of ventilation after exercise in the following space:

_____ breaths/min

Enter the total minute volume and the oxygen consumption per minute after exercise in the data table.

7. Calculate the *percent increase* after exercise for total minute volume and oxygen consumption. This is the difference between the exercise and resting measurements, divided by the resting measurement, and multiplied by 100%. Enter these values in the data table in your laboratory report.

B. Effect of Exercise on the Rate of CO_2 Production

Normally the rate of ventilation is matched to the rate of CO_2 production by the tissues, so that the carbonic acid, bicarbonate, and H^+ concentrations in the blood remain in the normal range. If hypoventilation occurs, however, the carbonic acid levels will rise above normal and the pH will fall below 7.35. This condition is called **respiratory acidosis.** Hyperventilation, conversely, causes an abnormal decrease in carbonic acid and a corresponding rise in blood pH. This condition is called **respiratory alkalosis.** Respiratory acidosis or alkalosis thus occurs when the blood CO_2 level (measured as the partial pressure, P_{CO_2}, in mm of mercury [Hg]) is different from the normal value (40 mm Hg) as a result of abnormal breathing patterns.

Clinical Significance

Hypoventilation results in the retention of carbon dioxide and thus in the excessive accumulation of carbonic acid; this produces a fall in blood pH called respiratory acidosis. Hyperventilation results in the excessive elimination of CO_2 and thus in low carbonic acid and high pH. This differs from the normal hyperpnea (increased total minute volume) that occurs during exercise, where increased respiration matches increased CO_2 production and where the arterial CO_2 levels and pH remain in the normal range.

Procedure

1. Fill a beaker with 200 ml of distilled water, and add 5.0 ml of $0.10N$ NaOH and a few drops of phenolphthalein indicator. This indicator is pink in alkaline solutions and clear in neutral or acidic solutions. Divide this solution into two beakers.
2. While sitting quietly, exhale through a glass tube or straw into the solution in the first beaker. Note the time it takes to turn the solution from pink to clear, and record this time in your laboratory report.
3. Exercise vigorously for 2 to 5 minutes by running up and down stairs or by doing jumping jacks. Exhale through a glass tube or straw into the second beaker, and again note the time it takes to make the pink solution clear.

Effects of Exercise on the Respiratory System

Read the assigned sections in the textbook before completing the laboratory report.

A. Total Minute Volume and Oxygen Consumption

1. Enter the total minute volume and rate of oxygen consumption during rest and exercise in the table below.
2. Calculate and enter the percent increases in your measurements after exercise, and enter these values in the table below.

Measurement	Resting	Exercise	% Increase
Total minute volume			
Oxygen consumption			

3. Why did the rate of oxygen consumption increase as a result of exercise?

4. Why did the total minute volume increase as a result of exercise? (Explain the physiological mechanisms that may be involved.)

5. High total minute volume during exercise is called *hyperpnea.* How does this differ from hyperventilation?

6. Define the term *oxygen debt*. Explain why hyperpnea continues for a time after exercise has stopped.

B. Effect of Exercise on the Rate of CO₂ Production

1. Enter your data in the spaces below:

    Time for color to change at rest _____

    Time for color to change after exercise _____

2. Explain your results in the space below:

3. Define the following terms:

    a. Acid _____

    b. Base _____

    c. Acidosis _____

    d. Alkalosis _____

4. A solution with a H^+ concentration of 10^{-9} molar has a pH of _____ ; its OH^- concentration is

    _____ .

5. Hypoventilation produces respiratory _____ , whereas hyperventilation produces respiratory

    _____ .

6. Draw equations to show how hypoventilation affects the blood concentration of carbon dioxide, carbonic acid, H^+, and bicarbonate.

7. Intravenous infusions of sodium bicarbonate are often given to acidotic patients to correct the acidosis and to relieve the strain of rapid breathing. Write a chemical equation, and describe the reason bicarbonate is helpful in this situation.

Urinary System

Before coming to class, review the following sections in chapter 25 of the textbook: "Gross Structure of the Kidneys," "Microscopic Structure of the Kidneys," and "Ureters, Urinary Bladder, and Urethra."

Introduction

Urine formation begins when a filtrate of plasma enters the nephron tubules of the kidneys. The composition and volume of this filtrate is then altered as water and many solutes from the filtrate are returned to the blood and as some additional waste products in the blood are secreted into the filtrate. The final product, urine, is channeled out of the kidneys in the ureters and stored in the urinary bladder. Through their formation of urine from blood, the kidneys are significantly involved in the maintenance of the homeostasis of blood volume and composition.

Objectives

Students completing this exercise will be able to:

1. Identify the gross anatomical structures of the urinary system.
2. Identify the gross features of a kidney in a coronal section.
3. Identify many of the microscopic structures of the urinary system.
4. Discuss the process of urine formation and the functional role of the nephron.

Materials

1. Fresh or preserved kidneys (sheep or pig)
2. Preserved cats
3. Dissecting trays and instruments
4. Models and charts of the urinary system
5. Reference text
6. Atlas
7. Colored pencils

A. Gross Anatomy of the Urinary System

1. Identify and label the following structures in figure 40.1:

   Abdominal aorta
   Inferior vena cava
   Kidney
   Ureter
   Urethra
   Urinary bladder

B. Gross Structure of the Kidney

1. Identify and label the following structures in figure 40.2:

   Calyx (major and minor)
   Capsule
   Hilum
   Renal column
   Renal cortex
   Renal medulla
   Renal papilla
   Renal pelvis
   Renal pyramid
   Ureter

2. Using colored pencils, color-code the renal cortex in one color, the renal pyramids in another color, and the renal calyces, renal pelvis, and ureter in a third color.

C. Microscopic Structure of the Kidney

1. Identify and label the following structures in figures 40.3 and 40.4:

   Urinary tubules
       Collecting duct
       Distal convoluted tubule
       Glomerular (Bowman's) capsule
       Nephron loop (loop of Henle)
       Proximal convoluted tubule
   Vascular tubules
       Afferent arterioles
       Arcuate arteries
       Efferent arteriole
       Glomerulus
       Interlobar arteries
       Interlobular arteries
       Peritubular capillaries

Figure 40.1 The urinary system.

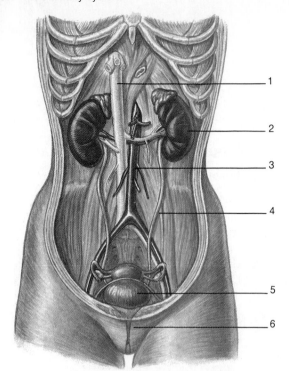

1. _____
2. _____
3. _____
4. _____
5. _____
6. _____

Figure 40.2 The internal structures of a kidney.

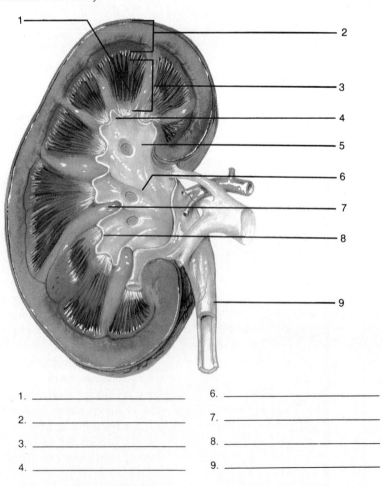

1. _____
2. _____
3. _____
4. _____
5. _____

6. _____
7. _____
8. _____
9. _____

Figure 40.3 A simplified illustration of blood flow from a glomerulus to an efferent arteriole, the peritubular capillaries, and the venous drainage of the kidneys.

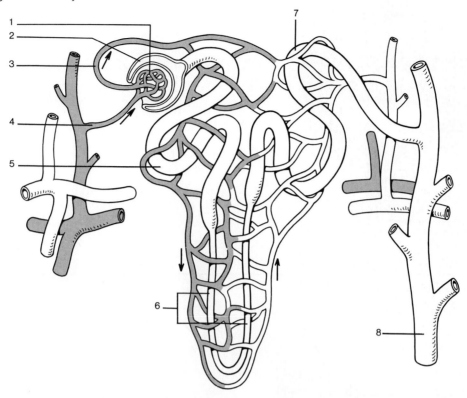

1. _____
2. _____
3. _____
4. _____

5. _____
6. _____
7. _____
8. _____

Figure 40.4 The vascular blood supply to the kidney.

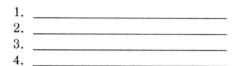

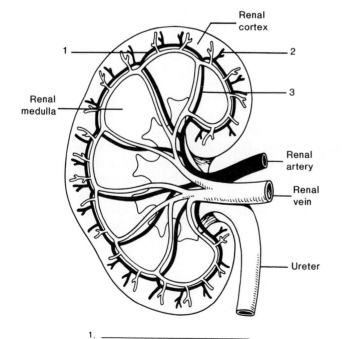

1. _____
2. _____
3. _____

2. Using colored pencils, color-code the renal cortex, renal medulla, and renal pelvis of figure 40.4 with three different colors.

D. Dissection of the Kidney

Obtain a fresh or preserved sheep or pig kidney. If it is a fresh kidney, it may be still embedded in protective adipose tissue. Carefully remove this fat, being careful not to damage the adrenal gland, which caps the superior border. Examine the concave hilum, and identify the renal vessels and the ureter. The shiny covering of the kidney is the *capsule*. Examine it by inserting a dissecting needle underneath it and gently lifting it from the *renal cortex*.

Make a longitudinal (coronal) cut through the kidney to produce dorsal and ventral halves. Rinse the kidney in tap water, and compare the structures of the kidney with the structures depicted in figure 40.2.

Figure 40.5   An anterior view of the urinary bladder, ureters, and urethra.

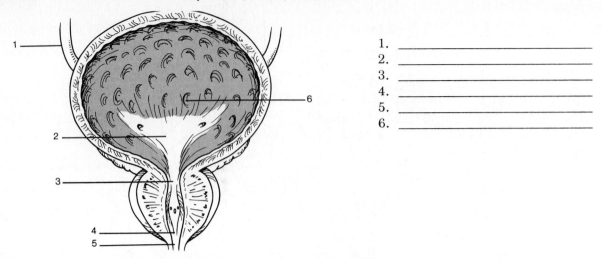

1. _____
2. _____
3. _____
4. _____
5. _____
6. _____

Figure 40.6   A coronal view of the kidney of a cat.

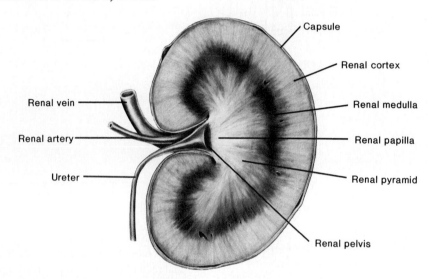

E. Ureters, Urinary Bladder, and Urethra

Identify and label the following structures in figure 40.5:

Ureter
Urethra
    External urethral sphincter
    Internal urethral sphincter
    Urethral orifice
Urinary bladder
    Rugae of urinary bladder
    Trigone

F. Urinary System of the Cat

Dissection of the cat urinary system is a simple procedure, and it may be conducted in conjunction with examination of the reproductive organs.

The urinary system consists of two *kidneys,* two *ureters,* the *urinary bladder,* and the *urethra.* The urinary system of the cat and that of the human are basically the same, but there are two differences. The left kidney in the human is superior to (or higher than) the right; in the cat, the right kidney is anterior to the left. In addition, six or more renal pyramids are found in each human kidney, while the cat usually has only one (fig. 40.6). It is not uncommon for a cat to have two renal veins draining the right kidney.

Laboratory Report 40

Name _____

Date _____

Section _____

Urinary System

Read the assigned sections in the textbook before completing the laboratory report.

1. Diagram and label a longitudinal view of a kidney.

2. Describe briefly the position of the kidneys.

3. Which microscopic structures are located within the renal cortex of the kidney, and which in the renal medulla?
   a. Renal cortex _____
   b. Renal medulla _____

4. Trace a drop of blood from an interlobular artery through a glomerulus and to an interlobular vein. List in order the vessels through which it would pass.

5. Trace the pathway of urine from the site of filtration at the glomerular capsule, through the nephron, and then outside of the body. List in order all the structures through which the urine would pass.

Renal Regulation of Fluid and Electrolyte Balance

Before coming to class, review the following sections in chapter 25 of the textbook: "Reabsorption of Salt and Water" and "Renal Control of Electrolyte Balance."

Introduction

Homeostasis is aided by the kidney's adjustments of urine volume and composition. Fluid balance is maintained by regulation of water reabsorption, and electrolyte balance is maintained by the filtration, reabsorption, and secretion of Na^+, K^+, Cl^-, HCO_3^-, and other ions. The kidneys are stimulated to reabsorb water by antidiuretic hormone (ADH), which is secreted by the posterior pituitary gland. Aldosterone, secreted by the adrenal cortex, stimulates the reabsorption of Na^+ and water as well as the secretion of K^+.

Objectives

Students completing this exercise will be able to:

1. Describe the roles of ADH and aldosterone in the regulation of fluid and electrolyte balance.
2. Calculate ion concentration in milliequivalents per liter (mEq/L).
3. Demonstrate and explain how the kidneys respond to water and salt loading by changing urinary volume, specific gravity, pH, and electrolyte composition.

Materials

1. Urine collection cups
2. Urinometers, droppers
3. pH paper (pH range 3–9), potassium chromate (20 grams per deciliter, or g/dl), silver nitrate (2.9 g/dl)
4. NaCl crystals or salt tablets

Milliequivalents

The concentrations of ions in body fluids are usually given in terms of mEq/L. To understand the meaning and significance of this unit of measurement, consider the chloride concentration in the urine:

Suppose that a urine sample had a chloride concentration of 610 mg/100 ml. How does the number of ions and number of charges compare with the number of other ions and charges that are present in the urine? To determine this, we must first convert the chloride concentration from mg/100 ml to millimoles (mM) per liter.

Example

The atomic weight of chloride is 35.5. Therefore,

$$\frac{610 \text{ mg of } Cl^-}{100 \text{ ml}} \times \frac{1g}{1000 \text{ mg}} \times \frac{1000 \text{ ml}}{1 \text{ L}} \times \frac{1 \text{ mole}}{35.5g}$$

$$= 0.171 \text{ M} \times \frac{1000 \text{ mM}}{1 \text{ M}} = 171 \text{ mM}$$

One mole of chloride has the same number of ions as 1 mole of Na^+ or 1 mole of Ca^{++} or 1 mole of SO_4^{-3} or 1 mole of anything else. One mole of Ca^{++}, however, has twice the number of charges (*valence*) as 1 mole of Cl^-. It requires 2 moles of Cl^-, therefore, to neutralize 1 mole of Ca^{++}. If this is taken into account by *multiplying the moles by the valence,* the product is termed the **equivalent weight** of an ion. One-thousandth of the equivalent weight dissolved in 1 liter of solution gives a concentration in mEq/L.

Example

171 mM × 1 (the valence of Cl^-)
= 171 mEq/L of Cl^-

The major advantage in expressing the concentrations of ions in mEq/L is that the total concentration of anions can easily be compared with the total concentration of cations. In an average sample of venous plasma, for example, the total anions and the total cations are each equal to 156 mEq/L. Chloride, the major anion, has a plasma concentration of 103 mEq/L. The chloride concentration in the urine is highly variable, ranging from 61 to 310 mEq/L.

Clinical Significance

Due in large part to the effects of ADH and aldosterone, the kidneys can vary their excretion of water and electrolytes to maintain homeostasis of the blood volume and composition. Abnormally low blood volume can produce *hypotension* (low blood pressure) and may result in *circulatory shock;* abnormally high blood volume contributes to *hypertension.* Renal regulation of Na^+-K^+ balance is also critical for health. Changes in blood Na^+ cause secondary changes in blood volume. Changes in blood K^+ affect the bioelectrical properties of all cells, but the effects on the heart are particularly serious. *Hyperkalemia* (high blood K^+) is usually fatal when the K^+ concentration rises from 4 mEq/L (normal) to over 10 mEq/L. This may be caused by a variety of conditions, including inadequate aldosterone secretion (in Addison's disease) or an excessive intake of potassium.

Procedure

1. Void and take samples of the urine at the beginning of the laboratory session. This first sample is used as the control (time zero).
2. All should drink 500 ml of water; one group should take 4.5 g of NaCl (salt tablets are easiest to take), and the rest should not take NaCl.
3. After drinking the solutions described in step 2, void every 30 minutes for 2 hours.
4. Analyze each of the five urine samples to determine pH, specific gravity, and chloride content in the following manner.
   a. *pH.* Determine the pH of the urine samples by dipping a strip of pH paper into the urine and matching the color that develops with a color chart. The urine normally has a pH between 5.0 and 7.5.
   b. *Specific gravity.* Determine the specific gravity of the urine samples by floating a urinometer in a cylinder nearly filled with the specimen (fig. 41.1). Read the specific gravity directly from the scale, making sure that the urinometer is not touching the bottom or the sides of the cylinder. The specific gravity is directly related to the concentration of solutes in the urine and ranges from 1.010 to 1.025.
   c. *Chloride concentration.* When Na^+ is reabsorbed by the renal tubules, Cl^- follows passively by electrical attraction.

Figure 41.1 Instruments for determining the specific gravity of urine: (*a*) a glass cylinder; (*b*) a urinometer float.

Determine the chloride concentration in the urine samples by the following method.

(1) Measure 10 drops of urine into a test tube (1 drop is approximately 0.05 ml).
(2) Add 1 drop of 20% potassium chromate solution with a second dropper.
(3) Add 2.9% silver nitrate solution, 1 drop at a time, using a third dropper, while shaking the test tube continuously. Count the minimum number of drops needed to change the color of the solution from yellow to brown.
(4) Determine the chloride concentration of the urine sample. Since each drop of 2.9% silver nitrate added in step 3 is equivalent to 61 mg of Cl^- per 100 ml of urine, simply multiply the number of drops added by 61 to obtain the chloride concentration of the urine in mg/100 ml.

Example

If 10 drops of 2.9% silver nitrate were required,
10×61 mg of Cl^-/100 ml = 610 mg Cl^-/100 ml.

5. Convert the chloride concentration to mEq/L, and enter your data in the appropriate table in the laboratory report.

Laboratory Report 41

Name _____

Date _____

Section _____

Renal Regulation of Fluid and Electrolyte Balance

Read the assigned sections in the textbook before completing the laboratory report.

1. Enter your data in the appropriate table below.
   a. Ingestion of water only:

Time	Volume (ml)	pH	Specific Gravity	Chloride (mEq/L)
0				
30				
60				
90				
120				

   b. Ingestion of water and NaCl:

Time	Volume (ml)	pH	Specific Gravity	Chloride (mEq/L)
0				
30				
60				
90				
120				

2. Name the hormone that stimulates the reabsorption of water and thus helps to produce a decrease in blood osmolality. _____

3. Name the hormone that is secreted in response to stimulation by angiotensin II. _____

4. Calcium is normally present in plasma at a concentration of about 0.1 g/L. Calculate the mEq/L of Ca^{++} in the plasma. (The atomic weight of calcium is 40.)

5. Imagine a dehydrated desert prospector and a champagne-quaffing party goer, each of whom drinks a liter of water at time zero and voids urine over a period of 3 hours. Using their urine samples, compare the probable differences in volume and composition. (*Hint*: alcohol inhibits ADH secretion.)

	Prospector	**Party Goer**
Urine Volume		
Specific Gravity		
Na^+ and Cl^- content		

6. Explain your answers to question 5.

7. Many diuretic drugs inhibit Na^+ reabsorption in the nephron loop. Predict these drugs' effects on the urinary excretion of Cl^- and K^+, and explain your answer.

Renal Plasma Clearance of Urea

Before coming to class, review the following sections in chapter 25 of the textbook: "Glomerular Filtration" and "Renal Clearance Rates."

Introduction

Urea and other waste products in the plasma are eliminated in the urine. The kidneys are thus said to *clear* the blood of these compounds. Urea is filtered through the glomeruli, but not all of the filtered urea is excreted in the urine; a variable amount (about 40%) of the filtered urea is reabsorbed. The *renal plasma clearance* for urea (about 75 ml/min) is thus less than the rate of plasma filtration (about 125 ml/min). The latter measurement is known as the *glomerular filtration rate,* or *GFR*. A renal plasma clearance for any substance that is less than the GFR always indicates that the substance is reabsorbed. Conversely, a renal plasma clearance that is higher than the GFR always indicates that the substance is secreted as well as filtered.

Clinical Significance

The plasma concentration of blood urea nitrogen (BUN) reflects both the rate of urea formation from protein and the rate of urea excretion by filtration through the glomeruli of the kidneys. In the absence of abnormal protein metabolism, therefore, a rise in BUN indicates abnormal kidney function, such as nephritis, pyelonephritis, or renal calculi (kidney stones).

Since urea is passively reabsorbed, the renal plasma clearance for urea is substantially less than the GFR. Because of this, the urea clearance is not a particularly good indicator of kidney function. More clinically useful indicators include the plasma clearance of exogenously administered inulin (a large polysaccharide), and plasma clearance of endogenous creatinine (a byproduct of creatine, a molecule found primarily in muscle). Since inulin is neither reabsorbed nor secreted by the nephron, its clearance equals the GFR. Creatinine is secreted to a slight degree by the renal nephron, so its clearance is 20%–25% greater than the true GFR (as defined by the inulin clearance test).

Objectives

Students completing this exercise will be able to:

1. Describe the chemical nature and physiological significance of urea.
2. Define the renal plasma clearance rate, and explain how it is measured.
3. Perform a renal plasma clearance rate measurement for urea, and explain the physiological significance of this measurement.
4. Explain how the renal plasma clearance rate for a solute is affected by filtration, reabsorption, and secretion.

Materials

1. Pipettes with mechanical pipettors (or Repipettes), mechanical microliter pipettes (capacity 20 microliters, or µl), disposable tips
2. Colorimeter, cuvettes
3. BUN reagents and standard (Sclavo Diagnostics, available through Curtin-Matheson Scientific, Inc.)
4. Sterile lancets, 70% alcohol
5. File
6. Microhematocrit centrifuge and heparinized capillary tubes
7. Container for the disposal of blood-containing objects.

Collection of Plasma and Urine Samples

1. Empty the urinary bladder, then drink 500 ml of water as quickly as is comfortable.
2. About 20 minutes after drinking the water, cleanse the fingertip with 70% alcohol and, using a sterile lancet, obtain a drop of blood from the fingertip. Fill a heparinized capillary tube at least halfway with blood, plug one end, and centrifuge in a microhematocrit centrifuge for 3 minutes (as in performing a hematocrit measurement).

☞ **Note:** *Always handle only your own blood and place all objects that have been in contact with blood in the container indicated by the instructor.*

3. Score the capillary tube lightly with a file at the plasma-cell junction, and break the tube at the scored mark. Expel the plasma into a small beaker or test tube.
4. Collect a urine sample 30 minutes after drinking the water. Measure the water (in ml) produced in 30 minutes, divide by 30, and enter the volume per minute of urine produced in the laboratory report.
5. Dilute the urine 1:20 with water. This can be done by adding 19 ml of water to 1 ml of urine, or by adding 1.9 ml of water to 0.10 ml (100 μl) of urine. Mix the diluted urine solution.

Measurement of Urea

1. Label four test tubes or cuvettes B (blank), S (standard), P (plasma), and U (urine).
2. Pipette 2.5 ml of Reagent A into each tube. Then pipette 20 μl of the following into the indicated tubes:

Tube B: 20 μl of distilled water
Tube S: 20 μl of urea standard (28 mg/dl)
Tube P: 20 μl of plasma from the capillary tube
Tube U: 20 μl of the 30-minute urine sample that was previously diluted to 1/20 of its original concentration

☞ **Note:** *Since 20 μl is a very small volume, pipetting must be carefully performed to be accurate. If an automatic microliter pipette is used, depress the plunger several times to wet the disposable tip thoroughly in the solution before withdrawing the 20-μl sample. When delivering the sample, depress the plunger to the first stop with the disposable tip against the wall of the test tube. Then pause several seconds before depressing the plunger to the second stop. Be sure to mix the test tube well so that the sample is washed from the test tube wall.*

3. Mix and incubate the tubes at room temperature for 10 minutes.
4. Pipette 2.5 ml of Reagent B into each tube and incubate for an additional 10 minutes.

5. Set the colorimeter at 600 nanometers (nm), standardize with the reagent blank, and record the absorbance values of the standard, plasma, and urine samples in the data table in your laboratory report.
6. Calculate the urea concentration of the plasma (using Beer's law, described in exercise 4), and enter this value in your laboratory report.
7. Calculate the urea concentration of the diluted urine sample in the same manner as in step 6. Then multiply your answer by the dilution factor of 20, and enter this corrected value of the urea concentration of the urine in your laboratory report.
8. Using your values for the volume of urine produced per minute (V), plasma urea concentration (P), and urine urea concentration (U), calculate your renal plasma clearance rate of urea. Enter this value in your laboratory report.

> The normal range of plasma urea is 5–25 mg/dl.

Example

Suppose V (from step 4) = 2.0 ml/min
P (from step 6) = 10 mg/dl
U (from step 7) = 375 mg/dl

$$\text{Clearance rate} = \frac{\dfrac{375\ \text{mg}}{\text{dl}} \times \dfrac{2.0\ \text{ml}}{\text{min}}}{\dfrac{10\ \text{mg}}{\text{dl}}}$$

$$= 75\ \text{ml/min}$$

> The normal range of urea renal plasma clearance is 64–99 ml/min.

Laboratory Report 42

Name _____

Date _____

Section _____

Renal Plasma Clearance of Urea

Read the assigned sections in the textbook before completing the laboratory report.

1. Enter the volume of urine produced per minute in the space below.

    $V =$ _____ ml/min

2. Enter your absorbance values in the table below:

Tube	Contents	Absorbance
P	Plasma	
U	Diluted urine	
S	Standard (28 mg/dl)	

3. Enter the urea concentration of the:

    plasma (P) _____ mg/dl

    urine (U) _____ mg/dl

4. Calculate your renal plasma clearance for urea:

    Clearance = _____ ml/min

5. Name a molecule in the plasma that is:

    a. filtered but neither reabsorbed nor secreted _____

    b. filtered and partially reabsorbed _____

    c. filtered and completely secreted _____

    d. filtered and only slightly secreted _____

    e. not filtered _____

6. Identify the substances that have the following clearance rates:

    a. The clearance rate is greater than zero but less than the GFR for _____ .

    b. The clearance rate is equal to the GFR for _____ .

    c. The clearance rate is slightly greater than the GFR for _____ .

    d. The clearance rate is equal to the total plasma flow rate to the kidneys for _____ .

7. Define the plasma clearance, and explain how it is measured.

8. Explain the mechanism by which each of the following conditions might produce an increase in the plasma concentration of urea: (a) increased protein catabolism, (b) decreased blood pressure (in circulatory shock), and (c) kidney failure.

Clinical Examination of the Urine

Before coming to class, review the following sections in chapter 25 of the textbook: "Glomerular Filtration," "Symptoms and Diagnosis of Urinary Disorder," and "Infections of Urinary Organs."

Introduction

The presence of abnormally large amounts of proteins and renal casts in the urine indicates glomerular damage. Renal casts are small cylindrical structures formed by the precipitation of protein into the renal tubules. The presence of bacteria and many white blood cells in the sediment indicates urinary tract infection. Abnormal concentrations of glucose, ketone bodies, bilirubin, and other plasma solutes may indicate that these molecules are present in abnormally high concentrations in the plasma, and thus are diagnostic of particular disease states.

A clinical examination of urine may provide evidence of urinary tract infection or kidney disease. Additionally, since urine is derived from plasma, an examination of the urine provides a convenient, nonintrusive means of assessing the composition of plasma and of detecting a variety of systemic diseases. A clinical examination of the urine includes an observation of its appearance, tests of its chemical composition, and a microscopic examination of urine sediment.

Objectives

Students completing this exercise will be able to:

1. Describe the physiological processes responsible for normal urinary concentrations of protein, glucose, ketone bodies, and bilirubin in the urine.
2. Explain the pathological processes that may produce abnormal urinary concentrations of solutes, and note their clinical significance.
3. Describe the normal constituents of urine sediment, and explain how the microscopic examination of urine sediment is clinically useful.

Table **43.1** Abnormal Appearance of Urine	
Color	**Cause**
Yellow-orange to brownish green	Bilirubin from obstructive jaundice
Red to red-brown	Hemoglobinuria
Smokey red	Unhemolyzed RBCs from urinary tract
Dark wine color	Hemolytic jaundice
Brown-black	Melanin pigment from melanoma
Dark Brown	Liver infections, pernicious anemia, malaria
Green	Bacterial infection (*Pseudomonas aeruginosa*)

Materials

1. Microscopes
2. Urine collection cups, test tubes, microscope slides, and cover slips
3. Albustix, Clinitest tablets, Ketostix, Hemastix, Ictotest tablets, or Multistix (all from Ames)
4. Sternheimer-Malbin stain
5. Centrifuge and centrifuge tubes
6. Transfer pipettes (droppers)

A. Appearance of the Urine

Obtain a urine sample in a clear plastic or glass cup. Observe the appearance of the urine while referring to table 43.1.

B. Test for Proteinuria

Since proteins are very large molecules (macromolecules), they are not normally present in measurable amounts in the glomerular ultrafiltrate or in the urine. The presence of protein in the urine may therefore indicate that the permeability of the glomerulus is abnormally increased. This may be caused by renal infections (glomerulonephritis) or by other diseases that secondarily affect the kidneys, such as diabetes mellitus, jaundice, or hyperthyroidism.

Procedure

Dip the yellow end of an Albustix[4] strip into a urine sample, and compare the color developed with the chart provided. Enter your observations in the data table in your laboratory report.

C. Test for Glycosuria

Although glucose is easily filtered by the glomerulus, it is not normally present in the urine because all of the filtered glucose should be reabsorbed from the renal tubules into the blood. This reabsorption process is *carrier mediated*—that is, the glucose is transported across the wall of the renal tubule by a carrier molecule.

When the concentration of glucose in the plasma and in the glomerular ultrafiltrate is within the normal limits (70–110 milligrams [mg] per 100 milliliters [ml]), there is a sufficient number of carrier molecules in the renal tubules to transport all the glucose back into the blood. However, if the blood glucose level exceeds a certain limit (called the **renal plasma threshold** for glucose, about 180 mg/dl), the glucose molecules in the glomerular ultrafiltrate will be greater than the number of available carrier molecules, and the untransported glucose will spill over into the urine.

The chief cause of glycosuria is diabetes mellitus, although other conditions, such as hyperthyroidism, hyperpituitarism, and liver disease may also have this effect. Glycosuria, therefore, is not a renal disease but a symptom of other systemic diseases that raise the blood sugar level.

Clinical Significance

When the kidneys are inflamed, the permeability of the glomerular capillaries may be increased, resulting in the leakage of proteins into the urine and the appearance of renal casts in the urine sediment. Since this represents a continuous loss of the solutes that produce the colloid osmotic pressure of plasma, fluid may accumulate in the tissues, resulting in *edema* together with the proteinuria.

Though the appearance of glucose in the urine suggests the presence of diabetes mellitus, a positive urine test alone is not sufficient for such diagnosis—a person may have hyperglycemia without glycosuria. This can occur if the plasma glucose concentration at the time of the test is not high enough to exceed the ability of the tubules to completely reabsorb glucose from the filtrate. If this *renal plasma threshold* for glucose is not exceeded, the urine will be free of glucose. A better test for diabetes is the oral glucose tolerance test.

---

[4] Albustix strips, Clinitest tablets, Ketostix strips, Hemastix strips, and Ictotest tablets are from Ames Laboratories.

Procedure

1. Place 10 drops of water and 5 drops of urine in a test tube.
2. Add a Clinitest tablet.
3. Wait 15 seconds, and compare the color developed with a color chart.
4. Record your observations in the data table in your laboratory report.

D. Test for Ketonuria

When there is carbohydrate deprivation, such as in starvation or high-protein diets, the body relies increasingly on the metabolism of fat for energy. This pattern is also seen in people with the disease diabetes mellitus, where lack of insulin prevents the body cells from utilizing the large amounts of glucose available in the blood. This occurs because insulin is necessary for the transport of glucose from the blood into the body cells.

The metabolism of fat proceeds in a stepwise manner: (1) triglycerides are hydrolyzed to fatty acids and glycerol; (2) fatty acids are converted into smaller intermediate compounds—*acetoacetic acid, beta-hydroxybutyric acid,* and *acetone;* and (3) these intermediates are utilized in aerobic cellular respiration. When the production of fatty acid metabolism intermediates (collectively known as **ketone bodies**) exceeds their utilization in respiration (when the body cannot use them in aerobic metabolism), these ketone bodies accumulate in the blood (*ketonemia*) and spill over into the urine (*ketonuria*).

Procedure

Dip a Ketostix strip into a urine sample; 15 seconds later, compare the color developed with the color chart. Record your observations in the data table in your laboratory report.

E. Test for Hemoglobinuria

Hemoglobin may appear in the urine when there is hemolysis in the systemic blood vessels (e.g., in transfusion reactions), rupture of the capillaries of the glomerulus, or hemorrhage in the urinary system. In the last condition, whole red blood cells may be found in the urine (*hematuria*), although the low osmotic pressure of the urine may cause hemolysis and the release of hemoglobin (hemoglobinuria) from these cells.

Procedure

Dip the test end of a Hemastix strip into the urine sample, wait 30 seconds, and compare the color developed with the color chart. Enter your observations in the data table in your laboratory report.

F. Test for Bilirubinuria

The fixed phagocytic cells of the spleen and bone marrow (*reticuloendothelial system*) destroy old red blood cells and convert the heme groups of hemoglobin into the pigment *bilirubin*. The bilirubin is secreted into the blood and carried to the liver, where it binds to (*conjugates* with) glucuronic acid, a derivative of glucose. Some of the conjugated bilirubin is secreted into the blood, and the rest is excreted in the bile as bile pigment that passes into the small intestine.

The blood normally contains a small amount of free and conjugated bilirubin. An abnormally high level of blood bilirubin may result from: (1) an increased rate of red blood cell destruction (hemolytic anemia), (2) liver damage (hepatitis, cirrhosis), and (3) obstruction of the common bile duct (gallstones). The increase in blood bilirubin results in *jaundice,* a condition characterized by a brownish yellow pigmentation of the skin, sclera of the eye, and mucous membranes.

The kidneys can usually excrete only bilirubin that is conjugated with glucuronic acid. Therefore, high urine bilirubin is often associated with jaundice due to liver disease or bile duct obstruction, but not normally with jaundice due to hemolytic anemia.

Procedure

1. Place 5 drops of urine on a square of the test mat.
2. Place an Ictotest tablet in the center of the mat.
3. Place 2 drops of water on the tablet.
4. Interpret the test as follows.
   a. Negative: Mat has no color or a slight pink to red color.
   b. Positive: Mat turns blue to purple. The speed and intensity of color development are proportional to the amount of bilirubin present.
5. Record your observations in the data table in your laboratory report.

G. Microscopic Examination of Urine Sediment

Microscopic examination of the urine sediment may reveal the presence of various cells, crystals, bacteria, and renal casts. Although a small number of casts are found in normal urine, a large number indicates renal disease such as glomerulonephritis and nephrosis. The casts may be noncellular, or they may be cellular and contain leukocytes, erythrocytes, or epithelial cells (fig. 43.1). The presence of large numbers of erythrocytes, leukocytes, or epithelial cells in the urine is indicative of renal disease.

Although a few crystals are present in normal urine, their presence in large numbers may suggest a tendency to form kidney stones, and a large number of uric acid crystals occur in gout (fig. 43.2).

Procedure

1. Fill a conical centrifuge tube three-quarters full of urine, and centrifuge at a moderate speed for 5 minutes.
2. Discard the supernatant, and place a drop of Sternheimer-Malbin stain on the sediment. Mix by aspiration with a transfer pipette.
3. Place a drop of the stained sediment on a clean slide, and cover with a cover slip.
4. Scan the slide with the low-power objective under reduced illumination, and identify the components of the sediment.

Figure 43.1 Components of urine sediment: (*a*) red blood cells; (*b*) white blood cells; (*c*) renal tubule epithelial cells; (*d*) urinary bladder epithelial cells; (*e*) urethral epithelial cells; (*f*) bacteria; (*g*) a hyaline renal cast; (*h*) a waxy renal cast.

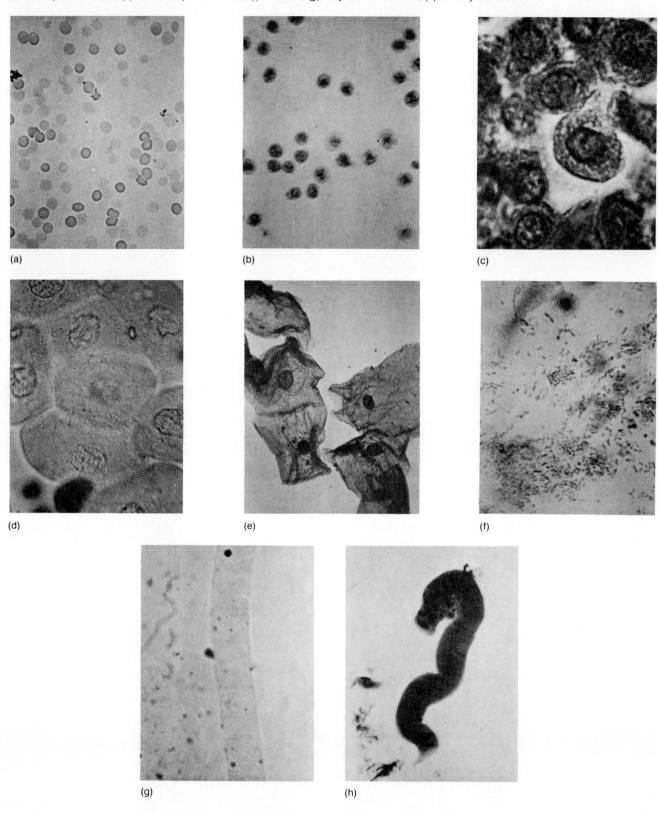

(a)

(b)

(c)

(d)

(e)

(f)

(g)

(h)

Figure 43.2   Crystals found in urine sediment.

Crystals

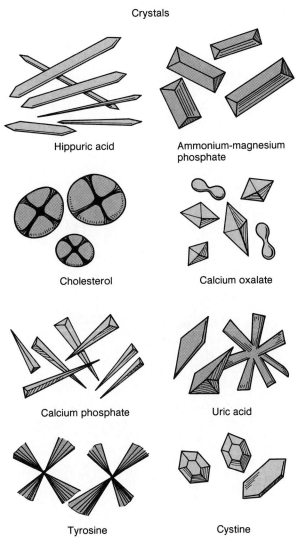

Hippuric acid

Ammonium-magnesium phosphate

Cholesterol

Calcium oxalate

Calcium phosphate

Uric acid

Tyrosine

Cystine

Laboratory Report 43

Name _____

Date _____

Section _____

Clinical Examination of the Urine

Read the assigned sections in the textbook before completing the laboratory exercise.

1. Record your data and the interpretations of your data in the table below:

Urine Test	Result of Exercise (Positive or Negative)	Physiological Reason for Negative Result	Clinical Significance of Positive Result
Proteinuria			
Glycosuria			
Ketonuria			
Hemoglobinuria			
Bilirubinuria			

2. Name a possible cause of each of the following conditions:
   a. Glycosuria _____
   b. Proteinuria _____
   c. Ketonuria _____
   d. Bilirubinuria _____

3. Explain the composition of renal casts and how they get into the urine.

4. Is it possible for someone to have an abnormally high plasma glucose concentration but not have glycosuria? Explain your answer.

5. Proteinuria and the presence of numerous renal casts in the urine are often accompanied by edema. What is the relationship between these symptoms?

Digestive System

Before coming to class, review the following sections in chapter 26 of the textbook: "Introduction to the Digestive System," "Mouth, Pharynx, and Associated Structures," "Esophagus and Stomach," "Small Intestine," "Large Intestine," and "Liver, Gallbladder, and Pancreas."

Introduction

The digestive system can be divided into a tubular *gastrointestinal (GI) tract,* or *alimentary canal,* and *accessory digestive organs.* Accessory digestive organs include the teeth, tongue, salivary glands, liver, gallbladder, and pancreas. The digestive system serves two primary functions: digestion and absorption. The process of *digestion* ultimately results in the conversion of food molecules into their subunits (such as glucose and amino acids) within the lumen of the GI tract. The process of *absorption* refers to the movement of these digestion products out of the lumen, through the mucosa of the small intestine, and into the blood or lymph.

Objectives

Students completing this exercise will be able to:

1. Identify the regions of the GI tract, and state the general function of each region.
2. Identify specific gross structures of the digestive system (e.g., flexures, sphincters, ducts).
3. Describe the general structure of the four tunics that compose the wall of the GI tract, and regional modifications that exist for performance of specific functions.
4. Discuss the location, structure, and function of the accessory digestive organs.
5. Identify histological structures of the various regions and organs of the digestive system.

Materials

1. Histological slides of specific regions of the GI tract and the accessory digestive organs
2. Models and charts of the digestive system
3. Tongue depressors, mirrors, and flashlights
4. Microscope
5. Reference text
6. Colored pencils
7. Preserved cats

A. Organization of the Digestive System

1. Identify and label the following structures in figure 44.1:

   Organs of the GI tract
       Esophagus
       Large intestine
       Oral (buccal) cavity
       Pharynx
       Small intestine
       Stomach
   Accessory digestive organs
       Liver
       Salivary glands: (a) sublingual,
       (b) submandibular, (c) parotid
       Teeth
       Tongue
       Pancreas
       Gallbladder

B. Serous Membranes of the Gastrointestinal Tract

1. Identify, label, and color-code the following serous membranes in figure 44.2:

   Greater omentum
   Lesser omentum
   Mesentery
   Parietal peritoneum
   Visceral peritoneum

Figure 44.1 The digestive system, including the GI tract and the accessory digestive organs.

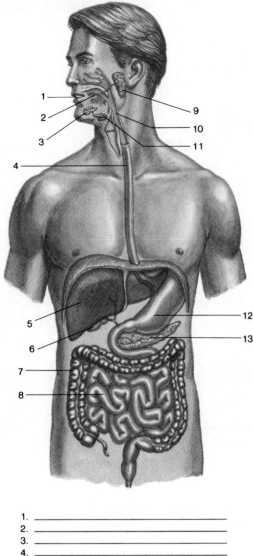

1. _____
2. _____
3. _____
4. _____
5. _____
6. _____
7. _____
8. _____
9. _____
10. _____
11. _____
12. _____
13. _____

C. Mouth, Pharynx, and Associated Structures

1. Identify and label the following structures in figures 44.3 through 44.7:

Oral (buccal) cavity
Lips
    Inferior labial frenulum
    Superior labial frenulum
    Vestibule

Figure 44.2 A midsagittal section of the abdominal cavity showing the peritoneum and various visceral organs.

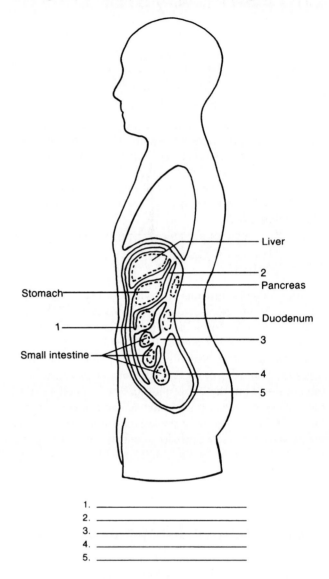

1. _____
2. _____
3. _____
4. _____
5. _____

Palate
    Hard palate
        Palatal rugae
    Soft palate
        Glossopalatine arch
        Pharyngopalatine arch
        Uvula
    Palatine tonsils
Pharynx
        Laryngopharynx
        Oropharynx
        Pharyngeal tonsil
Teeth
        Alveolus
        Canine tooth
        Cementum

Figure 44.3 The structures of the oral cavity as seen through the oral orifice (the left side of the roof of the mouth is dissected to show the muscular soft palate).

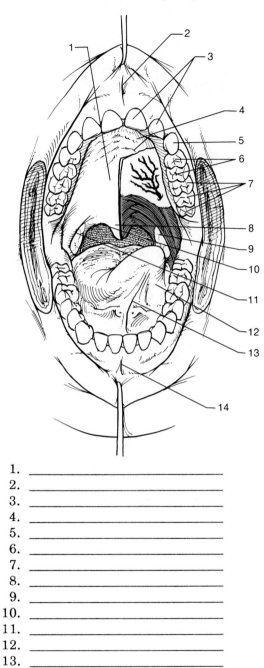

Figure 44.4 The dorsum of the tongue.

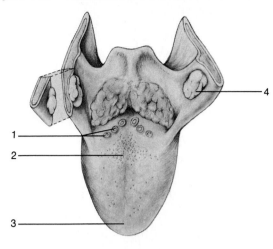

1. _____
2. _____
3. _____
4. _____
5. _____
6. _____
7. _____
8. _____
9. _____
10. _____
11. _____
12. _____
13. _____
14. _____

1. _____
2. _____
3. _____
4. _____

Cusp
Dentin
Enamel
Gingiva
Incisor teeth
Molar teeth
Neck
Premolar teeth
Root
Tongue
Lingual frenulum
    Papillae (containing
        taste buds)
        Filiform papillae
        Fungiform papillae
        Vallate papillae
Salivary glands
    Parotid gland
    Sublingual gland
    Submandibular gland

Figure 44.5 The structure of a tooth shown in a vertical section through a canine tooth.

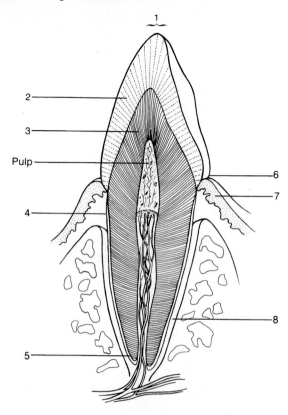

Figure 44.6 Salivary glands.

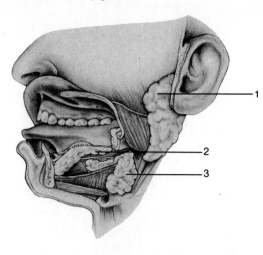

1. _____
2. _____
3. _____

1. _____
2. _____
3. _____
4. _____
5. _____
6. _____
7. _____
8. _____

Figure 44.7 A sagittal section of the facial region showing the oral cavity, nasal cavity, and pharynx.

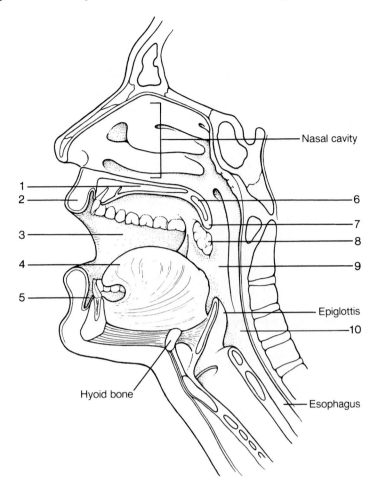

Nasal cavity

Epiglottis

Hyoid bone

Esophagus

1. _____
2. _____
3. _____
4. _____
5. _____
6. _____
7. _____
8. _____
9. _____
10. _____

2. Using a tongue depressor and a mirror, study your own mouth or examine the mouth of your lab partner. Identify the kinds and number of teeth. Are any wisdom teeth present? Observe the texture of the dorsum of the tongue, and note the median sulcus. At the rear of the mouth, the uvula can be seen suspended from the soft palate. Palatine folds of tissue and the palatine tonsils can be observed lateral to the uvula. With the tongue elevated, the lingual frenulum can be seen as it connects the tongue to the floor of the mouth. On both sides of the lingual frenulum, fleshy papillae form the openings of the submandibular ducts.

With your own tongue directed to your cheek opposite the second upper molar, feel the opening of the parotid duct. It will feel like a flap of skin high on the inner side of your cheek within the vestibule. Now direct your tongue to the roof of your mouth and feel the transverse rugae of the hard palate.

Figure 44.8   The major regions and structures of the stomach.

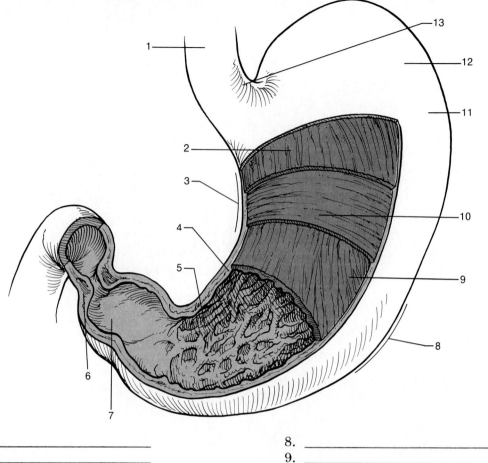

1. _____
2. _____
3. _____
4. _____
5. _____
6. _____
7. _____

8. _____
9. _____
10. _____
11. _____
12. _____
13. _____

D. Esophagus and Stomach

1. Identify and label the following layers, regions, and structures in figure 44.8:

   Esophagus
   Stomach
       Body of stomach
       Cardia of stomach
       Circular layer
       Fundus of stomach
       Gastric rugae of mucosa
       Greater curvature of stomach
       Lesser curvature of stomach
       Longitudinal layer

   Oblique layer
   Pyloric sphincter
   Pylorus of stomach
   Submucosa

2. Using colored pencils, color-code the three layers of smooth muscle in the stomach.

3. Examine a slide of an esophagus using the $10 \times$ objective lens; observe the mucosa, submucosa, muscularis externa (circular and longitudinal layers), and serosa. Use the photomicrographs in the text as a reference.

4. Examine a slide of the stomach using the $10 \times$ objective lens; identify the layers, the gastric pits, and the gastric glands. Use the photomicrographs in the text as a reference.

Figure 44.9 The regions of the small intestine.

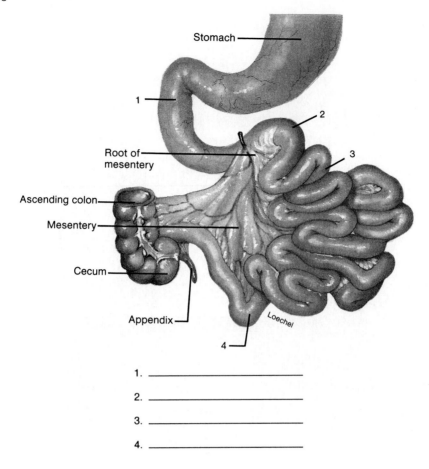

Stomach

1

2

Root of mesentery

3

Ascending colon

Mesentery

Cecum

Appendix

Loechel

4

1. _____

2. _____

3. _____

4. _____

E. Small Intestine

1. Identify and label the following regions and structures in figure 44.9:

   Duodenum
   Duodenojejunal flexure
   Ileum
   Jejunum

2. Examine a slide of the small intestine using the 10 × objective lens; identify the layers, villi, and intestinal crypts (crypts of Lieberkuhn).

F. Large Intestine

1. Identify and label the following regions and structures in figure 44.10:

   Anal canal
   Ascending colon
   Cecum
   Descending colon
   Epiploic appendage
   Haustrum
   Hepatic flexure
   Ileocecal valve
   Rectum

   Sigmoid colon
   Splenic flexure
   Taeniae coli
   Transverse colon
   Appendix

G. Liver, Gallbladder, and Pancreas

1. Identify and label the following structures in figure 44.11:

   Gallbladder
       Common bile duct
       Cystic duct
   Liver
       Caudate lobe of liver
       Falciform ligament
       Hepatic duct
       Left lobe of liver
       Porta
       Quadrate lobe of liver
       Right lobe of liver
   Pancreas
       Body of pancreas
       Pancreatic duct
       Tail

Figure 44.10   The large intestine.

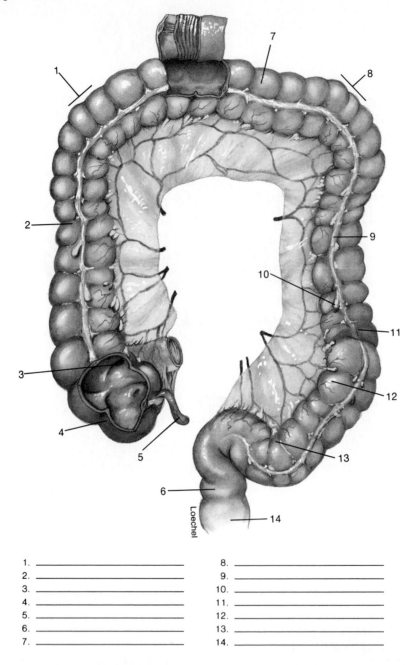

Loechel

1. _____   8. _____
2. _____   9. _____
3. _____   10. _____
4. _____   11. _____
5. _____   12. _____
6. _____   13. _____
7. _____   14. _____

2. Using colored pencils, color-code the liver, the structures that store or transport bile, and the pancreas and pancreatic duct.

3. Examine a slide of the liver using the $10 \times$ objective lens; observe the liver lobules, hepatic plates, sinusoids, portal triads, and central veins. Use the photomicrographs in the text as a reference.

4. Examine a slide of the pancreas using the $10 \times$ objective lens; identify the pancreatic acini and the pancreatic islets. Use the photomicrographs in the text as a reference.

H. Digestive System of the Cat

Refer to figure 44.12 and identify as many organs of the digestive system of an embalmed cat as possible. If the oral, thoracic, and abdominal cavities have not yet been opened, refer to the descriptions of these regions in the cat dissection information in exercises 30 and 37.

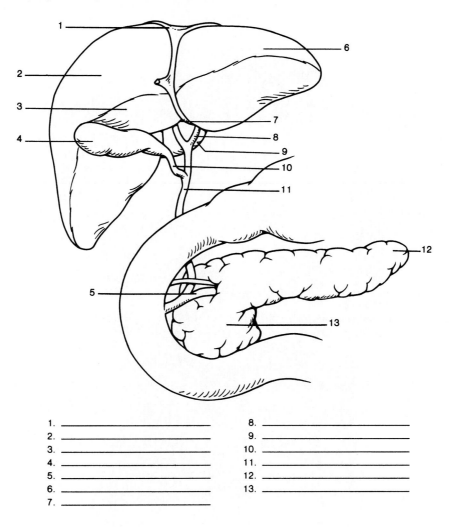

1. _____ 8. _____
2. _____ 9. _____
3. _____ 10. _____
4. _____ 11. _____
5. _____ 12. _____
6. _____ 13. _____
7. _____

Figure 44.12 A ventral view of the thorax and abdomen of the cat showing the organs of the digestive system.

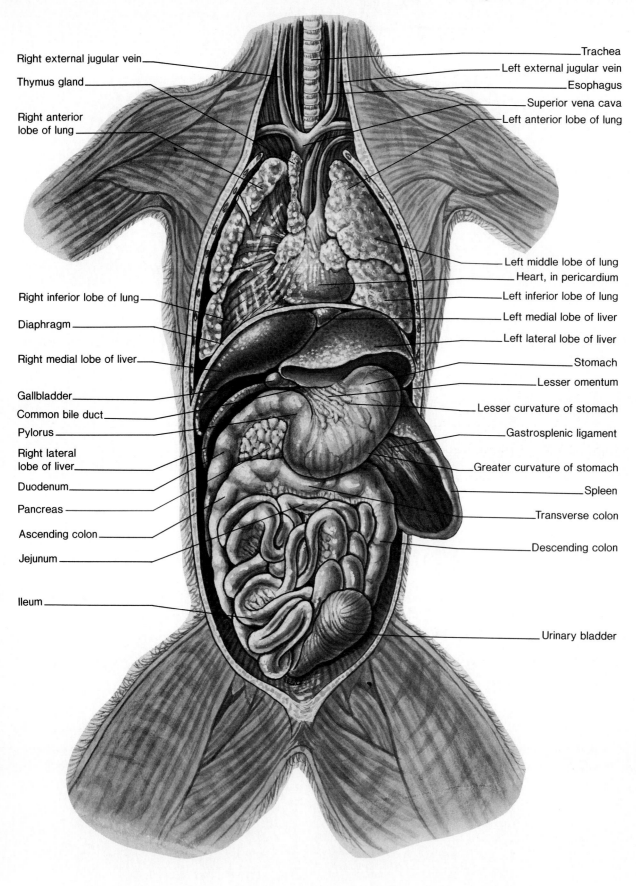

Right external jugular vein

Thymus gland

Right anterior lobe of lung

Right inferior lobe of lung

Diaphragm

Right medial lobe of liver

Gallbladder

Common bile duct

Pylorus

Right lateral lobe of liver

Duodenum

Pancreas

Ascending colon

Jejunum

Ileum

Trachea

Left external jugular vein

Esophagus

Superior vena cava

Left anterior lobe of lung

Left middle lobe of lung

Heart, in pericardium

Left inferior lobe of lung

Left medial lobe of liver

Left lateral lobe of liver

Stomach

Lesser omentum

Lesser curvature of stomach

Gastrosplenic ligament

Greater curvature of stomach

Spleen

Transverse colon

Descending colon

Urinary bladder

Laboratory Report 44

Name _____

Date _____

Section _____

Digestive System

Read the assigned sections in the textbook before completing the laboratory report.

A. Organization of the Digestive System

1. Trace a morsel of food through the GI tract. List the cavities and regions it traverses from the mouth to the anus.

2. List the layers, or tunics, of the GI tract, starting with the layer that faces the lumen. Indicate the type of tissue that each layer comprises.

B. Serous Membranes of the Gastrointestinal Tract

1. Describe the structure and location of each of the following:
   a. Lesser omentum

   _____

   b. Greater omentum

   _____

   c. Mesenteries

   _____

C. Mouth, Pharynx, and Associated Structures

1. The oral cavity and pharynx are lined with a stratified squamous epithelium. What is the advantage of this?

2. Place your fingers flat against the underside of your chin, and then feel and describe the action of the tongue and floor of the mouth as you swallow. Now place your fingers on your larynx (Adam's apple) and feel and describe its action as you swallow. What is the significance of these movements?

D. Esophagus and Stomach

1. Compare the mucosa and muscularis externa of the esophagus with those of the stomach. What functions do these layers serve?

2. Identify the following:

   a. Gastric rugae

   _____

   b. Gastric glands

   _____

E and F. Small Intestine and Large Intestine

1. Discuss the structure and function of the villi in the small intestine.

2. Identify each of the following structures and its location:

   a. Plicae circulares

   _____

   b. Ileum

   _____

   c. Cecum

   _____

   d. Haustra

   _____

   e. Taenia coli

   _____

G. Liver, Gallbladder, and Pancreas

1. Describe the structure of a liver lobule. What structures are contained in a portal triad?

2. Trace the flow of blood and bile in a liver lobule. Do blood and bile mix? Explain.

3. Distinguish between the hepatic duct and the cystic duct.

4. Describe the exocrine and endocrine structures of the pancreas.

Digestion of Starch by Salivary Amylase

Before coming to class, review "Enzymes as Catalysts" and "Control of Enzyme Activity" in chapter 4 of the textbook, and "Digestion and Absorption of Carbohydrates" in chapter 26.

Introduction

Starch digestion begins in the mouth, where starch mixes with saliva, which contains the enzyme *salivary amylase,* or *ptyalin.* Starch, a long chain of repeating glucose subunits, is hydrolyzed first into shorter polysaccharide chains, and eventually into the disaccharide *maltose,* which consists of two glucose subunits. Maltose, as well as glucose and other monosaccharides, is known as a **reducing sugar.**

In this exercise, the effects of pH and temperature on the activity of ptyalin will be tested by monitoring the disappearance of substrate (starch) and the appearance of product (maltose) at the end of an incubation period. The appearance of maltose in the incubation medium will be determined by the **Benedict's test,** where an alkaline solution of cupric ions (Cu^{++}) is reduced to cuprous ions (Cu^+), which then forms a yellow-colored precipitate of cuprous oxide (Cu_2O).

Clinical Significance

Starch digestion begins in the mouth with the action of salivary amylase (ptyalin). This is usually of minor importance in digestion (unless one chews excessively), because most of the degradation of polysaccharides and complex sugars to monosaccharides occurs in the small intestine.

The action of salivary amylase may help prevent the accumulation of carbohydrates between the gums (gingiva) and the teeth, and it may help protect against the growth of harmful bacteria that result in dental cavities (*caries*).

Objectives

Students completing this exercise will be able to:

1. Describe the action of salivary amylase, and explain how its enzymatic activity was demonstrated.
2. Explain how enzymatic activity is affected by changes in pH and temperature, and describe the pH and temperature optima of salivary amylase.

Materials

1. Water bath (set at 37° C), Bunsen burners, test tubes, test-tube clamp, graduated cylinders
2. Starch solution: dissolve 1.0 g per 100 ml over heat
3. Iodine (Lugol's reagent): dissolve 1.0 g iodine and 2.0 g potassium iodide in 300 ml of water
4. Benedict's reagent: dissolve 50.0 g sodium carbonate, 85.0 g sodium citrate, and 8.5 g copper sulfate in 5.0 liters of water

Procedure (see fig. 45.1)

1. Number four clean test tubes.
2. Obtain 10 ml of saliva (use a small, graduated cylinder). Salivation can be aided by chewing a piece of paraffin. If only 5 ml of saliva is obtained, dilute the saliva with an equal volume of distilled water.
3. Add 3.0 ml of distilled water to tube 1.
4. Add 3.0 ml of saliva to tubes 2 and 3.
5. Add 3 drops of concentrated HCl to tube 3.
6. Boil the remaining saliva by placing it in a Pyrex test tube and passing it through the flame of a Bunsen burner. (Use a test-tube clamp, and keep the tube at an angle, pointing it away from yourself and others.) Add 3.0 ml of this boiled saliva to tube 4.
7. Add 5.0 ml of cooked starch (provided by the instructor) to each of the four tubes.

Figure 45.1 The chart outlining the procedure in exercise 45.

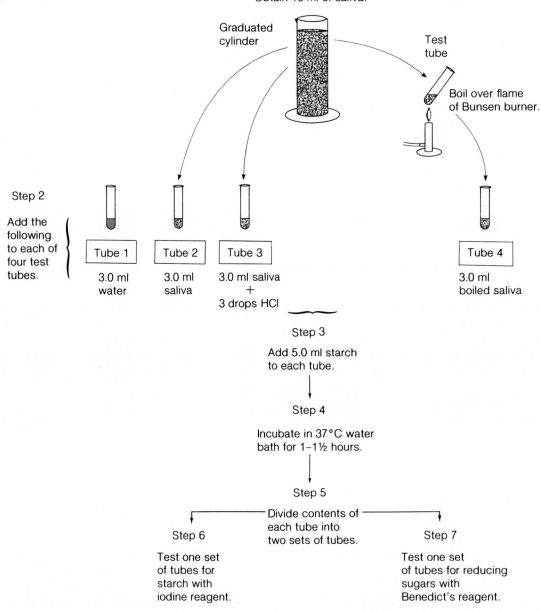

Step 1

Obtain 10 ml of saliva.

Graduated
cylinder

Test
tube

Boil over flame
of Bunsen burner.

Step 2

Add the
following
to each of
four test
tubes.

| Tube 1 | Tube 2 | Tube 3 | Tube 4 |

3.0 ml
water

3.0 ml
saliva

3.0 ml saliva
+
3 drops HCl

3.0 ml
boiled saliva

Step 3

Add 5.0 ml starch
to each tube.

Step 4

Incubate in 37°C water
bath for 1–1½ hours.

Step 5

Divide contents of
each tube into
two sets of tubes.

Step 6

Test one set
of tubes for
starch with
iodine reagent.

Step 7

Test one set
of tubes for reducing
sugars with
Benedict's reagent.

8. Allow the tubes to incubate for 1–1½ hours in a 37°C water bath.
9. Divide the contents of each sample in half by pouring into four new test tubes.
10. Test one set of solutions for starch by adding a few drops of Lugol's reagent. A positive test is indicated by the development of a purplish black color.
11. Test the other set of solutions for reducing sugars in the following way:

    a. Add 5.0 ml of Benedict's reagent to each of the four test tubes, and immerse them in a rapidly boiling water bath for 2 minutes.

    b. Remove the tubes from the boiling water with a test-tube clamp, and rate the amount of reducing sugar present according to the following scale:

    blue
    green   +
    yellow   ++
    orange +++
    red   ++++

12. Enter your results in the data table in your laboratory report.

Laboratory Report 45

Name _____

Date _____

Section _____

Digestion of Starch by Salivary Amylase

Read the assigned sections in the textbook before completing the laboratory report.

1. Enter your data in the table below, using the rating method described in the procedure.

Contents before Incubation	Starch after Incubation	Maltose after Incubation
1. Starch + distilled water		
2. Starch + saliva		
3. Starch + saliva + HCl		
4. Starch + boiled saliva		

2. Which tube(s) contained the most starch following incubation? Which tube(s) contained the most sugar? What conclusions can you draw from these results?

3. What conclusions can you draw if the test for starch *and* the test for sugar are positive for a particular tube? What might be the results if you let the tubes incubate for a longer period of time?

4. Reviewing your data, predict what would happen to salivary amylase activity once saliva is swallowed. Explain.

5. What effect does cooking have on enzyme activity? Explain why this effect occurs.

Digestion of Egg Albumin by Pepsin

Before coming to class, review "Enzymes as Catalysts" and "Control of Enzyme Activity" in chapter 4 of the textbook. Also review "Digestion and Absorption of Proteins" and "Regulation of Gastric Function" in chapter 26 of the textbook.

Introduction

Gastric juice has a pH less than 2 owing to secretion of **hydrochloric acid (HCl)** by the parietal cells of the gastric mucosa. This strong acidity coagulates proteins in the chyme and promotes the conversion of *pepsinogen,* secreted by the chief cells of the mucosa, into **pepsin.** The active enzyme pepsin, which partially digests the proteins in chyme, is maximally active (has a pH optimum) at the pH of gastric juice. In this exercise, the ability of pepsin to digest egg white albumin is assessed as a function of pH and temperature.

Clinical Significance

The stomach does not normally digest itself. When regions of the mucosa of the stomach or duodenum are digested by the strongly acidic gastric juice, a **peptic ulcer** (gastric or duodenal) is present. Although the etiology of peptic ulcers is not entirely known, it is believed that ulcers are caused by the influx of acid (H+) from the gastric lumen into the mucosa. Duodenal ulcers may be caused by excessive gastric acid secretion, but the stomach itself is normally protected from acid by the tight junctions between adjacent epithelial cells; in addition, cell division renews the gastric epithelium every three days. Since gastric acid secretion is stimulated by *histamine* released by mast cells within the connective tissue of the stomach, drugs that specifically block histamine-induced acid secretion (by blocking histamine receptors) are commonly used to treat peptic ulcers.

Objectives

Students completing this exercise will be able to:

1. Describe the action of pepsin and the factors that stimulate its secretion.
2. Describe the pH and temperature requirements of pepsin.
3. Explain why the stomach does not normally digest itself, and how peptic ulcers may be formed.

Materials

1. Water bath (set at 37° C), test tubes, droppers
2. Freezer or ice bath
3. Pepsin (5 g/100 ml), 2*N* HCl, 10*N* NaOH
4. White of hard-boiled eggs

Procedure (see figure 46.1)

1. Number five clean test tubes.
2. Using a sharp scalpel or razor blade, cut slices of egg white that are about the size of a fingernail and as thin as possible. This is critical; the slices should be very thin and uniform. Place an equally sized slice of egg white in each of the five test tubes.
3. Add 1 drop of distilled water to tube 1. Add 1 drop of concentrated HCl to tubes 2, 3, and 4. Add 1 drop of concentrated (10*N*) NaOH to tube 5.
4. Add 5.0 ml of pepsin solution to tubes 1, 2, 3, and 5. Add 5.0 ml of distilled water to tube 4.
5. Place tubes 1, 2, 4, and 5 in a 37° C water bath. Place tube 3 in a freezer or ice bath.
6. After 1–1½ hours, remove the tubes (thaw the one that was frozen), and record the appearance of the egg white in the data table in your laboratory report.

Figure 46.1 The chart outlining the procedure in exercise 46.

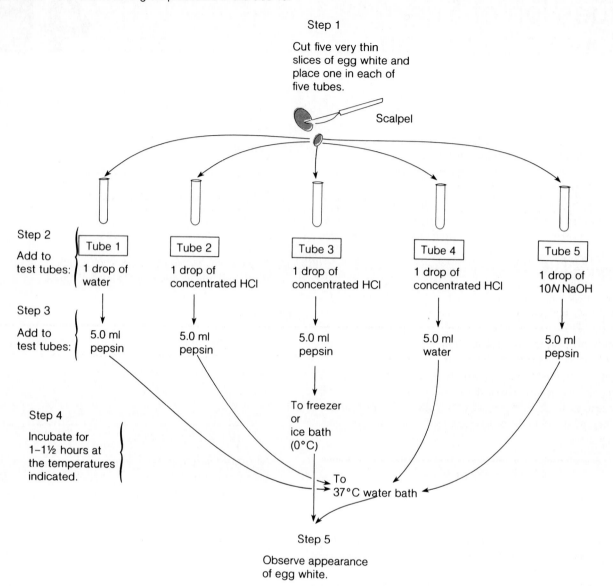

Step 1

Cut five very thin slices of egg white and place one in each of five tubes.

Scalpel

Step 2

Add to test tubes:

| Tube 1 | Tube 2 | Tube 3 | Tube 4 | Tube 5 |
| 1 drop of water | 1 drop of concentrated HCl | 1 drop of concentrated HCl | 1 drop of concentrated HCl | 1 drop of 10N NaOH |

Step 3

Add to test tubes:

5.0 ml pepsin

5.0 ml pepsin

5.0 ml pepsin

5.0 ml water

5.0 ml pepsin

To freezer or ice bath (0°C)

Step 4

Incubate for 1–1½ hours at the temperatures indicated.

To 37°C water bath

Step 5

Observe appearance of egg white.

Laboratory Report 46

Name _____

Date _____

Section _____

Digestion of Egg Albumin by Pepsin

Read the assigned sections in the text before completing the laboratory report.

1. Enter your observations in the data table below:

Conditions of Incubation	Appearance of Egg White after Incubation
1. Protein + pepsin at 37°C	
2. Protein + pepsin + HCl at 37°C	
3. Protein + pepsin + HCl at 0°C	
4. Protein + HCl at 37°C	
5. Protein + pepsin + NaOH at 37°C	

2. Which test tube showed the most digestion of egg albumin? What can you conclude about the pH optimum of pepsin?

3. Compare the effects of HCl on protein digestion by pepsin with the effects of HCl on starch digestion by salivary amylase (exercise 45). Explain the physiological significance of these effects.

4. Using the results of this exercise, explain why food that is frozen keeps longer than food at room temperature.

5. Why doesn't the stomach normally digest itself? Why doesn't gastric juice normally digest the duodenum?

Digestion of Fat by Pancreatic Juice and Bile

Before coming to class, review the following sections in chapter 26 of the textbook: "Liver, Gallbladder, and Pancreas" and "Digestion and Absorption of Lipids."

Introduction

Pancreatic juice contains a lipase enzyme that can hydrolyze the bonds joining fatty acids with glycerol, and thus convert triglycerides into free fatty acids and monoglycerides. Because lipase is water soluble, it is in the aqueous phase and can work only to hydrolyze a fat droplet at the interface of the fat droplet and water. Emulsification of the fat creates a finer suspension of fat droplets and increases their surface area, so that the hydrolysis reactions catalyzed by lipase can occur more rapidly. Emulsification of fat is promoted by the detergent action of bile salts, which are a constituent of bile. The bile is released into the duodenum by a reflex response to the arrival of fatty chyme.

Clinical Significance

The formation of **gallstones** is believed to be due, in part, to an excessive concentration of cholesterol in the bile. Blockage of the bile duct by a gallstone can cause inadequate flow of bile to the small intestine, which results in obstructive jaundice and *steatorrhea* (fat in the feces because of inadequate digestion and absorption). Steatorrhea is associated with a deficiency in the fat-soluble vitamins A, D, E, and K. Since vitamin K is necessary for normal blood clotting, this condition can be serious. *Obstructive jaundice* occurs when blood levels of the bile pigment *bilirubin* increase because the bile duct is blocked; high bilirubin levels then result in yellowish discoloration of the skin, the sclera of the eyes, and the mucous membranes.

Objectives

Students completing this exercise will be able to:

1. Define the term *emulsification,* and explain how this process aids in the digestion of fat.
2. Describe the action of lipase, and explain why the pH of the solutions tested in this exercise changed.
3. Describe how the secretion of pancreatic juice and bile is regulated.
4. Describe how fat is digested, absorbed, and transported in the body.

Materials

1. Water bath (set at 37° C), test tubes, droppers
2. pH meter or short-range pH paper
3. Pancreatin solution (1 g/dl), bile salts
4. Cream or vegetable oil

Procedure

1. Number three test tubes, and add the following to the indicated test tubes.
   Tube 1: 3.0 milliliters (ml) of cream or vegetable oil + 5.0 ml of water + a few grains of bile salts
   Tube 2: 3.0 ml of cream or vegetable oil + 5.0 ml of pancreatin solution
   Tube 3: 3.0 ml of cream or vegetable oil + 5.0 ml of pancreatin solution + a few grains of bile salts
2. Incubate the tubes at 37° C for 1 hour, checking the pH of the solutions at 20-minute intervals with a pH meter or with short-range pH paper.
3. Record your data in the table in your laboratory report.

Laboratory Report 47

Name _____

Date _____

Section _____

Digestion of Fat by Pancreatic Juice and Bile

Read the assigned sections in the textbook before completing the laboratory report.

1. Record your data in the table below:

pH

	1. Cream or oil + bile salts	2. Cream or oil + pancreatin	3. Cream or oil + bile salts + pancreatin
0 minutes			
20 minutes			
40 minutes			
60 minutes			

2. Explain why fat digestion affects the pH of the solution.

3. Does bile digest fat? Explain.

4. In which tube did fat digestion occur most rapidly? Explain the reason for this.

5. A person with gallstones has jaundice and an abnormally long clotting time. Explain the relationships between these observations.

6. How does the absorption of fat differ from the absorption of glucose and amino acids?

Caloric Balance and Body Weight[5]

Before coming to class, review the following sections in chapter 27 of the textbook: "Metabolism of Carbohydrates, Lipids, and Proteins," "Nutritional Requirements," and "Metabolic Regulation by Adrenal Hormones, Thyroxine, and Growth Hormone."

Introduction

Both the chemical energy consumed in foods and the metabolic energy expended by the cell are measured in kilocalories (kcal) or large calories (Calories). The major sources of food calories are *carbohydrate, fat (lipid)*, and *protein.* When allowances are made for inefficiency in the assimilation of each nutrient, one gram (g) of each of the three nutrients provides the body with approximately the following number of calories:

$$1 \text{ g carbohydrate} = 4.0 \text{ kcal}$$
$$1 \text{ g fat} = 9.0 \text{ kcal}$$
$$1 \text{ g protein} = 4.0 \text{ kcal}$$

The primary carbohydrates in food are the **sugars** such as glucose, fructose, and sucrose, and **complex carbohydrates** such as starches and dietary fiber. To meet the energy requirements of children and adults, it is recommended that more than half (about 55%) of the calories consumed per day come from carbohydrate sources in the diet. Emphasis should be on the increased consumption of complex carbohydrates, especially dietary fibers found in fruits, vegetables, legumes, and whole-grain cereals. In addition to providing a source of calories, dietary fiber has been associated with improving overall health by promoting normal stool elimination, enhancing satiety, and lowering plasma cholesterol levels.

---

5. Courtesy of Dr. Lawrence G. Thouin, Jr., Pierce College.

Lipids are generally divided into **triglycerides, phospholipids,** and **sterols (steroids).** The digestion, emulsification, and absorption of lipids also facilitates the absorption of **fat-soluble vitamins A, D, E, and K** and the **essential fatty acids.** Three unsaturated fatty acids are considered essential and must be present in the diet to maintain health: *linoleic acid, arachidonic acid,* and *linolenic acid.*

Clinical Significance

Triglycerides are the major lipid components of foods and the most concentrated source of energy (9 kcal/g). The average American currently derives about 36% of the daily total dietary calories from fats. High dietary fat and cholesterol intakes have been associated with the increased risk of cardiovascular disease and cancer. The Food and Nutrition Board's Committee on Diet and Health currently recommends: (1) the fat content of the U.S. diet be lowered to not exceed 30% of the caloric intake; (2) less than 10% of fat calories be provided from saturated fatty acids; and (3) dietary cholesterol be less than 300 milligrams mg/day (National Research Council, 1989).

Objectives

1. Describe the different nutrient classes, and list the calories per gram for carbohydrate, lipids, and fat.
2. Define *dietary record,* and demonstrate how one is used to assess food and fluid consumption.
3. Define *basal metabolic rate (BMR)* and demonstrate two different methods for estimating BMR.
4. Define *activity factor (AF)* and demonstrate how estimations are made of calories burned for various activities.

Materials

1. Home scale or physicians height-weight scale
2. Calorie-counting guide such as the U.S. Department of Agriculture Handbook, cookbooks, or popular diet books

A. Energy Intake and the Three-Day Dietary Record

U.S. Department of Agriculture (USDA) surveys find that, for the average American, about 15% of the total food energy intake is derived from protein. About 65% of this protein is from animal sources, primarily meat and dairy products, and about 20% from cereal grains. Despite increased protein requirements for certain populations, such as growing children, pregnant or lactating females, and the elderly, the typical American diet normally meets or exceeds the requirements. Interestingly, there is little evidence that physical exercise increases the need for protein, other than that required during the initial conditioning period. Therefore, no adjustment to the recommended allowance for protein is made for those eating a typical U.S. diet.

Nutrients that do not contribute energy to the body but nevertheless are required to maintain body functions are **vitamins, minerals,** and **water.** The *fat-soluble vitamins* (A, D, E, and K) are absorbed from the intestine with other food lipids and are concentrated to some degree in adipose tissue. The *water-soluble vitamins* are C, thiamine, riboflavin (B_2), niacin (B_1), pyridoxine (B_6), folate, cyanocobalamin (B_{12}), biotin, and pantothenic acid. Most water-soluble vitamins serve as *coenzymes* that assist enzymes in the regulation of metabolism. The *major minerals,* required in higher quantities, are calcium, phosphorus, magnesium, and the *electrolytes* (sodium, chloride, and potassium). Minerals required in lesser quantities are *minor (trace) elements* and include iron, zinc, iodine, selenium, copper, manganese, fluoride, chromium, and molybdenum. Recommended quantities of these nutrients are normally met when a variety of foods are consumed. For the average person eating a typical U.S. diet, therefore, no supplementation is recommended. Indeed, high intakes of these nutrients as supplements can be toxic.

Water is also an essential nutrient that must be consumed in the diet. Although assessments of adequate water intake involve many complex factors, the general recommendation is a minimum of 1.0 milliliters (ml) of water per kilocalorie of energy expended per day. Therefore, an individual expending 2000 kcal per day should consume at least 2 liters of water. With adequate water intake, the electrolyte requirements are normally met.

Procedure

1. Complete the 3-day *dietary record* sheets (provided in the laboratory report) over three consecutive days, making an attempt to include at least one weekend day. Try to weigh yourself under consistent conditions (comparable clothing, same scale, etc.) at the same time of day during this 3-day period.

2. Record all **foods** and **fluids** consumed each day in the appropriate columns provided, noting the approximate **time** of day. Estimate food quantities by weight (e.g., ounces) and fluids by volume (e.g., cups), depending upon the units listed in your calorie guide. Record the total volume of fluids consumed at the bottom of the column.

3. Look up the estimated number of **calories** for each food or fluid consumed in your diet record (look for calories provided on labeled items). Record the total caloric intake at the bottom of the column.

4. Comment on **where** you consumed the item. Do you always eat sitting down, at a table, in a quiet, relaxing environment? Or are you in the car, between classes, in front of the TV, in bed studying, etc.?

5. Comment on **why** you ate or drank the item. Do you always eat because you are hungry? Or is it because you were bored, the food was there, someone else paid for it, you were falling asleep, etc.?

6. Leave the **activity factor** box empty for now.

B. Energy Output: Estimates of the BMR and Activity

The total energy expended each day includes the energy required at rest and during physical activity. For most people, the calories expended at rest make up most of the total daily energy expenditure. This energy is used to pump blood, inflate the lungs, transport ions, and carry on the other functions of life. The measurement of this resting energy expenditure shortly after awakening and at least 12 hours after the last meal is known as the **basal metabolic rate (BMR).** With all other factors equal, BMR is influenced most by the amount of actively metabolizing tissues, or *lean body mass.* The BMR is higher in younger, more muscular people, and in males (who have a higher average muscle mass than females). The BMR is also influenced by various hormones, primarily thyroxine. People who are *hypothyroid* have a low BMR, and those that are *hyperthyroid* have a high BMR.

While most of our caloric output is spent at rest, most people are physically active to some degree, expending calories beyond the BMR. The additional number of activity calories expended will vary with the individual, and with the duration, intensity, and types of activities performed. This increased caloric output can be estimated by multiplying an **activity factor (AF)** by the BMR. In general, the average sedentary person raises the total number of calories burned per day to 130% (AF = 1.3) of the estimated BMR. Moderately active persons may raise daily expenditures 150% (AF = 1.5) above the BMR estimates; while top athletes may come close to doubling the BMR estimates (AF = 2.0). **Aerobic activities** such as running, swimming, bicycling, dancing, and others burn more calories than **anaerobic activities** such as weight lifting.

Procedure

1. Estimate your BMR using two different methods:

   Method 1: Your weight in kilograms

   (2.2 lbs/kg) _____ (kg):

   Female BMR = 0.7 kcal/kg/hour
   Male BMR = 1.0 kcal/kg/hour

   Use the above conversion factors to calculate your kilocalories per hour. Then multiply this figure by 24, and enter your answer in the space below.
   In one day (24 hours), your BMR is approximately

   _____ kcals.

   Method 2: Estimate your Ideal Body Weight (IBW) in pounds.

   Female IBW = 100 lbs for the first 5 feet in height
        +   5 lbs per inch above 5 feet in height
   Male   IBW = 106 lbs for the first 5 feet in height
        +   6 lbs per inch above 5 feet in height

   Your IBW is approximately _____ lbs; next, multiply your IBW × 10 for your daily estimated BMR.
   In one day (24 hours), your BMR is

   approximately _____ kcals.

2. Since these methods are only estimates, a difference between the two values is to be expected. Select the one BMR estimate you feel is most accurate and write that number in the space provided on the dietary record in the laboratory report.

3. For each day in your 3-day dietary record, select one *activity factor* that best reflects your total activity for that 24-hour period, and write that number in the appropriate box on the daily dietary record.

Activity Factors

Lying in bed all day, equal to the BMR — 1.0
Mild activity, normal routine, no exercise — 1.3
Moderate activity, 1 hour of aerobic exercise — 1.5
Heavy activity, 2–4 hours of aerobic exercise — 1.7
Top athletic training — 2.0

4. Calculate the total number of calories expended each day as follows:

Total calories expended = AF × BMR (method #1 or #2)

5. Write this total in the space provided (under AF) at the bottom of each dietary record page.

6. Subtract total calories expended from total calories consumed to determine the caloric balance lost or gained that day. Write the number of excess calories in the space provided in the bottom of the diet record and circle either "lost" or "gained."

7. Assuming that one pound of body tissue (not just fat) that is gained or lost represents approximately 3500 kcal, convert the excess calories from line 4 into body weight:

   Body weight = _____ (kcal)

   (lost/gained)   3500 (kcal/lb)

   = _____ (lb)

Write the pounds lost/gained that day in the space provided in the diet record.

Laboratory Report 48

Name _____

Date _____

Section _____

Caloric Balance and Body Weight

THREE-DAY DIETARY RECORD

Day

Your Weight (lb) _____

Time am/pm	Food-Units (e.g., oz.)	Fluid-Units (e.g., cup)	Calories	Where?	Why?
	Total	Total			

1. Total calories *consumed* (intake): _____ kcal.

2. Estimated BMR at rest: _____ kcal. (Part B, Step 1, Method #1 or #2)

3. Total calories *expended* (output): _____ kcal.

   Multiply line 2 times **Activity Factor (AF)** in box.

   AF

   ☐

4. Subtract line 3 from line 1 for the caloric balance today.

   Total kcal consumed _____ (−) total kcal expended _____ =

   today's caloric balance: _____ kcal **gained** or **lost** (circle one).

5. Given that one pound of body weight (not just fat) is equal to approximately *3500 kcal* gained or lost, how many pounds of body weight was gained or lost today? _____ lb

6. Enter values from today's report into the Dietary Record Evaluation form following these daily reports.

Day   Your Weight (lb) _____

Time am/pm	Food-Units (e.g., oz.)	Fluid-Units (e.g., cup)	Calories	Where?	Why?
		Total	Total		

1. Total calories *consumed* (intake): _____ kcal.
2. Estimated BMR at rest: _____ kcal. (Part B, Step 1, Method #1 or #2)
3. Total calories *expended* (output): _____ kcal.

   AF

   Multiply line 2 times **Activity Factor (AF)** in box.

   [ ]

4. Subtract line 3 from line 1 for the caloric balance today.

   Total kcal consumed _____ (–) total kcal expended _____ =

   today's caloric balance: _____ kcal **gained** or **lost** (circle one).

5. Given that one pound of body weight (not just fat) is equal to approximately *3500 kcal* gained or lost, how many pounds of body weight was gained or lost today? _____ lbs

6. Enter values from today's report into the Dietary Record Evaluation form following these daily reports.

Day   Your Weight (lb) _____

Time am/pm	Food-Units (e.g., oz.)	Fluid-Units (e.g., cup)	Calories	Where?	Why?
		Total	Total		

1. Total calories *consumed* (intake): _____ kcal.
2. Estimated BMR at rest: _____ kcal. (Part B, Step 1, Method #1 or #2)
3. Total calories *expended* (output): _____ kcal.

   Multiply line 2 times **Activity Factor (AF)** in box.

   AF
   ☐
4. Subtract line 3 from line 1 for the caloric balance today.

   Total kcal consumed _____   (−)   total kcal expended _____   =

   today's caloric balance: _____ kcal **gained** or **lost** (circle one).
5. Given that one pound of body weight (not just fat) is equal to approximately *3500 kcal* gained or lost, how many pounds of body weight was gained or lost today? _____ lbs
6. Enter values from today's report into the Dietary Record Evaluation form following these daily reports.

Three-Day Dietary Record Evaluation

1. Record the total number of calories **consumed** over three days:

   Day 1 _____ Day 2 _____ Day 3 _____ = _____ kcals

2. Record the total number of calories **expended** over three days:

   Day 1 _____ Day 2 _____ Day 3 _____ = _____ kcals

3. Subtract the total kcal expended from the total kcal consumed, and indicate if the difference represents a gain (+) or a loss (−).

   _____ kcal gained/lost (circle one) over 3 days

4. Divide your total from step 3 by 3500 kcal/lb (to convert kcal to pounds). Indicate if this represents a weight gain or loss.

   _____ lb gained/lost (circle one)

5. Determine your measured weight change by subtracting your weight on day 3 from your weight on day 1. Compare your calculated weight gain/loss with your actual weight gain/loss, and explain why the numbers may differ.

6. Study your diet record considering the six nutrient classes. Make a list of at least five specific suggestions for improving this diet.

   a.

   b.

   c.

   d.

   e.

7. Based on the recommendation of at least 1.0 ml of water intake per kilocalorie of energy expended each day, did you meet this requirement ? If not, propose a plan that would allow you to meet this requirement.

General Questions

1. A student wishes to lose 20 pounds before graduation ceremonies 10 months away. How many kilocalories *per week* would the student have to reduce intake or increase expenditure in order to lose 20 pounds?

2. How many calories *per day* would this student have to lose over the 10-month period? Is this a practical weight loss plan? Explain.

3. If this student wants to lose all of this weight by removing only fat calories from foods consumed, how many grams of fat *per day* must be removed from the diet? Is this a practical weight loss plan? Explain.

4. Define BMR and explain the impact physical activity (exercise) has on maintaining body weight.

Continuance of the Human Species

The following exercises are included in this unit:

49. Reproductive System
50. Steroid Hormones
51. Pregnancy Test
52. Patterns of Heredity

These exercises are based upon information presented in the following chapters of *Concepts of Human Anatomy and Physiology,* 4th ed.:

28. Male Reproductive System
29. Female Reproductive System
30. Human Development, Inheritance, and Aging

The reproductive system is unique among the body systems in its functional purpose. Whereas the other body systems exist to maintain the homeostasis of the individual organism, the reproductive system exists for the preservation of the *species,* not the individual. All higher organisms reproduce sexually, because sexual reproduction allows for far greater genetic diversity than does asexual reproduction. Genetic diversity, as shown by evolutionary biology, promotes the continuation of the species in the face of changing environments over time.

In this unit, you will study the organs of the reproductive system and some of the sex steroid hormones secreted by the gonads. A hormone secreted by the developing placenta is the basis of the pregnancy test, which will be performed on urine. Finally, the basic laws of inheritance discovered by Gregor Mendel will be used to predict the pattern of inheritance of particular human traits.

Reproductive System

Before coming to class, review the anatomy of the male and female reproductive systems in chapters 28 and 29 of your textbook.

Introduction

All body systems, except the reproductive system, sustain the individual. The reproductive system is specialized to perpetuate the species and to pass genetic material from generation to generation. This requires the production of haploid gametes through meiotic cell division and the participation of a mature male and female in sexual intercourse to produce a fertilized, diploid egg cell called a zygote. Through numerous mitotic cell divisions and specializations of the daughter cells into different tissues and organs, this zygote develops into a human embryo and fetus.

Objectives

Students completing this exercise will be able to:
1. List the organs and structures that compose the male and female reproductive systems, and discuss the functions of each.
2. Distinguish between primary and secondary sex characteristics, and identify them in both male and female.
3. Trace spermatozoa through the course of development and through the various ducts during ejaculation.
4. List homologous reproductive structures in the male and female.

Materials

1. Microscopes
2. Prepared slides of organs of the male and female reproductive systems—testis, epididymis, and ovary
3. Fresh or preserved reproductive organs from domestic animals
4. Models and charts of the male and female reproductive systems
5. Reference text
6. Colored pencils
7. Dissecting trays and instruments
8. Preserved cat

A. Male Reproductive System

1. Identify and label the following structures in figures 49.1, 49.2, and 49.3:
   Bulbourethral gland
   Ductus (vas) deferens (pl. ductus deferentia)
   Ejaculatory duct
   Epididymis: (a) body, (b) head, (c) tail (pl. epididymides)
   Penis
       Corpora cavernosa penis
       Corpus spongiosum penis
       Dorsal vein
       Glans penis
       Prepuce
   Prostate
   Scrotum
       Scrotal septum
   Seminal vesicle
   Spermatic cord
   Testes (sing. testis)
       Efferent ductules
       Rete testis
       Seminiferous tubules
       Septum
       Tunica albuginea
       Urethra
2. Using colored pencils, color-code the different structures illustrated in figures 49.2 and 49.3.
3. Examine a preserved or fresh testis from a bull or a ram. If the epididymis and ductus deferens are still intact, note their position relative to the testis. Make a cross-sectional cut into the testis, and observe the seminiferous tubules. Cut into the epididymis, and examine for spermatozoa.
4. Observe a histological slide of a testis under low- and high-power magnifications. Identify the seminiferous tubules and the interstitial tissue.

Figure 49.1 Organs of the male reproductive system: (*a*) a sagittal view; (*b*) a posterior view.

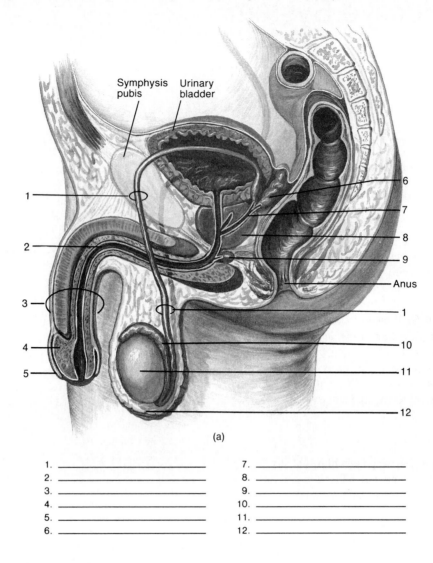

Symphysis pubis

Urinary bladder

6

7

8

9

Anus

1

10

11

12

1

2

3

4

5

(a)

1. _____   7. _____
2. _____   8. _____
3. _____   9. _____
4. _____   10. _____
5. _____   11. _____
6. _____   12. _____

B. Female Reproductive System

1. Identify and label the following structures in figures 49.4 through 49.7:

Breast
    Areola
    Lactiferous duct
    Mammary duct
    Mammary glands
    Nipple
External genitalia (vulva)
    Clitoris
    Hymen
    Labia majora (sing. labium majora)
    Labia minora (sing. labium minora)
    Mons pubis
    Perineum

Internal organs
    Ovary
    Uterine (fallopian) tubes
        Fimbriae
        Follicle
    Uterus
        Body
        Cervix
        Endometrium
        Fundus
        Myometrium
        Perimetrium
    Vagina
        Fornix
        Vaginal rugae (not shown)
        Vaginal orifice

Figure 49.1   Continued.

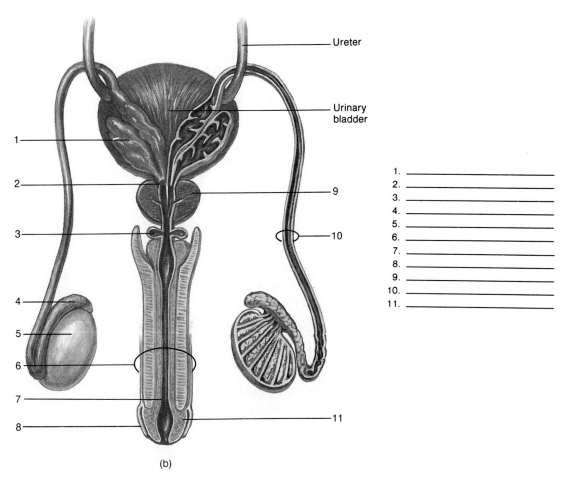

Ureter

Urinary
bladder

1. _____
2. _____
3. _____
4. _____
5. _____
6. _____
7. _____
8. _____
9. _____
10. _____
11. _____

(b)

Figure 49.2   Structural features of the testis and epididymis (longitudinal view).

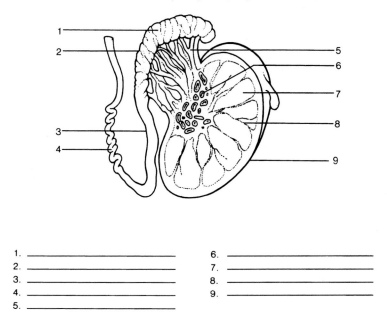

1. _____   6. _____
2. _____   7. _____
3. _____   8. _____
4. _____   9. _____
5. _____

Figure 49.3 The internal structure of the penis.

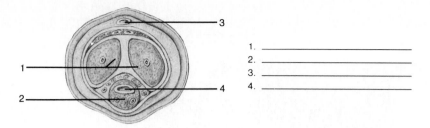

1. _____
2. _____
3. _____
4. _____

Figure 49.4 The female reproductive organs (an anterior view).

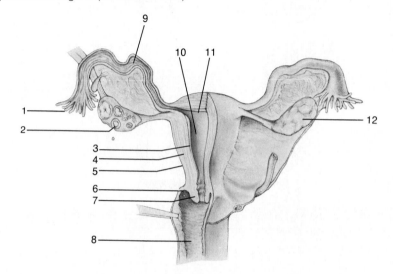

1. _____
2. _____
3. _____
4. _____
5. _____
6. _____

7. _____
8. _____
9. _____
10. _____
11. _____
12. _____

Figure 49.5 The female reproductive organs (a sagittal view).

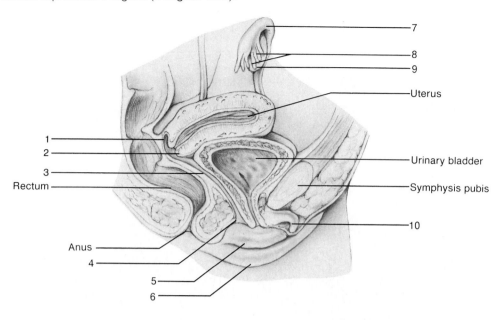

Labels in figure: 7, 8, 9, Uterus, 1, 2, 3, Rectum, Anus, 4, 5, 6, Urinary bladder, Symphysis pubis, 10

1. _____
2. _____
3. _____
4. _____
5. _____

6. _____
7. _____
8. _____
9. _____
10. _____

2. Color-code the endometrium and myometrium of the uterus in figures 49.4 and 49.5. Use different colors for the ovary and the uterine tubes.
3. Observe a slide of the ovary, and identify the following structures of an ovarian follicle:
   Antrum
   Corona radiata
   Cumulus oophorus
   Theca interna
   Ovum

C. Reproductive System of the Cat

Male Reproductive System

To expose the testes of a cat, an incision must be made in the scrotum. A scrotal septum separates each testis into its own compartment. Each testis is covered with a tough layer of peritoneum, the *tunica albuginea*. The *epididymis* is on the anterolateral surface of the testis. Trace the *ductus* (*vas*) *deferens* cranially through the inguinal canal. The ductus deferens, testicular vessels, lymph vessels, and nerves constitute the *spermatic cord*.

The *penis* is ventral to the testes (fig. 49.8). The *prepuce* of the penis will have to be dissected to expose the *glans penis*. A careful longitudinal cut through the penis of the cat will expose the *baculum* (os penis). Bacula are bones that many mammals have to facilitate copulation. A cut through the pelvic muscles and the symphysis pubis is necessary to expose the *urethra* and accessory reproductive organs of the male cat. Take care to avoid cutting the urethra as the symphysis pubis is split. Trace the urethra from the urinary bladder through the penis. Identify the *prostate* at the neck of the urinary bladder and the urethra. The paired *bulbourethral glands* connect to the urethra near the base of the penis.

Figure 49.6 The structure of the breast and mammary glands: (*a*) a sagittal section; (*b*) an anterior view partially sectioned.

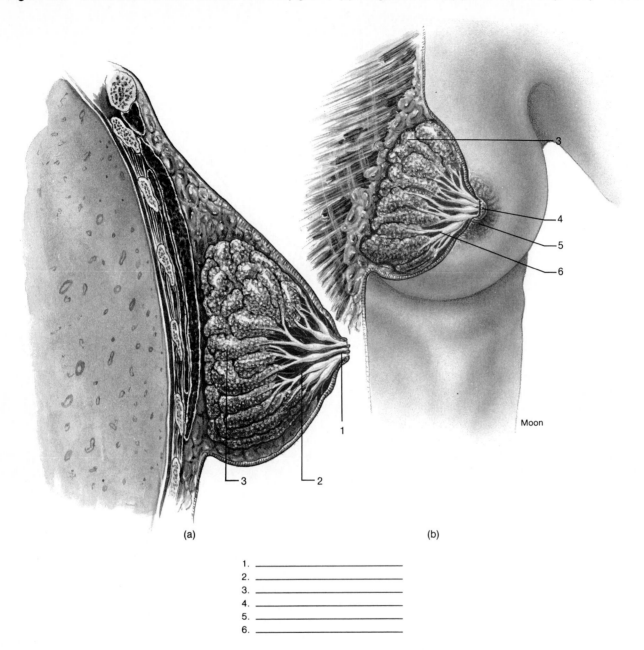

(a)

(b)

Moon

1. _____
2. _____
3. _____
4. _____
5. _____
6. _____

Figure 49.7 The external genitalia (vulva) of the female.

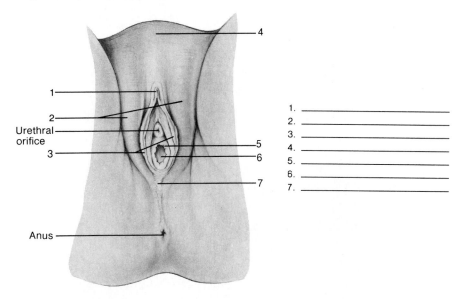

1.
2.
Urethral orifice
3.

4

5
6

7

Anus

1. _____
2. _____
3. _____
4. _____
5. _____
6. _____
7. _____

Female Reproductive System

The *ovaries* of the cat are small, whitish organs posterior to the kidneys. An *ovarian ligament* connects each ovary to its *uterine tube.* The *ostium* should be apparent at the free end of the uterine tube near the ovary. A number of fringed, fingerlike processes called *fimbriae* spread from the uterine tube over the lateral surface of the ovary. Each uterine tube (oviduct) is delicately convoluted and extends from the *infundibulum* to the *uterine horn* (fig. 49.9). The horns of the uterus are the sites for implantation, and they enable cats to have large litters. The uterine horns unite medially to become the *body of the uterus.* The uterine horns and body of the uterus are supported by the *broad ligament* (mesometrium) and the *round ligament.*

A longitudinal cut through the pelvic muscles and the symphysis pubis must be made to expose the lower portion of the body of the uterus and the vagina. Care should be taken to avoid damaging the underlying reproductive and urinary organs. The *vagina* will be seen as a chamber between the body of the uterus and the opening of the urethra, an area called the *urogenital sinus* (vestibule).

The urogenital sinus is a common passageway for both the urinary and reproductive systems. The external opening of the urogenital sinus is called the *urogenital orifice.* The *external genitalia,* or *vulva,* consist of the longitudinal folds of tissue called the *labia majora* and the erectile tissue called the *clitoris.*

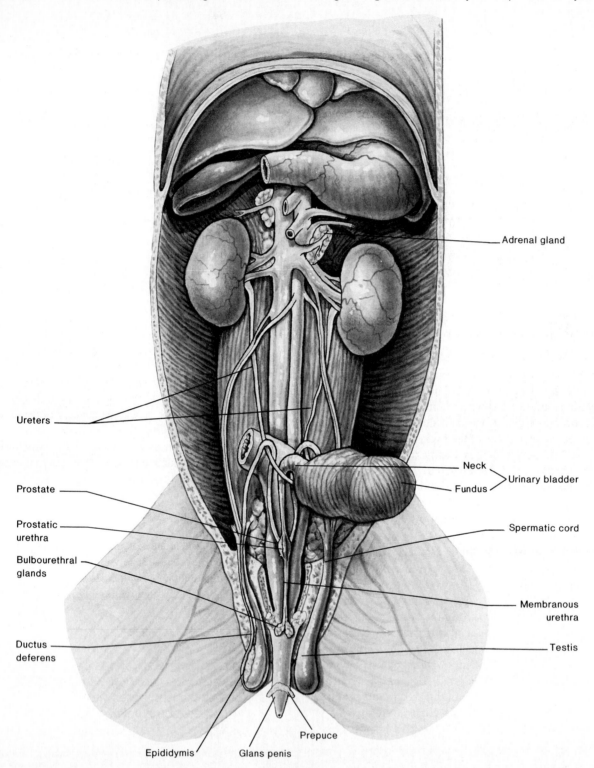

Adrenal gland

Ureters

Neck

Urinary bladder

Fundus

Prostate

Spermatic cord

Prostatic urethra

Bulbourethral glands

Membranous urethra

Ductus deferens

Testis

Prepuce

Epididymis   Glans penis

Figure 49.9 A ventral view of the pelvic region of a female cat showing the organs of the urinary and reproductive systems.

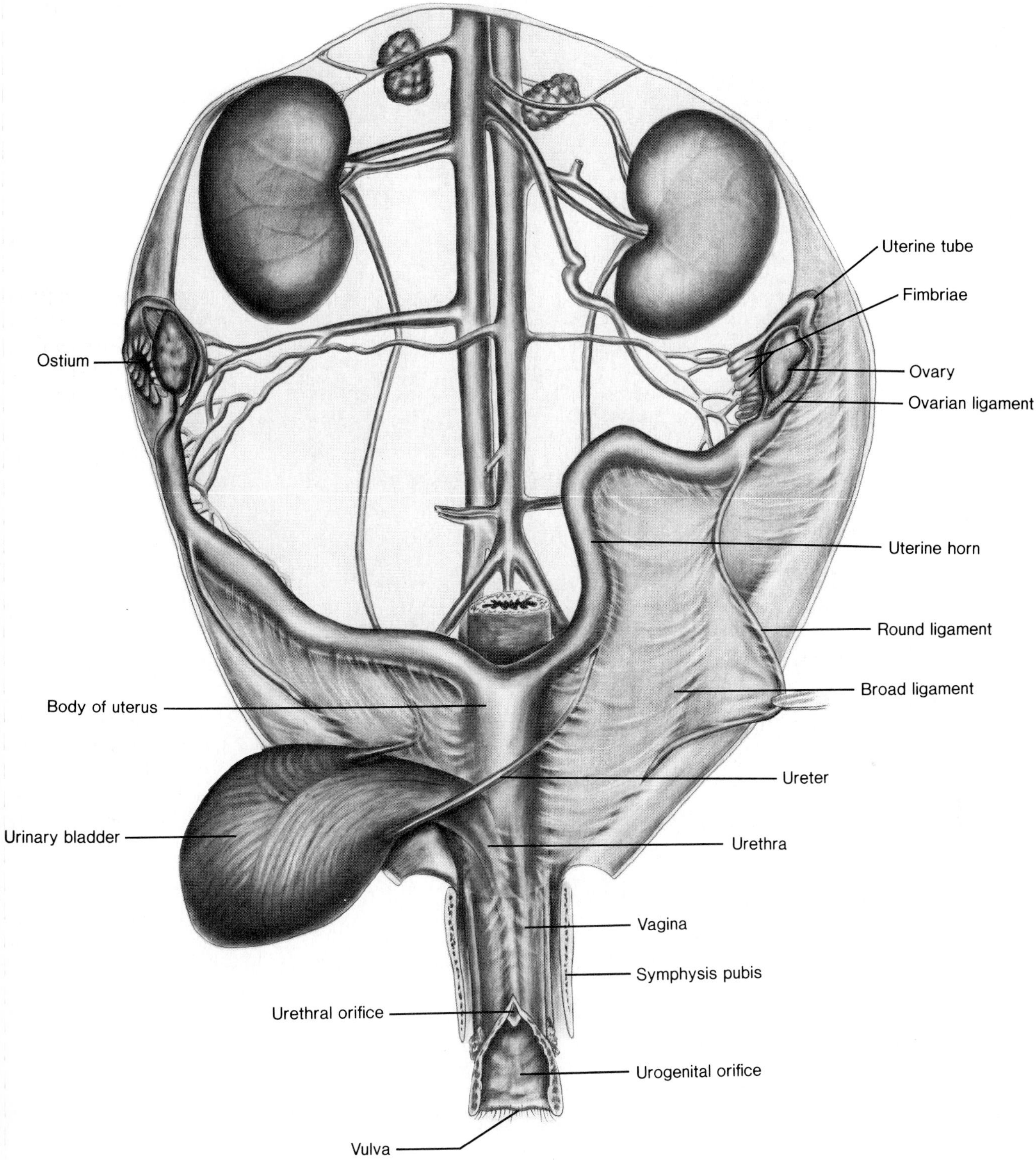

Ostium

Uterine tube

Fimbriae

Ovary

Ovarian ligament

Uterine horn

Round ligament

Broad ligament

Body of uterus

Ureter

Urethra

Urinary bladder

Vagina

Symphysis pubis

Urethral orifice

Urogenital orifice

Vulva

Laboratory Report 49

Name _____

Date _____

Section _____

Reproductive System

Read the assigned sections in the textbook before completing the laboratory report.

A. Male Reproductive System

1. Trace the passage of spermatozoa from the seminiferous tubules to ejaculation.

2. What structures are altered in a vasectomy? What effect does a vasectomy have on the secretion of testosterone? Explain why.

3. Describe the embryonic and fetal development of the testes and how the testes descend into the scrotum.

B. Female Reproductive System

1. Which male and female reproductive organs are homologous?

2. Describe the tissue layers of the uterus, and identify the significance of these layers during the menstrual cycle, pregnancy, and parturition (childbirth).

3. Describe the changes that occur in the ovarian follicles during the course of a menstrual cycle.

Steroid Hormones

Before coming to class, review "Chemical Classification of Hormones" in chapter 19 of the textbook, "Testes" in chapter 28, and "Ovaries and the Ovarian Cycle" in chapter 29.

Introduction

Steroid hormones are secreted by the adrenal cortex and gonads. These hormones can be grouped by function into **mineralocorticoids,** which regulate electrolyte balance; **glucocorticoids,** which regulate glucose balance, and **sex steroids,** which regulate different aspects of the reproductive system. Hydrocortisone, cortisone, and corticosterone are glucocorticoids; aldosterone and deoxycorticosterone are mineralocorticoids. These hormones are secreted by the adrenal cortex and thus are known as corticosteroids. Estradiol and progesterone are female sex steroids. Estradiol is secreted by ovarian follicles and the placenta, and progesterone is secreted by the corpus luteum of the ovaries and by the placenta. Testosterone is the male sex steroid secreted by the testes.

The differences in the biological effects of these hormones are related to the slight differences in their chemical structures (fig. 50.1). The differences in chemical structure can be used to separate a mixture of steroids and to identify unknown molecules.

Objectives

Students completing this exercise will be able to:

1. Identify the major classes of steroid hormones and the glands that secrete them.
2. Describe the primary differences between different functional classes of steroid hormones.
3. Demonstrate the technique of thin-layer chromatography, and explain how this procedure works.

Materials

1. Thin-layer plates (silica gel, F-254), chromatography developing chambers, capillary tubes

2. Driers (chromatography or hair driers), ultraviolet viewing box (short wavelength), rulers or spotting template (optional)
3. Steroid solutions in absolute methanol, 1.0 mg/ml of testosterone, of hydrocortisone, of cortisone, of corticosterone, and of deoxycorticosterone; 5 mg/ml of estradiol
4. Unknown steroid solution containing any two of the previous steroids
5. Developing solvent: 60 ml tolvene plus 10 ml ethyl acetate plus 10 ml acetone, or a volume containing a comparable 6:1:1 solvent ratio

Thin-Layer Chromatography

In this exercise, you will attempt to identify two unknown steroids that are present in the same solution. To do this, you must first *separate* and then *identify* these steroids by comparing their behavior with that of known steroids.

Since each steroid has a different structure, each will have a different *solubility* (ability to be dissolved) in a given solvent. These differences will be used to separate and identify the steroids on a *thin-layer plate.*

The thin-layer plate consists of a thin layer of porous material (in this procedure, silica gel) that is coated on one side of a plastic, glass, or aluminum plate. The solutions of steroids are applied on different spots of the plate (a procedure called "spotting"), and the plate is placed in a solvent bath with the spots above the solvent.

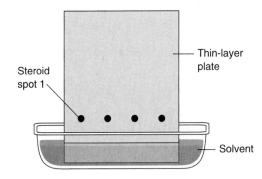

Figure 50.1 Structures of representative steroid hormones.

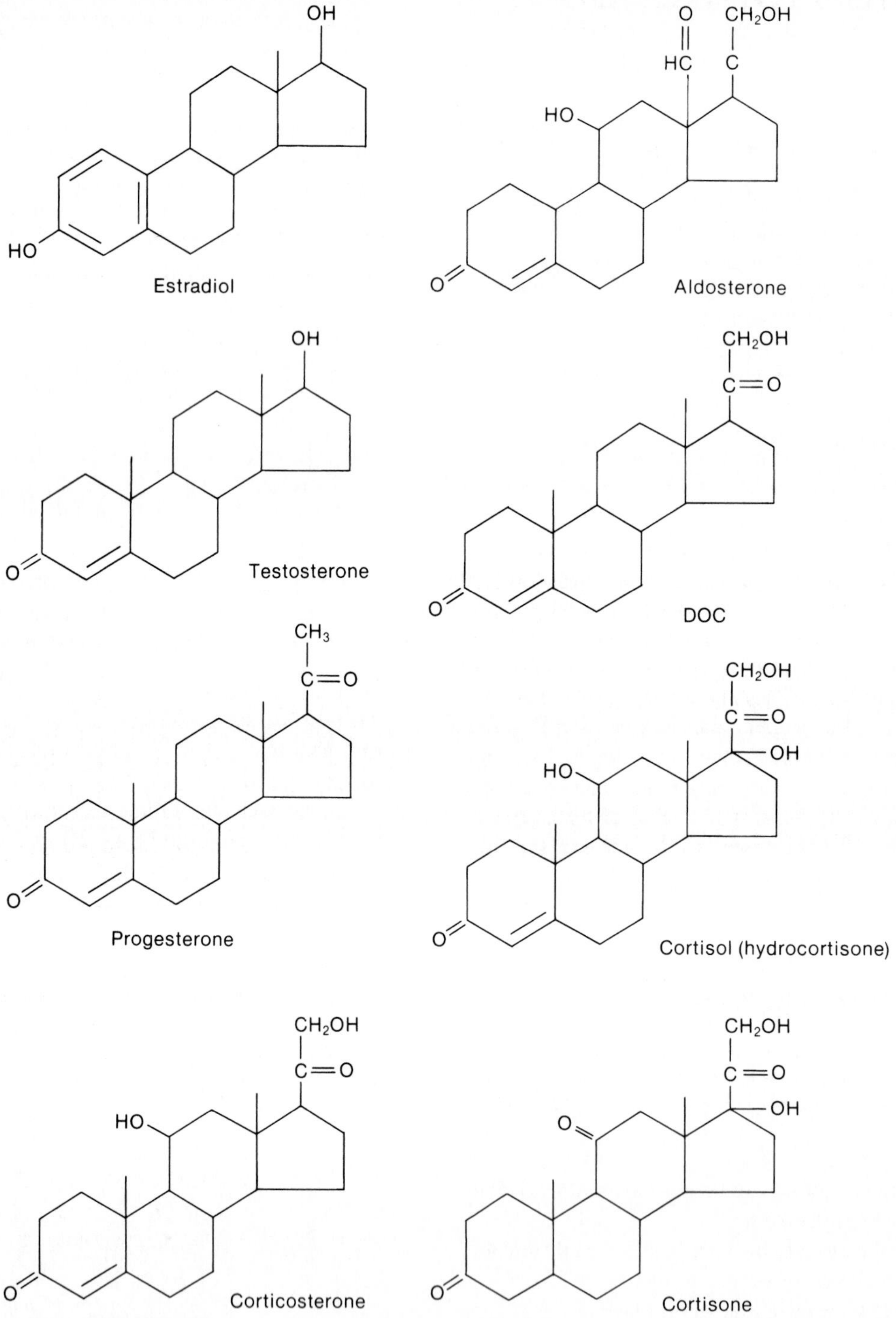

Estradiol

Aldosterone

Testosterone

DOC

Progesterone

Cortisol (hydrocortisone)

Corticosterone

Cortisone

As the solvent creeps up the plate by capillary action, it will wash the steroids off their original spots (the *origin*) and carry them upward toward the other end of the plate. Because the solubility of each steroid is different, it takes longer for the solvent to wash and carry some compared to others. If the process is halted before all the steroids have been washed off the top of the plate, some will have migrated farther from the origin than others.

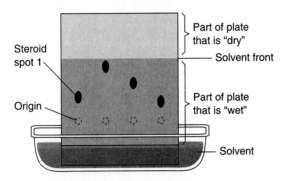

If this chromatography were repeated using the same steroids and the same solvent, the final pattern (*chromatogram*) would be the same as obtained previously. In other words, the distance that a given steroid migrates in a given solvent, relative to the *solvent front,* can be used as an identifying characteristic of that steroid. We can give this identity a numerical value by calculating the distance the steroid traveled relative to the front (the R_f value) as follows:

$$R_f = \frac{\text{distance from origin to steroid spot } (D_s)}{\text{distance from origin to solvent front } (D_f)}$$

We can identify the unknown steroid by comparing its R_f value in a given solvent with the R_f values of known steroids in the same solvent.

Clinical Significance

The chromatographic separation and identification of steroid hormones have revealed much about endocrine physiology that is clinically useful. It was learned, for example, that the placenta secretes estrogens that are more polar (water soluble) than the predominant ovarian estrogen, estradiol. These polar placental estrogens, *estriol* and *estetrol,* are now measured clinically during pregnancy to assess the health of the placenta.

Chromatography of androgens recovered from their target tissues (such as the prostate) has revealed that these tissues convert testosterone into other products. Further, these products appear to be more biologically active (more androgenic) than testosterone itself. Thus, testosterone secreted by the testes is a *prehormone* that is enzymatically converted in the target tissue into more active products, such as *dihydrotestosterone (DHT)*, in many tissues. Males who have a congenital deficiency in 5α-reductase, the enzyme responsible for this conversion, show many symptoms of androgen deficiency even though their testes secrete large amounts of testosterone.

Procedure

1. Using a pencil, make a tiny notch on the left margin of the thin-layer plate, approximately 1½ inches from the bottom. The origin of all the spots will lie on an imaginary line extending across the plate from this notch.
2. Using a capillary pipette, carefully spot steroid solution 1 (estradiol) about one-half inch in from the left-hand margin of the plate, along the imaginary line. Repeat this procedure, using the same steroid at the same spot two more times. Allow the spot to dry between applications.
3. Repeat step 2 with each of the remaining steroid solutions (solution 2, testosterone; solution 3, hydrocortisone; solution 4, cortisone; solution 5, corticosterone; solution 6, deoxycorticosterone; solution 7, unknown), spotting each steroid approximately one-half inch to the right of the previous steroid, along the imaginary line.
4. Observe the steroid spots at the origin under an ultraviolet (UV) lamp. (**Caution:** *Do not look directly at the UV light.*)
5. Place the thin-layer plates in a developing chamber filled with solvent (tolvene/ethyl acetate/acetone, 6:1:1), and allow the chromatogram to develop for 1 hour.
6. Remove the thin-layer plate, dry it, and observe it under the UV light. Using a pencil, outline the spots observed under the UV light.
7. In the laboratory report, record the R_f values of the known steroids, and determine the steroids present in the unknown solution.

Laboratory Report 50

Name _____

Date _____

Section _____

Steroid Hormones

Read the assigned sections in the textbook before completing the laboratory report.

1. Record your data in the table below, and calculate the R_f value of each spot:

Steroid	Distance to Front	Distance to Spot	R_f
1. Estradiol			
2. Testosterone	same		
3. Hydrocortisone	———		
4. Cortisone	———		
5. Corticosterone	———		
6. Deoxycorticosterone	———		
7. Unknown 1	———		
Unknown 2			

2. The chief estrogenic hormone is _____; the most potent androgen is _____.
3. Name the following steroids:

   a. 18-carbon sex steroid _____

   b. 19-carbon sex steroid _____

   c. 21-carbon sex steroid _____

4. Progesterone is secreted by the _____ and the _____ .
5. The most potent mineralocorticoid is _____ ; the major

   glucocorticoid is _____.

6. Review the R_f values of the different steroids in this exercise; by comparing these, describe the relative solubilities of the steroids in the solvent used.

7. Using an outline or flowchart, indicate the categories and subcategories of steroid hormones, including their glands of origin.

Pregnancy Test

Before coming to class, review the following sections in chapter 30 of the textbook: "Implantation" and "Extraembryonic Membranes."

Introduction

Shortly after fertilization occurs, embryonic cells called *trophoblast cells,* which will form the chorionic membrane, secrete a hormone called **human chorionic gonadotropin, or hCG.** Because this hormone is secreted by the embryo and not by the mother, laboratory tests for the presence of this hormone in a woman's urine or blood provide an accurate assessment for pregnancy.

The presence of hCG will be assayed by means of an antigen-antibody reaction. An *antigen* is capable of stimulating lymphocytes to produce *antibodies,* which are proteins capable of bonding to specific antigens. The antigen-antibody reaction in the body is termed the *immune response,* and an assay that makes use of this reaction is an *immunoassay.*

In this immunoassay for hCG, antibodies are stuck onto tiny white latex particles. If urine containing hCG (the antigen) is added, chemical bonds form between the antigens and antibodies, causing the latex particles to *agglutinate.* Since antibodies are relatively specific in their reaction with antigens, no agglutination reaction will occur if hCG is absent from the urine. See figure 51.1 for examples of positive and negative agglutination reactions.

Clinical Signifance

The immunoassay of hCG in urine is a very accurate test for pregnancy, and in one form or another, this is now the most widely used type of pregnancy test. If this test is performed too soon after conception, however, the hCG level may be below the sensitivity of the assay and may produce false negative results. These tests should therefore be performed at least two weeks following a missed period. False positive tests are less common and may be due to tumors that secrete hCG (such as choriocarcinomas).

Figure 51.1 A positive pregnancy test: (*a*) no agglutination when urine is added to a control solution containing latex particles coated with rabbit gamma globulin (from a rabbit not sensitized to hCG); (*b*) agglutination of latex particles when urine is added to latex coated with antibodies from a rabbit sensitized to hCG. (Note the appearance of agglutinated latex particles.)

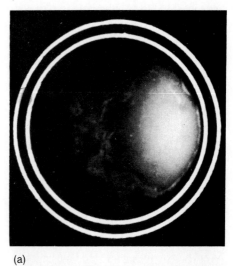

(a)

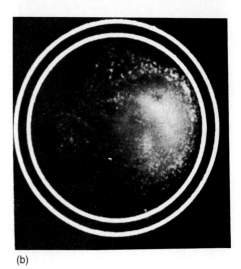

(b)

Objectives

Students completing this exercise will be able to:

1. Describe how a corpus luteum is formed, and explain what happens to the corpus luteum at the end of a nonfertile cycle.
2. Describe the function of the corpus luteum during pregnancy.
3. Describe the source of hCG, and explain the function of this hormone during pregnancy.
4. Explain how pregnancy can be detected by the test performed in this laboratory exercise.

Materials

1. Urine collection cup
2. Pregnancy kit (DAP test, Wampole test, or similar test)

Procedure

1. Allow the urine sample to reach room temperature.
2. Fill the plastic reservoir provided in the pregnancy kit with urine, and insert the filtering attachment.
3. Squeeze the reservoir to expel the urine onto two circles on the disposable slide.
4. Shake the bottle of reagent (latex particles with antibodies against hCG), and add 1 drop to the first circle. Mix the urine and reagent with an applicator stick by spreading the mixture over the entire circle.
5. Shake the latex control (latex particles with gamma globulin), and add 1 drop to the second circle. Mix as before.
6. Rock the slide gently for 1 minute, and look for agglutination. In a negative test, the solution will remain milky, whereas in a positive test it will appear grainy (fig. 51.1).

Laboratory Report **51**

Name _____

Date _____

Section _____

Pregnancy Test

Read the assigned sections in the textbook before completing the laboratory report.

1. The structure that secretes estrogen and progesterone for the first 10 weeks of pregnancy is the _____.

2. Human chorionic gonadotropin is secreted by the _____.

3. The action of hCG is similar to that of _____ , which is a hormone secreted by the anterior pituitary gland.

4. Describe the formation, function, and fate of the corpus luteum during the nonfertile menstrual cycle. What happens to the corpus luteum when fertilization occurs?

5. Why are most pregnancy tests not valid unless performed at least a couple of weeks after a missed period?

6. Why is this pregnancy test called an immunoassay? Explain how this test works to detect pregnancy.

Patterns of Heredity

Before coming to class, review the following sections in chapter 30 of the textbook: "Inheritance" and "Clinical Considerations."

Introduction

The inheritance of many aspects of body structure and function follows a pattern that can be understood by applying relatively simple concepts. A knowledge of these patterns of heredity is needed for proper understanding of anatomy and physiology, and for clinical purposes, such as the genetic counseling of prospective parents.

Clinical Significance

Most of the concepts of heredity discussed in this exercise were discovered in the 1860s by an Austrian monk named Gregor Mendel; consequently, these patterns of heredity are often called *simple Mendelian heredity.* A proper knowledge of these patterns is obviously needed for genetic counseling of carriers of genetic diseases. If both parents are carriers of such diseases as sickle-cell anemia, Tay–Sachs disease, phenylketonuria (PKU), and others that are inherited as autosomal recessive traits, it is important they understand there is a 25% chance their children will get the disease. If only one parent is a carrier, they should know their children will definitely not get the disease. Further, they should be cautioned that, regardless of the number of children they have, the probability that their next child will get the disease always remains the same.

Objectives

Students completing this exercise will be able to:

1. Define and explain the terms *dominant, recessive, homozygous,* and *heterozygous.*
2. Describe and distinguish between autosomal and sex-linked inheritance.
3. Explain the nature of sickle-cell anemia and its inheritance.
4. Explain the inheritance of hemophilia and color blindness.

Materials

1. PTC paper
2. Sickle-Sol tube test (Dade)
3. Ishihara color blindness charts
4. Lancets and 70% ethyl alcohol

A. Sickle-Cell Anemia

Sickle-cell anemia is an autosomal recessive disease affecting 8%–11% of the African-American population of the United States. In this disease, a single base change in the DNA, through the mechanisms of transcription and translation, results in the production of an abnormal hemoglobin (*hemoglobin S*), which differs from the normal hemoglobin (*hemoglobin A*) by the substitution of one amino acid for another (valine for glutamic acid) in one position of the protein.

A person inherits two sets of genes controlling every trait, one from the mother and one from the father (assuming these genes are *autosomal*—that is, not located on the sex chromosomes). If both genes are identical, the person is said to be **homozygous** for that trait. A person who is homozygous for normal hemoglobin, for example, has the genotype *AA,* whereas a person who is homozygous for hemoglobin S has the genotype *SS.*

If a person inherits the gene for hemoglobin A from one parent and the gene for hemoglobin S from the other parent, the person is said to be **heterozygous** for that trait and has the genotype *AS.* This person has the sickle-cell *trait* but does not have sickle-cell *disease.* The phenotype (in this case the absence of sickle-cell disease) is the same for the heterozygote as it is for the person who is homozygous normal. Thus, the gene for hemoglobin A is **dominant** to the gene for hemoglobin S (or the gene for hemoglobin S is **recessive** to the gene for hemoglobin A). Although the heterozygote does not display the phenotype of sickle-cell disease, this person is a carrier of the sickle-cell trait, since one-half of the gametes will contain the gene for hemoglobin A and one-half will contain the gene for hemoglobin S. If this individual mates with one who is homozygous *AA,* the probability is 50% that an offspring will be homozygous *AA,* and 50% that an offspring will be heterozygous *AS.*

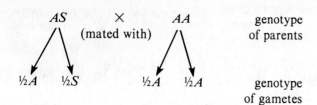

$$\frac{1}{2}A \times \frac{1}{2}A = \frac{1}{4}\ AA \left.\right\} \ 2/4\ AA$$
$$\frac{1}{2}A \times \frac{1}{2}A = \frac{1}{4}\ AA$$
$$\frac{1}{2}S \times \frac{1}{2}A = \frac{1}{4}\ AS \left.\right\} \ 2/4\ AS$$
$$\frac{1}{2}S \times \frac{1}{2}A = \frac{1}{4}\ AS$$

If two individuals who are both heterozygous *AS* mate, one-fourth of the progeny will have the genotype *AA*, one-fourth will have the genotype *SS*, and one-half will have the genotype *AS*. Although individuals with the homozygous genotype *AA* and the heterozygous genotype *AS* are healthy, the chances are *one in four* that a child from this mating will have the phenotype of sickle-cell disease (genotype *SS*).

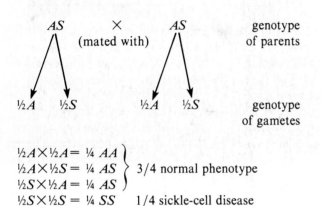

$$\frac{1}{2}A \times \frac{1}{2}A = \frac{1}{4}\ AA \left.\right\}$$
$$\frac{1}{2}A \times \frac{1}{2}S = \frac{1}{4}\ AS \left.\right\} \ 3/4\ \text{normal phenotype}$$
$$\frac{1}{2}S \times \frac{1}{2}A = \frac{1}{4}\ AS \left.\right\}$$
$$\frac{1}{2}S \times \frac{1}{2}S = \frac{1}{4}\ SS \qquad 1/4\ \text{sickle-cell disease}$$

A rapid test for sickle-cell anemia is based upon the fact that, under conditions of reduced oxygen tension, hemoglobin S is less soluble than hemoglobin A and tends to make a solution turbid, or cloudy (fig. 52.1a).

Procedure

1. Fill a calibrated capillary tube with blood up to the line, and blow the blood into a test tube containing 2.0 milliliters of test reagent (contains sodium dithionite, which produces low oxygen tension).

2. If the solution does not become cloudy within 5 minutes, the test is negative.
3. Record your data in the laboratory report.

B. Inheritance of PTC Taste

The ability to taste PTC paper (phenylthiocarbamide) is inherited as a dominant trait. Therefore, if T is taster and t is nontaster, then tasters have the genotype *TT* or *Tt*, and nontasters have the genotype *tt*.

Procedure

Taste the PTC paper by leaving a strip of it on the tongue for a minute or so. If the paper has an unpleasantly bitter taste, you are a taster. Determine the number of tasters and nontasters in the class, and enter this data in your laboratory report.

C. Sex-Linked Traits: Inheritance of Color Blindness

The sex of an individual is determined by one of the 23 pairs of chromosomes inherited from the parents—these are the sex chromosomes, X and Y. The female has the genotype *XX,* and the male has the genotype *XY.* Traits that are determined by genes located on the *X chromosome,* as opposed to the other (autosomal) chromosomes, are called *sex-linked traits* (the Y chromosome apparently carries very few genes).

Unlike the previously considered patterns of heredity, where the genes are carried on autosomal chromosomes, the inheritance of genes carried on the X chromosome follows a different pattern for males than for females. This is because the male inherits only one X chromosome (and thus only one set of sex-linked traits) from his mother, whereas the female inherits an X chromosome from both parents.

The genes for color vision and for some of the blood-clotting factors are carried on the X chromosome, where the phenotypes for color blindness and hemophilia are recessive to the normal phenotypes. A normal female may have either the homozygous dominant or the heterozygous ("carrier") genotypes, whereas a male must have either the normal or the affected phenotypes.

Suppose a normal man mates with a woman who is a carrier for color blindness (*C* is normal, *c* is color-blind).

Figure 52.1 (a) A turbidity test for sickle-cell anemia; (b) sickled cells under a light microscope; (c) normal red blood cells under a scanning electron microscope; (d) sickled red blood cells under a scanning electron microscope.

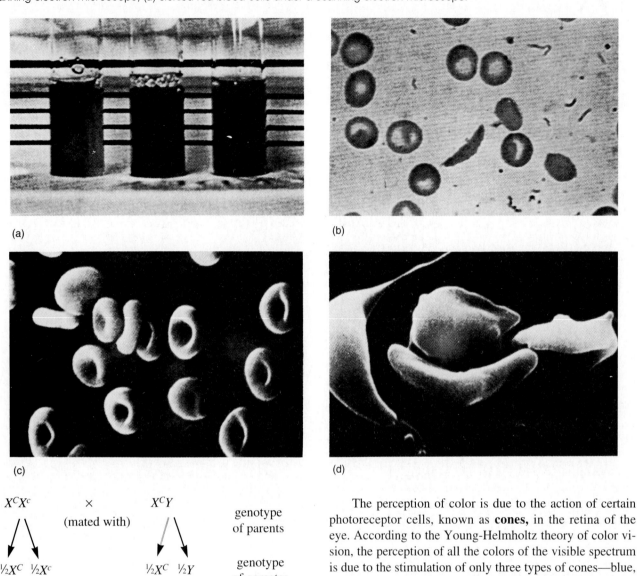

(a)

(b)

(c)

(d)

$$X^C X^c \quad \times \quad X^C Y \qquad \text{genotype of parents}$$

(mated with)

$$\tfrac{1}{2}X^C \quad \tfrac{1}{2}X^c \qquad \tfrac{1}{2}X^C \quad \tfrac{1}{2}Y \qquad \text{genotype of gametes}$$

$$\tfrac{1}{2}X^C \times \tfrac{1}{2}X^C = \tfrac{1}{4}X^C X^C$$
$$\tfrac{1}{2}X^C \times \tfrac{1}{2}Y = \tfrac{1}{4}X^C Y$$
$$\tfrac{1}{2}X^c \times \tfrac{1}{2}X^C = \tfrac{1}{4}X^C X^c$$
$$\tfrac{1}{2}X^c \times \tfrac{1}{2}Y = \tfrac{1}{4}X^c Y$$

genotypes of progeny

The probability that a child formed from this union will be color-blind is 25%, and the probability that this child will be male is 100%. All female children formed from this union will, of course, have the normal phenotype, but the probability that a given female child will be a carrier for color blindness is 50%.

The perception of color is due to the action of certain photoreceptor cells, known as **cones,** in the retina of the eye. According to the Young-Helmholtz theory of color vision, the perception of all the colors of the visible spectrum is due to the stimulation of only three types of cones—blue, green, and red cones. These three names refer to the regions of the spectrum at which each type of cone is maximally stimulated. When one of these three types of cones is defective owing to the inheritance of a sex-linked recessive trait, the ability to distinguish certain colors is diminished.

Procedure

In the Ishihara test, colored dots are arranged in a series of circles in such a way that a person with normal vision can see a number embedded within each circle, whereas a color-blind person will only see an apparently random array of colored dots.

Laboratory Report *52*

Name _____

Date _____

Section _____

Patterns of Heredity

Read the assigned sections in the textbook before completing the laboratory report.

A. Sickle-Cell Anemia

1. Was your test positive or negative? _____

B. Inheritance of PTC Taste

1. Are you a taster? _____
2. Enter the number of tasters and nontasters in your class in the table below:

Number in Class	
Tasters	
Nontasters	

3. Calculate the proportion of tasters (the number of tasters divided by the total number of students). Enter this value in the space below:

   _____

C. Sex-Linked Traits: Inheritance of Color Blindness

1. Are you color-blind? If so, what type of color blindness do you have? _____

General Questions

1. A student discovers that he and his mother are PTC tasters and his father is a nontaster. What are the genotypes of these three people?

2. If a man with sickle-cell disease marries a woman with sickle-cell trait, what is the probability that their children will have (a) sickle-cell trait and (b) sickle-cell disease?

3. A normal man marries a woman who is a carrier for hemophilia (a sex-linked trait). What is the probability that their first child will have hemophilia? What is the sex of the child? What is the probability that their next child will have hemophilia? Explain.

Sources of Equipment and Solutions

American Scientific Products, Division of American Hospital Supply Corporation

Atlanta
American Scientific Products
1750 Stoneridge Drive
Stone Mountain, Georgia 30083
(404)943–4070
(800)232–3550 (GA)
(800)241–6640 (Out of state)

Boston
American Scientific Products
20 Wiggins Avenue
Bedford, Massachusetts 01730
(617)275–1100
(800)842–1208 (MA)
(800)225–1642 (Out of state)

Charlotte
American Scientific Products
8350 Arrowridge Boulevard
Charlotte, North Carolina 28210
(704)525–1021
(800)432–6997 (NC)
(800)438–1234 (Out of state)

Chicago
American Scientific Products
1210 Waukegan Road
McGaw Park, Illinois 60085
(312)689–8410
(800)942–4591 (IL)
(800)323–4515 (Out of state)

Cleveland
American Scientific Products
3201 East Royalton Road
Broadview Heights, Ohio 44147
(216)526–2430
(800)362–9111 (OH)

Columbus
American Scientific Products
2340 McGaw Road
Obetz, Ohio 43207
(614)491–0050
(800)848–9670 (OH)
(800)282–9640 (Out of state)

Dallas
American Scientific Products
210 Great Southwest Parkway
Grand Prairie, Texas 75050
(214)647–2000
(800)492–4820 (TX)
(800)527–6230 (Out of state)

Denver
American Scientific Products
4910 Moline Street
Denver, Colorado 80239
(303)371–0565
(800)332–1241 (CO)
(800)525–1251 (Out of state)

Detroit
American Scientific Products
30500 Cypress
Romulus, Michigan 48174
(313)729–6000
(800)482–3740 (MI)
(800)521–0757 (Out of state)

Honolulu
American Scientific Products
274 Puuhale Road
Honolulu, Hawaii 98619
(808)847–1585

Houston
American Scientific Products
4660 Pine Timbers
Houston, Texas 77041
(713)462–8000
(800)392–2054 (TX)

Kansas City
American Scientific Products
1118 Clay Street
North Kansas City, Missouri 64116
(816)221–2533
(800)892–2433 (MO)
(800)821–2006 (Out of state)

Los Angeles
American Scientific Products
17111 Red Hill Avenue
P.O. Box C19505
Irvine, California 92713
(714)540–5320
(800)432–7141 (CA)

Miami
American Scientific Products
1900 N.W. 97th Avenue
Miami, Florida 33152
(305)592–4620

Minneapolis
American Scientific Products
13505 Industrial Park Boulevard
Minneapolis, Minnesota 55441
(612)553–1171
(800)642–3220 (MN)
(800)328–7195 (Out of state)

New Orleans
American Scientific Products
155 Brookhollow Esplanade
P.O. Box 23628
Harahan, Louisiana 70183
(504)733–7571
(800)452–8738 (LA)
(800)535–7333 (Out of state)

New York
American Scientific Products
100 Raritan Center Parkway
Edison, New Jersey 08817
(201)494–4000
(212)964–3500 (New York City)
(215)925–3983 (Philadelphia)
(800)526–7510 (Out of state)

Ocala
American Scientific Products
601 S.W. 33rd Avenue
Ocala, Florida 32670
(904)732–3480
(800)342–0191 (FL)

Philadelphia
American Scientific Products
2550 Boulevard of Generals
Valley Forge, Pennsylvania 19482
(215)631–9300

Phoenix
American Scientific Products
602 West 22nd Street
Tempe, Arizona 85282
(602)968–3151
(800)352–1431 (AZ)
(800)528–4471 (Out of state)

Puerto Rico
American Scientific Products
G.P.O. 2796
San Juan, Puerto Rico 00936
(809)788–1200

Rochester
American Scientific Products
2 Town Line Circle
Rochester, New York 14623
(716)475–1470
(716)856–0114 (Buffalo)
(315)242–0747 (Syracuse)
(800)462–5673 (NY state)

St. Louis
American Scientific Products
10888 Metro Court
Maryland Heights, Missouri 63043
(314)569–2960
(800)392–4234 (MO)
(800)325–4520 (Out of state)

Salt Lake City
American Scientific Products
P.O. Box 27568
Salt Lake City, Utah 84125
(801)972–3032
(800)453–4690, 4691

San Francisco
American Scientific Products
255 Caspian Drive
Sunnyvale, California 94086
(408)743–3100
(800)538–1670 (N.CA)
(800)672–8610 (S.CA)

Seattle (Office Handling Alaska)
American Scientific Products
3660 148th Avenue N.E.
Redmond, Washington 98052
(206)885–4131
(800)562–8060 (WA)
(800)426–2950 (ID, MT, OR)
(800)426–6360 (AK)

Washington, D.C.
American Scientific Products
8855 McGaw Road
Columbia, Maryland 21045
(301)992–0800
(800)638–2813

Bio-Analytic Laboratories, Inc.
3473 Palm City School Road
P.O. Box 388
Palm City, Florida 34990
(407)287–3340
(800)327–8282

Carolina Biological Supply Company
Main Office
Burlington, North Carolina 27215
(800)334–5551
West Coast Customers
Powell Laboratory Division
Gladstone, Oregon 97027
(800)547–1733

Curtin-Matheson Scientific, Inc., A Coulter Subsidiary Company

Atlanta
2140 Newmarket Parkway
Marietta, Georgia 30067
(404)424–0500

Boston
110A Commerce Way
Woburn, Massachusetts 01888
(617)935–8888

Chicago
1850 Greenleaf Avenue
Elk Grove Village, Illinois 60007
(312)439–5880

Cincinnati
12101 Centron Place
Cincinnati, Ohio 45246
(513)671–1200

Cleveland
4540 Willow Parkway
Cleveland, Ohio 44125
(216)883–2424

Dallas
1103–07 Slocum Street
Dallas, Texas 75207
(214)747–2503

Denver
12950 East 38th Avenue
Denver, Colorado 80239
(303)371–5713

Detroit
1600 Howard Street
Detroit, Michigan 48216
(313)964–0310

Honolulu (Sales Office)
99–1169 Iwaena Street
Aiea, Hawaii 96701
(808)487–7220

Houston
4220 Jefferson Avenue
Houston, Texas 77023
(713)923–1661

Indianapolis (Sales Office)
2511 East 46th, Suite V–1
Indianapolis, Indiana 46205
(317)546–5401

Kansas City
6111 Deramus Road
Kansas City, Missouri 64120
(816)241–5000

Los Angeles
2750 Saturn Street
Brea, California 92621
(714)996–1310

Miami (Sales Office)
795 West 83rd Street
Hialeah, Florida 33014
(305)558–0851

Minneapolis
2218 University Avenue S.E.
Minneapolis, Minnesota 55414
(612)378–1110

New Orleans
627 Distributors Row
Harahan, Louisiana 70123
(504)733–7763

New York
357 Hamburg Turnpike
Wayne, New Jersey 07470

Orlando
7524 Currency Drive
Orlando Central Park
Orlando, Florida 32809
(305)859–8281

St. Louis
11526 Adie Road
Maryland Heights, Missouri 63043
(314)872–8100

San Francisco
470 Valley Drive
Brisbane, California 94005
(415)467–1040

Seattle
1177 Andover Park West
Tukwila, Washington 98188
(206)575–0575

Tulsa
6550 East 42nd Street
Tulsa, Oklahoma 74145
(918)622–1700

Washington, D.C.
10727 Tucker Street
Beltsville, Maryland 20705
(301)937–5950

Fisher Scientific Company

Atlanta
2775 Pacific Drive
P.O. Box 829
Norcross, Georgia 30091
(404)449–5050

Baton Rouge
4334 South Sherwood Forest
Boulevard
Baton Rouge, Louisiana 70816
(504)293–8801

Boston
461 Riverside Avenue
P.O. Box 379
Medford, Massachusetts 02155
(617)391–6110

Chicago
1600 West Glenlake Avenue
Itasca, Illinois 60143
(312)773–3050

Cincinnati
5418 Creek Road
Cincinnati, Ohio 45242
(513)793–5100

Cleveland
Building A
3355 Richmond Road
Beachwood, Ohio 44122
(216)292–7900

Dallas
4301 Alpha Road
Dallas, Texas 75234
(214)387–0850

Denver
14 Inverness Drive East
P.O. Box 3129
Englewood, Colorado 80155
(303)741–3440

Detroit
34401 Industrial Road
Livonia, Michigan 48150
(313)261–3320

Ft. Lauderdale
1815 East Commercial Boulevard
Ft. Lauderdale, Florida 33308
(305)491–7360

Houston
10700 Rockley Road
P.O. Box 1307
Houston, Texas 77001
(713)495–6060

Los Angeles
2761 Walnut Avenue
Tustin, California 92680
(714)832–9800

Louisville
1900 Plantside Drive
Louisville, Kentucky 40299
(502)491–7384

Memphis
4403 Delp Street
P.O. Box 181150
Memphis, Tennessee 38118
(901)362–3444

New York
52 Fadem Road
Springfield, New Jersey 07081
(201)379–1400

Orlando
7464 Chancellor Drive
P.O. Box 13430
Orlando, Florida 32809
(305)857–3600

Parkersburg
703 Rayon Drive
P.O. Box 3322
Parkersburg, West Virginia 26101
(304)485–1751

Philadelphia
191 South Gulph Road
King of Prussia, Pennsylvania 19406
(215)265–0300

Pittsburgh
585 Alpha Drive
Pittsburgh, Pennsylvania 15238
(412)781–3400

Raleigh
3315 Winton Road
P.O. Box 11666
Raleigh, North Carolina
(919)876–2351

Richmond
Seaboard Building, Room 434
3600 West Broad Street
Richmond, Virginia 23230
(804)359–1301

Rochester
15 Jet View Drive
P.O. Box 8740
Rochester, New York 14624
(716)464–8900

San Francisco
2170 Martin Avenue
Santa Clara, California 95050
(408)727–0660

St. Louis
1241 Ambassador Boulevard
P.O. Box 12405
St. Louis, Missouri 63132
(314)587–7000

Washington, D.C.
7722 Fenton Street
Silver Spring, Maryland 20910
(301)587–7000

Medical Analysis Systems, Inc.
Lincoln Technology Park
542 Flynn Road
Camarillo, California 933012
(800)582–3095
(805)987–7891

Stanbio Laboratory, Inc.
2930 East Houston Street
San Antonio, Texas 78202
(800)531–5535
(512)222–2108

VWR Scientific Inc., Subsidiary of Univar

Southern Region

Houston
P.O. Box 33348
Houston, Texas 77033
(713)641–0681

Dallas
P.O. Box 35106
Dallas, Texas 75235
(214)631–0261

New Orleans
5717 Salmen Street
Harahan, Louisiana 70123
(504)733–4181

Nashville
1100 Elm Hill Pike
Nashville, Tennessee 37210
(615)327–1327

Atlanta
P.O. Box 1307 Sta. K
Atlanta, Georgia 30324
(404)262–3141

Miami
P.O. Box 520127
Miami, Florida 33152
(305)625–7181

Eastern Region

Rochester
P.O. Box 1050
Rochester, New York 14603
(716)247–0610

Pittsburgh
147 Delta Drive
Pittsburgh, Pennsylvania 15238
(412)782–4230

Baltimore
6601 Amberton Drive
Baltimore, Maryland 21227
(301)796–8500

Philadelphia
P.O. Box 8188
Philadelphia, Pennsylvania 19101
(215)467–3333

Boston
P.O. Box 232
Boston, Massachusetts 02101
(617)964–0900

New York City
P.O. Box 999
South Plainfield, New Jersey 07080
(201)756–8030

Columbus
P.O. Box 855
Columbus, Ohio 43216
(614)445–8281

Detroit
3140 Grand River Avenue
Detroit, Michigan 48208
(313)833–7800

Midland
P.O. Box 2210
Midland, Michigan 48640
(517)496–3930

Western Region

Seattle
P.O. Box 3551
Seattle, Washington 98124
(206)575–5100

Anchorage
1301 East First Avenue
Anchorage, Alaska 99501
(907)272–9507

Portland
P.O. Box 14070
Portland, Oregon 97214
(503)234–9272

San Francisco
P.O. Box 3200
San Francisco, California 94119
(416)468–7150

Honolulu
P.O. Box 29697
Honolulu, Hawaii 96820
(808)833–9544

Denver
P.O. Box 39398
Denver, Colorado 80239
(303)371–0970

Salt Lake City
P.O. Box 1678
Salt Lake City, Utah 84110
(801)486–4851

Los Angeles
P.O. Box 1004
Norwalk, California 90650
(213)921–0821

San Diego
P.O. Box 80962
San Diego, California 92138
(714)297–4851

Phoenix
P.O. Box 29027
Phoenix, Arizona 85038
(602)269–7511

Central Region

Chicago
2619 Congress Street
Bellwood, Illinois 60104
(312)647–3900

Davenport
P.O. Box 2827
Davenport, Iowa 52809
(319)322–6223

Milwaukee
16675 West Glendale Drive
New Berlin, Wisconsin 53151
(414)786–9400

St. Louis
P.O. Box 13320
St. Louis, Missouri 63157
(314)231–9770

Kansas City
P.O. Box 23037
Kansas City, Missouri 64161
(816)842–9536

Minneapolis
1124 Stinson Boulevard
Minneapolis, Minnesota 55413
(612)331–4850

Warren E. Collins, Inc.
220 Wood Road
Braintree, Massachusetts 02184
(617)843–0610

Credits

PHOTOGRAPHS

Exercise 2

2.1: Carl Ziess, Inc. Thornwood, N.Y.

Exercise 4

4.1: Courtesy Bausch and Lomb, Analytical Systems Division

Exercise 7

7.3: From R. G. Kessel and R. H. Kardon: Tissues & Organs: A Text Atlas of Scanning Electron Microscopy © W. H. Freeman and Company, 1979

Exercise 8

8.9A, 8.9B, 8.9C, 8.9D, 8.9E, 8.9F, 8.9G, 8.9H, 8.9I, 8.9J: © Edwin A. Reschke

Exercise 9

9.3: © Dr. Kerry Openshaw

Exercise 11

11.5A, 11.5B: Courtesy of Utah Valley Hospital, Department of Radiology

Exercise 12

Ex. 1, Ex. 2, Ex. 3, Ex. 4: Courtesy of Eastman Kodak

Exercise 14

14.1, 14.2: Narco Biology Systems; **14.3A, 14.3B, 14.4A, 14.4B, 14.5A, 14.5B, 14.5C:** Stuart Ira Fox

Exercise 15

15.1, 15.2, 15.3: Stuart Ira Fox (author)

Exercise 17

17.5A, 17.5B, 17.5C, 17.5D: Kent M. Van De Graaff (author)

Exercise 18

18.4, 18.8A, 18.8B, 18.12A, 18.12B, 18.12C, 18.12D: Kent M. Van De Graaff

Exercise 20

20.8, 20.9, 20.10, 20.11: Courtesy of John D. Lee

Exercise 21

21.4A, 21.4B, 21.4C, 21.4D, 21.4E: Kent M. Van De Graaff

Exercise 26

26.A, 26.B: Stuart Ira Fox (author)

Exercise 30

30.3, 30.4, 30.5, 30.6, 30.7: Courtesy of John D. Lee

Exercise 32

32.4: Stuart Ira Fox

Exercise 34

34.2: Stuart Ira Fox

Exercise 38

38.1: Courtesy of Warren E. Collins, Inc., Braintree, MA; **38.2:** Courtesy of Intelitool, Inc.; **38.3, 38.4, 38.5, 38.6, 38.7:** Stuart Ira Fox

Exercise 39

39.1: Stuart Ira Fox

Exercise 43

43.1A, 43.1B, 43.1C, 43.1D, 43.1E, 43.1F, 43.1G, 43.1H: Courtesy of Abbott Laboratories

Exercise 52

52.1A, 52.1B, 52.1C, 52.1D: From McCurdy, P. R. "Sickle Cell Disease" © Medcom, Inc. 1973. Reprinted by permission.

Color Plates

1–24: © Wm. C. Brown Communications/Ralph Stevens, Photographer; **26A, 26B:** Stuart Ira Fox

LINE ART

Exercise 22

22.2: From John W. Hole, Jr., *Human Anatomy and Physiology,* 6th ed. Copyright © 1993 Wm. C. Brown Communications, Inc., Dubuque, Iowa. All Rights Reserved. Reprinted by permission.

Exercise 24

24.1: From John W. Hole, Jr., *Human Anatomy and Physiology,* 6th ed. Copyright © 1993 Wm. C. Brown Communications, Inc., Dubuque, Iowa. All Rights Reserved. Reprinted by permission.

Exercise 25

25.1: From John W. Hole Jr., *Human Anatomy and Physiology,* 6th ed. Copyright © 1993 Wm. C. Brown Communications, Inc., Dubuque, Iowa. All Rights Reserved. Reprinted by permission.

Exercise 26

26.2: From John W. Hole, Jr., *Human Anatomy and Physiology,* 4th ed. Copyright © 1987 Wm. C. Brown Communications, Inc., Dubuque, Iowa. All Rights Reserved. Reprinted by permission. **26.3 & 26.5:** From John W. Hole, Jr., *Human Anatomy and Physiology,* 6th ed. Copyright © 1993 Wm. C. Brown Communications, Inc., Dubuque, Iowa. All Rights Reserved. Reprinted by permission.

Exercise 31

31.2: From John W. Hole, Jr., *Human Anatomy and Physiology,* 6th ed. Copyright © 1993 Wm. C. Brown Communications, Inc., Dubuque, Iowa. All Rights Reserved. Reprinted by permission.

Exercise 34

34.3 & page 297: Source: Hewlett-Packard Company, HP Medical Products Group, Andover, MA.

Mons pubis, 400
Motor cerebral cortex, 197, 198f
Motor nerve(s), stimulation of, 112, 113f
Motor neurons, 185
Mouth, 362–65, 363f–65f, 371–72
Multifidus spinae muscle, of cat, 161f
Muscle(s)
  of abdominal wall, 143, 144f, 147t
  antagonistic, 123
  of brachium, 131, 132f–34f, 133t
    of cat, 164–65, 165f–66f, 166t, 170
  contraction, 117
  cramps, 118
  of facial expression, 123, 124f, 125t
  fatigue, 117–22
    in frog gastrocnemius muscle, 117,
      118f, 121
  of hand movement, 131
  of head and neck, 123–29
    of cat, 160, 160f–61f, 162t, 169
  of hip, 146f, 147, 147f–48f, 157
    of cat, 165–67, 168f, 168t, 170
  of iris, 204
  of lower extremity, 143–58, 152f–54f, 156t,
    158
  of pectoral girdle and upper extremity,
    131–42
  of shoulder, of cat, 160–63, 161f, 162t, 169
  synergistic, 123
  that move wrist, hand, and fingers, 138,
    138f, 139t, 142
  of thigh, 148f, 149, 149f, 157
    of cat, 165–67, 167f, 168t, 168f, 170
  of trunk, 143–58
  of vertebral column, 143, 145f
Muscle degeneration, 112
Muscle reflexes, tests for, 187
Muscle spindles, 187
Muscular dystrophy, 118
Muscular tissue, 49, 53f, 58
Musculocutaneous nerve, test (biceps brachii
    jerk), 189, 189f
Myasthenia gravis, 112
Myelencephalon, 173, 178
Mylohyoid muscle, 124, 127t
  of cat, 160, 160f, 162t
Myocardial ischemia, 305
  ECG findings in, 287, 288f
Myocardium, 253
Myograph recording, 276–77, 277f
Myometrium, 400
Myopia, 202–3, 205

Nail(s), 61–62, 63f
  body of, 62
Nasal bone, 79
Nasal cavity, 309, 365f
  of cat, 313
Nasal conchae, 73, 309
Nasalis muscle, 123, 125t
Nasolacrimal canal, 79
Nasopharynx, 311
  of cat, 314f
Near point of vision, 204
Nearsightedness, 202
Neck, 3
  arteries of, 254
    of cat, 257, 257f
  dental, 363
  of femur, 93

muscles of, 123–24, 126f, 127t
  anterior, 131, 132f
    of cat, 160, 160f–61f, 162t, 169
  posterior, 124
  of radius, 91
  veins of, 263
    in cat, 267–71, 269f
Negative feedback, 15–17
  in constant-temperature water bath, 15, 17
  control of pulse rate, 15–16, 18
Neopallium, 180f
Nephron loop, 339
Nerve, 54f
Nerve fibers, of pituitary, 219
Nervous tissue, 49–50, 53f, 58
Neural tube, 173
Neuroglial cells, 50
Neurohypophysis, 219
Neuron, 50
Neutrophil leukocytosis, 238
Neutrophils, 237
Nicotine, effects on frog heart, 279
Nipple, 400
Normal values, 16, 19
Nostril, 309
Nuclear membrane, cell, 33
Nucleolus, cell, 33
Nucleus
  in cardiac muscle, 49
  cell, 33
  in nervous tissue, 50
  in skeletal muscle, 49
  in smooth muscle, 49
Nutrient foramen, of bone, 67
Nutrient foramina, of humerus, 91
Nutrient vessel, of bone, 67
Nystagmus, 204

Objective lenses, of microscope, 9–10, 10f
Oblique layer, of stomach, 366
Obstructive lung disorders, 325
Obturator foramen, 92
Occipital artery, of cat, 257f
Occipital bone, 73
Occipital condyles, 73
  superior articular surface for, 80
Occipitalis muscle, 123, 125t
Occipital lobes, cerebral, 174, 182f
Ocular lens, of microscope, 10, 10f
Oculomotor nerve, 175, 180f
Odontoid process, 80
Oil-immersion objective, of microscope, 10–11
Olecranon, of ulna, 91
Olecranon fossa, of humerus, 91
Olfactory bulbs, 175, 179, 180f–81f
Olfactory nerves, 179
Olfactory tract, 175, 180f
Omentum, 3
Omohyoid muscle, 124, 127t
Ophthalmic nerve, 175
Ophthalmoscope, examination of eye with, 205
Optic chiasma, 175, 177, 180, 180f–81f
Optic disc, 202, 205
Optic foramen, 74
Optic nerve, 180, 180f, 201
Oral cavity, 361–65, 363f–65f
  of cat, 313, 314f
Orbicularis oculi muscle, 123, 125t
Orbicularis oris muscle, 123, 125t
Organ of Corti, 209, 211f
Organ of Ruffini, 195

Oropharynx, 311, 362
  of cat, 314f
Os coxa, 92, 95f
Osmolal, 44
Osmolality, 44
Osmometers, 44
Osmosis, 1, 43, 43f, 47
  across red blood cell membrane, 44–45,
    47–48
  across synthetic semipermeable membrane,
    43–44, 47
Osmotic pressure, 44
Os penis, of cat, 403
Osseous tissue. See Bone tissue
Osteocyte, 49, 67
Osteon, 55f, 67, 69f
Ostium, of uterine tube, of cat, 405, 407f
Otitis media, 210
Otosclerosis, 210
Outer compact bone, 67
Oval window, 81f, 209
Ovarian arteries, 254
  of cat, 258, 259f
Ovarian follicle, 220, 224f
Ovarian ligament, of cat, 405, 407f
Ovarian veins, 265
  in cat, 269
Ovaries, 220, 223f–24f, 226–27, 400
  of cat, 405, 407f
Ovum, 220, 224f
Oxygen consumption, 331–33, 335–36
Oxygen debt, 331
Oxygen transport, 231–36

Pacemaker region, of heart, 275
Pacinian corpuscles, 195
Pain receptors, 195
Palatal rugae, 362
Palate, 362
Palatine bone, 79
Palatine process, of maxilla, 79
Palatine tonsils, 311
  of cat, 314f
Palmaris longus muscle, 138, 139t
Palmar surface, 3
Pancreas, 3, 220, 221f, 266f, 361, 362f, 367–68,
    369f, 373
  body of, 220, 367
  tail of, 220, 367
Pancreatic duct, 367
Pancreatic islets, 23, 220, 221f, 226
Pancreatic juice, digestion of fat by, 383–86
Pancreaticoduodenal vein(s)
  anterior, of cat, 271, 271f
  posterior, of cat, 271, 271f
Pancreatic vein, 265
Papillae, of tongue, 363
Papillary muscle(s), 249, 253
Papilledema, 205
Paranasal sinuses, 77f, 309
Parasympathetic nerve(s), control of
    heartbeat, 15
Parasympathomimetic drugs, 278
Parathyroid glands, 219, 221f, 225
Parietal bone, 74
Parietal lobes, cerebral, 174
Parietal pericardium, 313f
Parietal peritoneum, 361
Parietal pleura, 312
Parietooccipital fissure, cerebral, 174
Parotid gland(s), 361, 363
  of cat, 314f